UNIVERSITY OF LIVERPOOL
013563330

WITHDRAWN
FROM STOCK

D1555927

Fractal Imaging

NING LU

LIMITED WARRANTY AND DISCLAIMER OF LIABILITY

Fractal Imaging

NING LU

ACADEMIC PRESS
San Diego London Boston
New York Sydney Tokyo Toronto

This book is printed on acid-free paper. ♾

ACADEMIC PRESS
525 B Street, Suite 1900, San Diego, CA 92101-4495, USA
1300 Boylston Street, Chestnut Hill, MA 02167, USA
http://www.apnet.com

Academic Press Limited
24–28 Oval Road, London NW1 7DX, UK
http://www.hbuk.co.uk/ap/

Library of Congress Cataloging-in-Publication Data

Lu, Ning, 1963-
Fractal imaging / Ning Lu.
p. cm.
Includes bibliographical references and index.
ISBN 0-12-458010-6 (alk. paper)
1. Image processing—Digital techniques. 2. Fractals. 3. Image compression. I. Title.
TA1637.L8 1997
620.36'7—dc21 97-6616
CIP

Printed in the United States of America

97 98 99 00 01 IC 9 8 7 6 5 4 3 2 1

To My Father,
子林

Endless Sky
and
Placid Lake
Share
The Same Color

Contents

Here is one leaf reserved for me,
From all thy sweet memorials free;
And here my simple song might tell
The feelings thou must guess so well.
. . .

— Verses Written in an Album

Thomas Moore

Preface

For the last six years I have been dedicated to *fractal computer imaging,* and the experience has been magnificent. Fractal imaging, a mathematical wonder, has been transformed into commercial success and has truly fulfilled the hopes and dreams of many. Iterated Systems's leading products, and those who developed them, have proven the reality of fractal imaging. However, this is not the only reason for my writing this book.

My vision, from time to time, shines like a hanging dewdrop, sparkles like a spraying splash, flows like a murmuring stream, or rushes like a roaring waterfall. Ultimately, it forms a river. The river runs, without looking back, shaped by its bank, led by its course, and is eventually engulfed by the ocean. The vision, like a river, is reforming, reorienting, and it joins with other thoughts to generate a higher wave. Because this insight is the most valuable thing one can offer to the world, it was the driving force while I was writing this book. But this is a technical book, so I tried to keep this question of the audience in mind while writing this book.

The main theme of this book is *fractal imaging using affine transformations.* The most important application covers *digital image compression.* This book will explore the following processes: generating fractal objects, modeling real world images, and storing and transmitting data and information efficiently in the compressed fractal form. A global view of the most current compression technologies will be presented for the first time in a unified way.

The number one requirement for a reader is a willingness to expand his or her knowledge of the book's subject matter. In this case, to understand the whole story, one needs to have a suitable technical background: primarily, a working knowledge of college-level mathematics and computer science. In addition, some previous experiences in digital signal and image processing will help the reader appreciate the many subtleties and insights found in this book.

1. About This Book

Chapter 1 begins with the magic of the spleenwort fern, describes and defines what a fractal is, provides the history of fractal mathematics, and illustrates the fractal mechanism in the human visual system.

Chapter 2 presents a basic mathematical background for fractal and iterated function systems. The goal is to set up a clear, simple, minimal, and solid mathematical foundation for the rest of the book. Metric spaces, affine transformations, iterated function systems (IFS), and recurrent iterated function systems (RIFS) are introduced with ideas, intuitions, and examples. Discussions will primarily be restricted to real vector spaces.

Chapter 3 establishes a unified fractal imaging model of photographic images. As an example, the partitioned iterated function system (PIFS) theory is presented, which is the basis for building a basic fractal compressor. In detail, a C-implementation of the PIFS fractal image compression system is described to represent nature images in fractal transformation parameters.

Chapter 4 studies many aspects of image partitioning and demonstrates that in some cases, there is an advantage in using image overlapped covering rather than partitioning. One of the main factors in fractal image compression is contributed by image partitioning. The most popular and sound partitioning method, quadtree splitting, is presented as the central theme.

Chapter 5 provides a detailed study of spatial fractal parameters. A fractal transformation consists of two components: the spatial transform and the intensity transform. The spatial parameters behave quite redundantly—they are particularly inclined to stay locally nearby, enabling better compression techniques. In this chapter the seed image and universal code book technique is introduced as a special scenario of vector quantization.

Chapter 6 explores the color intensity parameters: brightness translation β-values and contrast rescaling γ-values. The intensity mean values are studied as alternative parameters of translation β-values. A JPEG DCT is presented as a specific combination of intensity orthogonal terms.

Chapter 7 exhibits a new block match-searching method by using a vector clustering technique. This method reduces a typical compression time from 8 hours to 25 seconds.

Chapter 8 is a collection of advanced decompression algorithms. The exquisite pixel-chaining algorithm and the fast pyramid-iteration algorithm are described as the main examples. Fractal image segmentation and fractal image enhancement are two major fractal-unique features covered in this chapter.

Chapter 9 introduces a new resolution refining compression scheme by further exploring fractal image enhancement technique. Subband decomposition and wavelet image compression are presented and linked as extensions of fractal multiresolution image compression theory.

Chapter 10 focuses on video compression techniques, particularly fractal transformations applied to the time axis, which are equivalent to motion compensation vectors. The digital video image standards, H.263 and MPEG-4, are discussed.

Chapter 11 presents techniques that deal with color images, including colorimetry, color decomposition, and various color image compression techniques.

Chapter 12 discusses various entropy coding techniques that are used in packing fractal parameters, as in Huffman coding, arithmetic coding, and runlength coding. In addition, some fast decoding algorithms are studied, and a new efficient entropy coding scheme is introduced: incremental coding scheme. These entropy coding theories are also displayed as fractal dynamic systems. Some well-known entropy distributions, e.g., Gaussian, Laplacian, Gamma, are analyzed.

Chapter 13 examines the difference between the current image measurement model and the human visual system. A new image quality assessment metric is defined.

Chapter 14 dissects digital image formats into sections of headers (i.e., sections in the beginning of a file), tailers (i.e., sections toward the end of a file), panels, and slices. The progressiveness of an image format is illustrated as a key property of digital imaging for Internet transmissions, now and in the future.

Chapter 15 contributes a new outlook for future image capturing and displaying devices: fractal optical cameras, fractal optical scanners, fractal copy machines, fractal television . . . .

2. ACKNOWLEDGMENTS

When I joined Iterated Systems in September 1990, I shifted my research area from 3-dimensional topology to fractal geometry, from manipulating the knots and braids to formulating the ferns and leaves. The reason was, in part, fascination with their mathematical, artistic, and aesthetic beauty; yet mainly I was driven by the job opportunity, or to use another word, the destiny. I have been working on fractal image compression research and production since then.

Foremost, my gratitude goes to Michael Barnsley, who guided me into the fractal realm, a place full of treasure, full of wonder, and full of magic. His vision, his enthusiasm, and his spirit will always inspire and enlighten me in a delightful way.

My appreciation goes to each of my colleagues: Alan Sloan, Steve Demko, John Elton, Els Withers, Hawley Rising, Lyman Hurd, Gang Liang, Li-Zhe Tan, Keshi Chen, John Muller, Steve Addison, Doug Hardin, Yaakov Shima, Jakko Kari, Brian Li, David Howard, Sylvia Williamson, Anca Deliu, Amir Said, Ruifeng Xie, and many others for inspiring me and sharing with me their insight on the beauty of fractals from one angle or another over the years.

My appreciation also goes to Iterated Systems's Still Imaging R&D team: Zhiwu Lu, Echeyde Cubillo, David Roy, David Knight, Mark Feldman, Tim Smith, Smadar Gefen, Eric Hyche, Bill Coleman, Yan Zhuang, Jean Hess, and others for the terrific job we completed together in showing fractal technology to the world.

My thanks extend to Christine Smith, Louisa Barnsley, Nicki Brown, Karen Morrione, Shawn Talley, Erika Buckman, and Susan Coleman for reading and commentary of some manuscripts, and to Jim Cavedo and Paul Rich for helping me in marketing materials.

I wish also to thank Iterated Systems for giving me the permission to use its product *Fractal Imager* as the attached CD-ROM and most importantly, for providing me with opportunity, knowledge, encouragement, excitement, challenge, success, and most of all, friendship.

I wish to thank Chuck Glaser, Bettina Burch, and Katy Tynan, of Academic Press, for their confidence in me and for their beautifully done work in publishing this book. I am also grateful to the copyeditor, David Kramer, and the compositor, Reuben Kantor, for their contribution in getting the book to its current form.

My gratitute extends to Robert Marlow, of Digital Photographic Restoration, who edited the early manuscript of this book and spent many long hours with me ironing out every detail.

There are many others who have shown me wonderful things in life and beyond. I'll keep those thanks deep in my heart.

Virtue, like a river, flows soundlessly in the deepest place.

Foreword

Written by Michael Barnsley. All rights to this material are reserved. However, permission is granted to Academic Press and Harcourt Brace Jovanovitch use this material as the forward to the book entitled Fractal Imaging, written by Ning Lu.

The unreasonable success of mathematics in providing key discoveries to science is clear. Historically, time and again, when an area of scientific study has seemed to be a mass of complication and eclectic detail, mathematical scientists have provided a clear sighted solution in the form of a unifying model.

For example, quantum mechanics was discovered from a collection of apparently inconsistent experimental data, observations, heuristics, and folklore concerning atomic spectra. The stunning realization of Schroedinger was that the dual nature of light and matter could be modeled with a differential equation. This equation relates the complex square root of a certain probability density function to the configuration of a physical system. It governs the shape and form of atoms and molecules and their interactions with each other and with electromagnetic radiation. The abstract machinery of complex numbers, Hilbert spaces, and differential equations provides highly accurate predictions of such real world objects as the spectra of hydrogen atoms. This is the basis of the science of physical chemistry. The central mystery of atoms and light is captured in the language of mathematics.

There are numerous other illustrations where mathematics has provided the critical model to an area of science: field theory, group theory, fluid mechanics, classical mechanics, the Lorenz transformation, $E=MC^2$, thermodynamics, information theory, and so on. Note, however, that before the mathematical breakthroughs are made there is typically much complication and uncertainty.

I believe that imaging science has reached the stage where there is a reasonable likelihood of it submitting to the powers of mathematical scientists. Specifically my hope is that a breakthrough will be made in the area of image complexity, resulting in a profound model for real world images.[1]

Such a discovery has not yet been made but all the signs are there that one will be. We have for the first time ready access to the data, which can be measured using high resolution digital image acquisition devices such as scanners. We have powerful computational platforms in the form of personal computers that are capable of storing and processing large images. Also, the area has caught the attention of mathematicians. There are many examples, including the work of Adelson[2] on affine video image segmentation, Mumford[3] on visual image understanding, the Geman brothers[4] on Markov random fields, Chui,[5] and Shapiro[6] on wavelet-based image compression, the work on fractal imaging by Jacquin,[7] Vrscay,[8] Levy-Vehel,[9] and work on fractal-wavelet hybrids by Berger[10] and Davis.[11]

Imaging science is deeply linked with the human vision system, which occupies sixty percent of the brain. It is concerned with what we perceive. Thus we have another sign that a breakthrough may occur, namely the existence of diverse models for how the brain works. When these models are resolved it is expected that a corresponding insight will be provided to imaging science.

What type of mathematical model is to be expected? Here are some examples that suggest the flavor of what we may hope for. One is provided by the work of John Hutchinson[12] who made a clean mathematical model, now called IFS theory, to unify and analyze a wide class of the fractal images. Might a similar but different theory encompass all real world images? Another is provided by the work of Mandelbrot[13] which is expressive of the hope for a unifying mathematical theory. Might some class of fractal distributions serve to capture the statistics of real world image data?

This book records observations made during the recent development of fractal image compression and representation technology at Iterated Systems. The experiments and results expressed here concern the commercial feasibility of using self-reference and affine transformations to represent digital real world images. In the course of the research not only were mathematical image algorithms found, but insights into the underlying science of images.

This book was written from the unique point of view of a mathematician who has a doctorate in geometry and topology and who now turns his hand to computational engineering problems. Ning Lu collaborated with me and other mathematicians on the development and the commercialization of IFS theory, the fractal transform, and associated ideas. The mathematicians, mainly from Georgia Tech, included: Alan Sloan, Stephen Demko, Andrew Harrington, John Elton, John Herndon, Anca Deliu, Els Withers, Lyman Hurd and Douglas Hardin. This period of research was intense, exciting, and fruitful.

The core of this book describes the development of a state-of-the-art fractal image compression system. As the system is developed, insights into the underlying science of images are revealed. Fractal compression involves a partition of an image into domain regions; and for each domain the selection and description of a transformation from the domain into the image. What is the best choice of image partition? What is the best class of transformations to use? How should one search for the transformation? What is the information cost of storing a given transformation? Each time such a question is answered one has not only more data about the nature of real world images, but also a growing awareness of what is not known.

I believe that a two-dimensional, geometrical information theory, appropriate to real world images, will be discovered. This theory will play an analogous role to that played by Shannon's work[14] with regard to one-dimensional data streams. I expect that fractal geometry and dynamical systems built up using affine transformations will play an important role. Why affine transformations? Simply because they feature in computer graphics, in human vision, in motion compensation and image segmentation, they are of low complexity, and because, as is explained most poetically and practically in this book, real world images contain significant affine redundancy.

REFERENCES

[1] M. F. Barnsley, L. P. Hurd, *Fractal Image Compression*, A.K. Peters Ltd., 1993.

[2] J. Y. Wang, E. H. Adelson, *Representing Moving Images with Layers*, IEEE Trans. of Image Processing, 3 (1994), 625–638.

[3] D. Mumford, *Mathematical Theories of Shape: do they model perception?*, Proc. Conference 1570, Soc. Photo-optical & Ind. Engineers, (1991) 2–10.

[4] S. Geman and D. Geman, *Stochastic relaxation, Gibbs distribution, and Bayesian restoration of images*, IEEE Trans. Patt. Anal. Mach. Intel., 6 (1984), 721–741.

[5] C. K. Chui, *An Introduction to Wavelets,* Academic Press, 1992.

[6] J. Shapiro, *Embedded Image Coding Using Zerotrees of Wavelet Coefficients,* IEEE Trans. On Signal Processing, 41 (1993) 3445–3462.

[7] A. E. Jacquin, *A Fractal Theory of Iterated Markov Operators with Applications to Digital Image Coding,* Ph.D Thesis, Georgia Tech, 1989.

[8] E. R. Vrscay and C. J. Roehrig, *Iterated Function Systems and the Inverse Problem of Fractal Construction Using Moments,* in *Computers and Mathematics* (E. Kaltofen and S. M. Watt, eds), Springer-Verlag, (1989) 250–259.

[9] J. Levy-Vehel, P. Mignot, *Multifractal Segmentation of Images,* Fractals, 2 (1994), 371–378.

[10] M. A. Berger, *Apparatus and Method for Encoding Digital Signals,* U.S. Patent No. 5497435, 1996.

[11] G. M. Davis, *A Wavelet-Based Analysis of Fractal Image Compression,* IEEE Trans. On Image Processing, (1996), 100–116.

[12] J. E. Hutchinson, *Fractals and Self-similarity,* Indiana Univ. Math. J. 3 (1981), 713–747.

[13] B. B. Mandelbrot, *The Fractal Geometry of Nature,* W. H. Freeman and Company, 1983.

[14] C. E. Shannon and W. Weaver, *The Mathematical Theory of Communication,* University of Illinois Press, 1949.

The most beautiful thing we can experience is the mysterious. It is the source of all true art and science.

...

When the solution is simple, God is answering.

— What I believe

Albert Einstein

Introduction: It's Magic

1

1.1 FERN AND LEAF

This story begins with magic.

> *Start with a piece of blank paper with an (x, y)-coordinate system marked and pick an arbitrary point on the paper; then find its coordinates. Randomly select one of the four affine transformations listed below:*

$$\begin{aligned} w_1: &\ (x,y) \mapsto (0.85\cdot x+0.04\cdot y, -0.04\cdot x+0.85\cdot y+40), \\ w_2: &\ (x,y) \mapsto (0.20\cdot x-0.26\cdot y, 0.23\cdot x+0.22\cdot y+40), \\ w_3: &\ (x,y) \mapsto (-0.15\cdot x+0.28\cdot y, 0.26\cdot x+0.24\cdot y+11), \\ w_4: &\ (x,y) \mapsto (0, 0.16\cdot y). \end{aligned}$$

> *Then apply the transformation to this point, and the coordinates of a new point are obtained. Notice the new point. Plot it in black on the paper. Again select randomly one of the above four transformations and apply it to this point to obtain the next new point. Notice the new point. Plot it in black on the paper. Again pick randomly one of the above four transformations and apply it to the point to obtain the next new point. Notice the new point. Plot it in black on the paper (you can repeat this process indefinitely). If you are patient and persistent enough, gradually the Iterated Systems trademark, a spleenwort fern (Figure 1.1.1*a*), will appear on the paper like magic.*

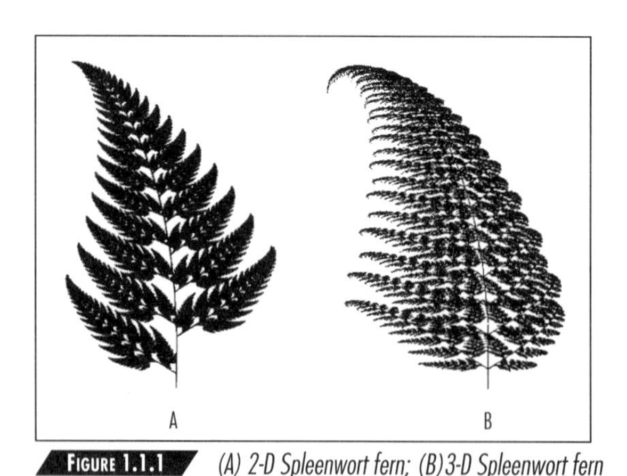

FIGURE 1.1.1 (A) 2-D Spleenwort fern; (B) 3-D Spleenwort fern

How does this magic happen? The process above shows that with the right mathematical model a perfect picture can be described with infinitely fine and marvelously rich textures in only 24 numbers:

85, 4, 0, –4, 85, 40; 20, –26, 0, 23, 22, 40; –15, 28, 0, 26, 24, 11; 0, 0, 0, 0, 16, 0.

These four affine transformations form an *iterated function system (IFS)*. The fern image created is an example of *a fractal*, which mathematically is the *attractor* of this IFS. The repeated drawing procedure is called *iteration*.

The magic of the fern is really a mathematically chaotic game. Once the procedure is known, wondrous objects can be generated. For example, the *dragon* in Figure 1.1.2*a* takes only 2 transforms, i.e., 12 numbers:

45, –50, 0, 40, 55, 0; 45, –50, 100, 40, 55, 0.

And the *Sierpinsky triangle* in Figure 1.1.2*b* is created using the following 18 numbers:

50, 0, 0, 0, 50, 0; 50, 0, 50, 0, 50, 0; 50, 0, 0, 0, 50, 50.

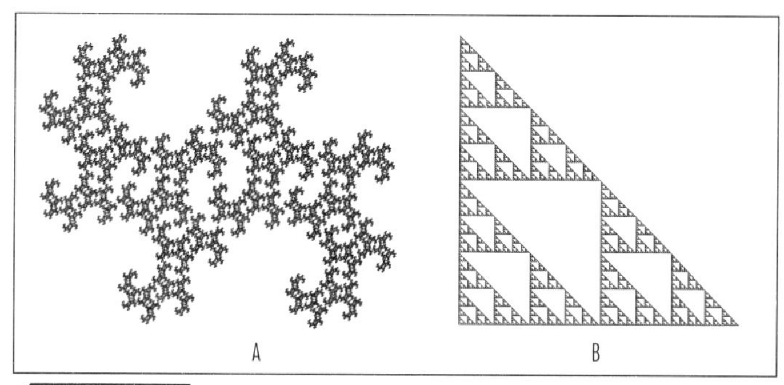

FIGURE 1.1.2 (A) Dragon; (B) Sierpinsky triangle

Planar affine transforms could be generalized to 3-dimensional space. Then each affine transform will be characterized by 12 coefficients. Consequently, more exciting fractals can be created. *The 3-D Sierpinsky gasket* in Figure 1.1.3 takes the following four affine transformations:

50, 0, 0, 0, 0, 50, 0, 0, 0, 0, 50, 0; 50, 0, 0, 50, 0, 50, 0, 0, 0, 0, 50, 0;

50, 0, 0, 25, 0, 50, 0, 50, 0, 0, 50, 0; 50, 0, 0, 25, 0, 50, 0, 25, 0, 0, 50, 50;

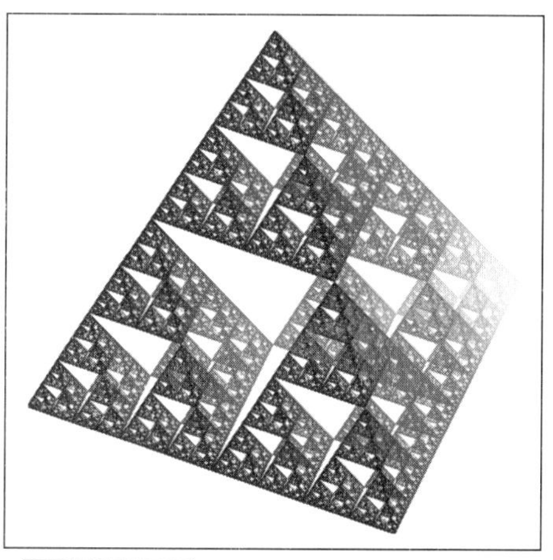

FIGURE 1.1.3 3-D Sierpinsky gasket

And the *3-D spleenwort fern* in Figure 1.1.1*b* is given by the following 4 affine transformations:

85, 0, 0, 0, 0, 85, 13, 81, 0, –9, 85, 0; 31, –34, 0, 0, 12, 21, 0, 24, 0, 0, 30, 0;
–29, 33, 0, 0, 12, 19, 0, 66, 0, 0, 30, 0; 0, 0, 0, 0, 0, 20, 0, 0, 0, 0, 0, 0.

Another excellent example of a fractal is the *maple leaf* shown in Figure 1.1.4:

49, 1, 25, 0, 62, –2; 27, 52, 0, –40, 36, 56;
18, –73, 88, 50, 26, 8; 4, –1, 52, 50, 0, 32.

FIGURE 1.1.4 *Maple leaf*

What is the definition of a *fractal*? A fractal is an image or picture that can be completely described by a mathematical algorithm in its infinitely fine texture and detail. In the most interesting cases those textures and details cannot be predicted using classical geometry.

1.2 FRACTAL

The *fractal*, as a mathematical phenomenon, dates back to the *Weierstrass nowhere differentiable continuous function* (Figure 1.2.1), to the *classic Cantor set* (Figure 1.2.2), to the *Hilbert space filling curve* (Figure 1.2.3), and even beyond.

1.2.1 WEIERSTRASS FUNCTIONS

Weierstrass functions were introduced in 1875 as a set of counterexamples in calculus to show that differentiability and continuity are two totally different concepts in mathematics.

The general formula of Weierstrass functions is given as

$$f(x)=\sum_{i=0}^{\infty}\lambda^{-si}\cos(\lambda^{i}x),\ \ 0<s<1 \ \text{ and } \ \lambda>1. \tag{1.2.1}$$

These functions, with parameters λ and s, are continuous but nowhere differentiable. Extending these functions to $s>1$, they become both continuous and differentiable. Clearly, the above property of the function is not obvious from the expression. However, it is visually clear from its graph, in Figure 1.2.1.

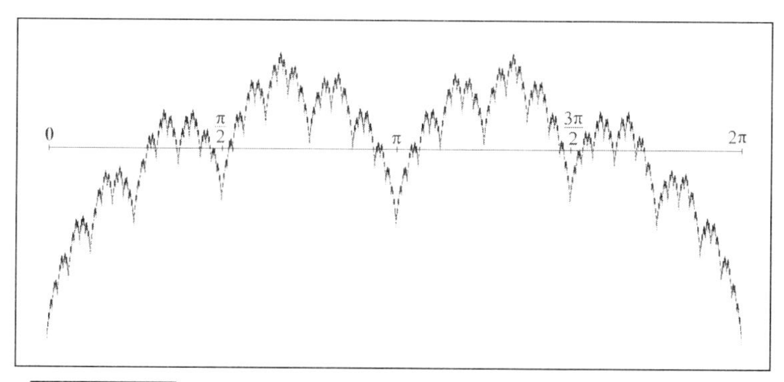

FIGURE 1.2.1 *Weierstrass function $\lambda = 2$ and $s = 0.5$*

Today it is known that the graphs of such functions are *thicker* than ordinary curves. The Weierstrass function has a fractional dimension $2-s$, which should be greater than or equal to 1. For example, when $\lambda=2$ and $s=0.5$, the function of dimension 1.5, can be described by the following:

$$f(\frac{x}{2})=\sum_{i=0}^{\infty}\left(\sqrt{2}\right)^{-i}\cos(2^{i}\frac{x}{2})=\frac{\sqrt{2}}{2}\left(f(x)+\cos\frac{x}{2}\right) \tag{1.2.2}$$

and

$$f(2\pi-\frac{x}{2})=\sum_{i=1}^{\infty}\left(\sqrt{2}\right)^{-i}\cos(2^{i}\left(2\pi-\frac{x}{2}\right))=f(\frac{x}{2})=\frac{\sqrt{2}}{2}\left(f(x)+\cos\frac{x}{2}\right), \tag{1.2.3}$$

for all $x\in[0,\ 2\pi)$.

Let us consider the interval $[0,\ 2\pi]$. For any function f on that interval, we can define a new function $W(f)$ using the above two formulas:

$$W(f)(x)=\begin{cases}\frac{\sqrt{2}}{2}\left(f(2x)+\cos x\right), & \text{if } x\in[0,\ \pi];\\ \frac{\sqrt{2}}{2}\left(f(4\pi-2x)+\cos x\right), & \text{if } x\in(\pi,\ 2\pi].\end{cases} \tag{1.2.4}$$

Consider the sequence

$$f, W(f), W^2(f) = W(W(f)), \cdots, W^k(f) = W(W^{k-1}(f)), \cdots.$$

The graph of its limit is indicated in Figure 1.2.1; it is the graph of Weierstrass function and is a fractal.

Just prior to the beginning of the twentieth century, mathematics was not in today's logical form. Weierstrass's discovery had caused a crisis, for the phenomenon was beyond everyone's intuitive comprehension. It was one of the critical motivations that led many scientists to formalize mathematics under a logical axiomatic system.

1.2.2 THE CANTOR SETS

The *Cantor set* was created in 1877 as a counterexample in measure theory. It was actually constructed as a fundamental byproduct of an investigation in Cantor's studies into the convergence of Fourier series (cf. [Z]).

Given the two functions

$$f_1(x) = \frac{x}{3}, \quad f_2(x) = \frac{x+2}{3} \tag{1.2.5}$$

defined on the unit interval [0, 1], we construct the sets

$$C_0 = [0, 1], \text{ and } C_{k+1} = f_1(C_k) \cup f_2(C_k) \text{ for } k = 0, 1, \ldots.$$

See Figure 1.2.2. The Cantor set C is the limit of the sequence $\{C_k\}$. Equivalently, $C = \bigcap_k C_k$, since $C_{k+1} \subset C_k$ for all k. It was declared as a nonmeasurable set because there is no measure that gives a value on the Cantor set and a value on its complement that add to exactly 1. And as is known, the measure of the whole interval [0, 1] is always 1.

Today, the Cantor set is known as a fractal geometric object of a dimension between 0 and 1, a fractional number: $\dim C = \frac{\ln 2}{\ln 3} \approx 0.63$.

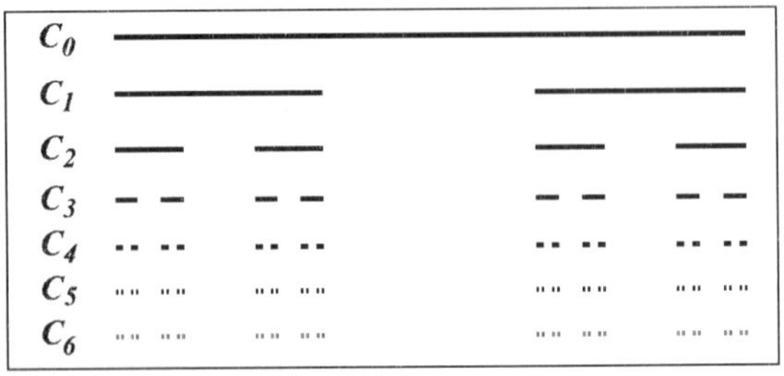

FIGURE 1.2.2 Cantor sets

1.2.3 THE HILBERT CURVE

Algebraically it is very easy to prove that *there are as many points on a line as in a plane*, however intuitively this is difficult to accept. Given a unit segment [0, 1] and a unit square [0, 1]×[0, 1], each point in them can be written in decimals as $0.d_1d_2d_3\ldots$ or $(0.a_1a_2a_3\ldots, 0.b_1b_2b_3\ldots)$, respectively. A bijection can be defined as follows:

$$\begin{array}{ccc} [0,\ 1] & \longleftrightarrow & [0,\ 1]\times[0,\ 1] \\ 0.d_1d_2d_3d_4d_5\cdots & \longrightarrow & (0.d_1d_3d_5\cdots, 0.d_2d_4d_6\cdots) \\ 0.a_1b_1a_2b_2a_3\cdots & \longleftarrow & (0.a_1a_2a_3\cdots, 0.b_1b_2b_3\cdots). \end{array}$$

To those who believe in only what they see, the proof seems unacceptable because it cannot be visualized. A geometric proof was needed. In 1890, the first geometric proof was given by Peano. Later, Hilbert introduced an easier, iterative construction. Hilbert filled up the unit square with a segment, as illustrated below, that now is called the *Hilbert space filling curve*. Considering the unit square [0, 1]×[0, 1], the following four transformations are chosen:

$$w_0\begin{pmatrix}x\\y\end{pmatrix}=\begin{pmatrix}0 & 0.5\\-0.5 & 0\end{pmatrix}\begin{pmatrix}x\\y\end{pmatrix}+\begin{pmatrix}0\\0.5\end{pmatrix},\quad w_1\begin{pmatrix}x\\y\end{pmatrix}=\begin{pmatrix}-0.5 & 0\\0 & 0.5\end{pmatrix}\begin{pmatrix}x\\y\end{pmatrix}+\begin{pmatrix}0.5\\0.5\end{pmatrix},$$

$$w_2\begin{pmatrix}x\\y\end{pmatrix}=\begin{pmatrix}0.5 & 0\\0 & 0.5\end{pmatrix}\begin{pmatrix}x\\y\end{pmatrix}+\begin{pmatrix}0.5\\0.5\end{pmatrix},\quad w_3\begin{pmatrix}x\\y\end{pmatrix}=\begin{pmatrix}0 & -0.5\\-0.5 & 0\end{pmatrix}\begin{pmatrix}x\\y\end{pmatrix}+\begin{pmatrix}1\\0.5\end{pmatrix}.$$

The transformation w_0 maps the whole square to the bottom left corner with a 90° rotation; w_1 maps the whole square to the top left corner with a vertical flip; w_2 shrinks the whole square to the top right corner; and w_3 maps the whole square to the bottom right corner with a diagonal flip.

Starting from the black segment in Figure 1.2.3*a* and applying the above four transformations, we obtain the black curve in Figure 1.2.3*b* except for a tiny portion by the end of the curve. Repeating the process, the curves obtained are as shown in Figures 1.2.3*c*, 1.2.3*d*, and 1.2.3*e*. The limit of these curves is called the *Hilbert curve H*. Actually, H is the attractor of the IFS formed by the four transformations w_0, w_1, w_2, and w_3.

It is interesting to see that all the fractals we have mentioned, created in various periods of time and in different areas of mathematics, played the same role—the role of altering conventional thinking—and actually providing the same key—the key to open a door to unfamiliar possibilities. It is not accidental that fractals will present once again more insights into our universe from a new angle with a new excitement.

By analyzing fractal objects, the following characteristics are identified and stated in Fisher's book [F]: *fractals have detail at every scale, fractals are self-similar,*

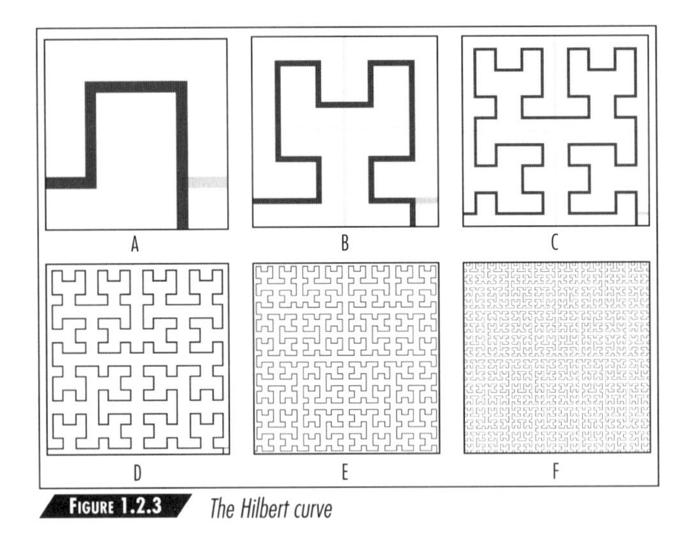

FIGURE 1.2.3 *The Hilbert curve*

fractals have fractional dimensions, and fractals can be described by a mathematical algorithm. Summarizing all the above, we see the key word: *similarity.* A fractal is self-similar because it is described by an algorithm. It can be generated by an algorithm because of the self-similarity on every scale. In essence, *a fractal is a mathematical model of the self-similar nature of the real world.*

1.2.4 FROM MANDELBROT TO BARNSLEY

Benoit Mandelbrot's observations in *The Fractal Geometry of Nature* [M] marked a new era of mathematics: *fractal geometry.* Today's computers not only enable us to view fractals as amazing mathematical objects but also furnish new ways to model the real world. Figure 1.2.4, a picture from Peitgen and Saupe's *The Science of Fractal Images* [PS], shows such an example of "Disney World–like" landscapes created by using fractal techniques. After finding fractals in mountains, clouds, and islands, Michael Barnsley started to see fractals everywhere (Figure 1.2.5). As he wrote in the very beginning of his work *Fractals Everywhere* [B]:

> *Fractal Geometry will make you see everything differently.... You risk the loss of your childhood vision of clouds, forests, galaxies, leaves, feathers, flowers, rocks, mountains, torrents of water, carpets, bricks, and much else besides. Never again will your interpretation of these things be quite the same.*

Why are we so excited about fractals? The answer comes from the heart of mathematics. What is mathematics? Mathematics is the attempt to model the universe with abstract simple rules: the elliptic Earth, the pyramidal tomb, the mythical golden section, the legendary π of a circle, the fascinating hexagon of a beehive, and the stunning icosahedron of a flu virus. The list goes on and on. Each of these findings had its own glorious moment in human civilization. But

Figure 1.2.4 *The Mandelbrot set in a landscape* (Courtesy of H. Peitgen and D. Saupe [PS])

Figure 1.2.5 *Fractal imitation of real world images. Left:* Sunflower; *Right:* Wolf (Courtesy of M. Barnsley [B])

we never thought of how simply mountains and clouds could be illustrated, or how clearly a green fern and a red leaf could be characterized, until we saw fractals. It was a great feeling of discovery, a deep appreciation of a miracle, and a wonderful experience of "ευρεκα!"

Having seen fractal imitations of the real world, Barnsley and his colleagues challenged themselves with the question, Can we give a fractal description to every real-world image? Jacobi once said, "*Man muss immer umkehren*" (one must always invert). By identifying the fractal self-similarities of images their efforts led to new scientific territory: *fractal image compression and representation.*

Two theories were developed in parallel in late 1980s: *the recurrent iterated function systems* (RIFS) and *the partitioned iterated function systems* (PIFS), by Barnsley, Sloan, Elton, Hardin, Jacquin, and others ([BEH], [BS], [BJ], [J3]).

Using RIFS theory, an interactive system named VRIFS was built on SUN workstations. Nicknamed a "graduate student system," VRIFS requires a huge labor-intensive effort to produce any photorealistic fractal representations. However, it proved the scientific concept and provided us some amazing footage.

The PIFS theory has been developed as the main method of fractal image compression. In 1988 Barnsley and Sloan founded Iterated Systems, a company devoting itself fully to fractal imaging, in particular, *fractal image compression*, based on their patent [BS2]. The major part of this book will record the progress Iterated Systems has made. We will show step by step how to write a basic, as well as an advanced, fractal image compression and decompression system, including how to capture, represent, and display a real-world image in the fractal format.

The image on the right is the original 256×256 8-bit grayscale digital image. The image on the left is the fractal representation created using Iterated Systems's product *Fractal Imager 1.0* in a 10:1 compression ratio.

FIGURE 1.2.6 Lena. *Left: compressed 6416 bytes; Right: original 65,540 bytes* (Reproduced by special permission of Playboy Magazine © 1972 Playboy)

From this book, you will recognize that the fractal model is quite different from other types of engineering models. Instead of modeling the unknown by smoothing, nullifying, and rectifying, we will model the unknown by magnifying, duplicating, and transforming from what we know. In other words, rather than cover missing data by stretching its surrounding data, clone that data from its neighborhood matching data. Viewing the classical engineering models as zero-order approaches, we will see that fractal modeling is a first-order approach.

1.3 FRACTALS AND SELF-SIMILARITY IN IMAGES

Stare at a real-world image. When we focus on a single spot, localization becomes an important process. Texture, color, brightness, and distance are among the first observations. It is widely believed that our visual cortical receptive fields sharpen at the focal point and decay quickly around it. Appropriate descriptions of this process have been modeled as *windowed eye-filter functions*. Our eyes are directly coordinated with our brains—both play crucial roles when processing an image. However, in comparison, the brain's role is far more important.

When a picture is viewed, the first thing we do is to *identify* the objects within the picture. The identification of objects is a process of segmenting the image and comparing these segments with similar objects in our memory. We can consider this process as finding similarities in the time dimension. We expand our view by identifying things and places spatially: for example, five people, two dogs, and three trees. After a while subtle details that were not noticed initially can be seen. All these processes can be classified as the *tempo-spatial process of searching for similarities*. Further discussion on human visual science will be presented in Chapter 13, where a new image metric is proposed.

Among all of the methods that describe an image, those that use global tempo-spatial similarities are by far the most effective. For example, by cutting the choice of color intensity values in half, i.e., from 256 colors to 128 colors, we save only 1 bit per value. However, if an image has a reflection, data can be conserved by half. Figures 1.3.1 and 1.3.2 show how much spatial redundancies and self-similarities there are in ordinary real-world images.

People generally communicate new experiences based upon past experiences that are commonly shared. The human language is made up entirely of codes of such shared experiences and objects. The fractal mechanism in human vision is extremely effective and powerful, for it follows the same procedure of understanding a current theme by referring it to what is known. Naturally, the next question is, Can we model real-world images with this fractal mechanism so that high image compression can be achieved?

FIGURE 1.3.1 *Spatial symmetry in an image*

FIGURE 1.3.2 *Self-similarity in a photo*

It is a great challenge for us to attempt to imitate the human brain. A good example is the human visual measurement system. Take a look at Figure 1.3.3. Notice that b appears to be far more similar to a than to c. Since a computer has to have a reference in its recognition and matching process, it would most likely find b more similar to c, because a differs from b and c in its explicit representation of a triangle, while the human eye notices the triangle implicit in b immediately.

In summary, fractal image compression and representation is based on a strong belief—a belief that there are tremendous similarities and redundancies

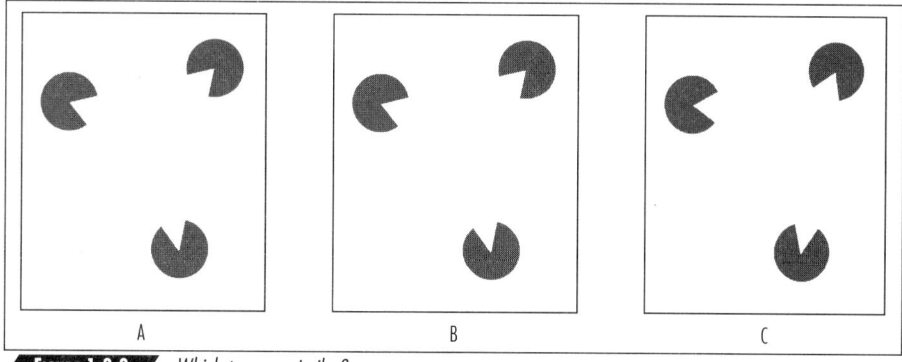

FIGURE 1.3.3 *Which two are similar?*

in most real-world images. Guided by this belief, research and development follows a sound strategy and a clear direction: *to extract image similarity information as much as possible in the tempo-spatial space and to characterize this information as efficiently as possible in a classified and unified way.*

This book will fully focus on this strategy and direction by asking, exploring, and modeling the following questions:

- *Are there good metrics to measure and recognize similarities?*
- *What are the effective methods for capturing and matching similarities?*
- *How should similarities be classified and indexed?*
- *What are the right forms for similarities to be described and represented?*
- *Are there better spaces for extracting and illustrating similarities?*
- *What are the interrelationships among similarities?*

So all their praises are but prophecies
Of this our time, all you prefiguring;
And, for they look'd but with divining eyes,
They had not skill enough your worth to sing:
For we, which now behold these present days,
Have eyes to wonder, but lack tongues to praise.

—When in the Chronicle of Wasted Time

William Shakespeare

MATHEMATICAL FOUNDATION 2

Before beginning this exciting journey, a few fundamentals need to be discussed regarding the mathematical background and history of *iterated function system (IFS) theory.* Barnsley's book *Fractals Everywhere* [B] provides a complete reference on this topic. This chapter will review some of the most interesting results of IFS theory. In addition, a new image model will be introduced. The material covered here is intended to provide a minimal and sufficient mathematical foundation for further reading.

2.1 METRIC SPACES

The spaces carrying fractals are mathematically called *metric spaces*. There are three types of metric spaces of interest that will be studied in this chapter: *the n-dimensional real vector space*, *the space of compact subsets in a real vector space*, and *the space of images*. These spaces will be used throughout the this book. A good reference for the more mathematical aspects of the metric spaces is Munkres's book *Topology—A First Course* [M6].

2.1.1 COMPLETE METRIC SPACE

Given a set X, a function

$$d : X \times X \longrightarrow [0, \infty) \tag{2.1.1}$$

is called a *metric*, or a *distance function*, if

1. (*identity*) $d(x, y) = 0$ if and only if $x = y$ for all $x, y \in X$,
2. (*symmetry*) $d(x, y) = d(y, x)$ for all $x, y \in X$,
3. (*triangle inequality*) $d(x, y) \leq d(x, z) + d(z, y)$ for all $x, y, z \in X$.

The pair (X, d) is called a *metric space*.

A sequence $\{x_i\}_{i=1,2,\cdots}$ in X *converges* to a point $x \in X$ in the metric space (X, d) if and only if the sequence $\{d(x_i, x)\}_{i=1,2,\cdots}$ of numbers converges to zero.

A sequence $\{x_i\}_{i=1,2,\cdots}$ in X is said to be *bounded* in the metric space (X, d) if there is a constant δ such that $d(x_i, x_j) \leq \delta$ for any integers i, j.

A sequence $\{x_i\}_{i=1,2,\cdots}$ in X is said to be a *Cauchy sequence* in the metric space (X, d) if for any positive number $\delta > 0$ there is an integer k such that $d(x_i, x_j) \leq \delta$ for any integers $i, j > k$. In other words,

$$d(x_i, x_j) \xrightarrow{i,j->\infty} 0. \tag{2.1.2}$$

A metric space (X, d) is said to be *complete* if any Cauchy sequence $\{x_i\}_{i=1,2,\cdots}$ in X converges to some point $x \in X$. Intuitively, it is saying, *no limit point in the space is missing.*

From now on, *all metric spaces used in our applications will be assumed to be complete.*

2.1.2 CONTRACTION MAPPING THEOREM

A map $f : X \longrightarrow X$ from the metric space (X, d) into itself is called a *transformation.* In general, in most applications, a transformation is expected to be *bijective*, that is, for any point x of X there is some unique point z of X to map into, $f(x) = z$, and there is also some unique point u of X to be mapped from, $f(u) = x$. Or equivalently, f is *invertible*, i.e., there is another transform $f^{-1} : X \longrightarrow X$, called the *inverse* of f, such that their composition is the identity:

$$f \circ f^{-1} = f^{-1} \circ f = \text{identity}. \tag{2.1.3}$$

A transformation $f : X \longrightarrow X$ on a metric space (X, d) is called *contractive* if there is a constant $0 \le s < 1$ such that

$$d(f(x), f(y)) \le s \cdot d(x, \ y) \tag{2.1.4}$$

for all $x, y \in X$. Any such number s is called a *contractivity factor* for the transformation f.

A point $a \in X$ is called a *fixed point* of the transformation f if $f(a) = a$.

THE CONTRACTION MAPPING THEOREM:

Let $f : X \longrightarrow X$ be a contractive transformation on a complete metric space (X, d). Then the transformation f possesses exactly one fixed point $a \in X$. Moreover, for any $x \in X$, the sequence

$$x, f(x), f^2(x) = f(f(x)), \cdots, f^k(x) = f(f^{k-1}(x)), \cdots \tag{2.1.5}$$

converges to the fixed point a, i.e.,

$$\lim_{k \to \infty} f^k(x) = a. \tag{2.1.6}$$

Proof: The proof of the theorem is straightforward. Let $x \in X$. Let $0 \le s < 1$ be the contractivity factor of the transformation f. Then for any $j \ge i \ge 0$,

$$
\begin{aligned}
d(f^i(x), f^j(x)) &\le s \cdot d(f^{i-1}(x), f^{j-1}(x)) \\
&\le s^i \cdot d(x, f^{j-i}(x)) \\
&\le s^i \cdot \sum_{k=1}^{j-i} d(f^{k-1}(x), f^k(x)) \\
&\le s^i \cdot \sum_{k=1}^{j-i} s^{k-1} \cdot d(x, f(x)) \\
&\le \frac{s^i}{1-s} d(x, f(x)),
\end{aligned}
\tag{2.1.7}
$$

from which it follows that the sequence $\{f^k(x)\}_{k=0,1,2,\cdots}$ is a Cauchy sequence. Since the space (X, d) is complete, there is a point $a \in X$ that is the limit of the sequence $\{f^k(x)\}_{k=0,1,2,\cdots}$.

Because

$$
\begin{aligned}
d(a, f(a)) &\le d(a, f^k(x)) + d(f^k(x), f(a)) \\
&< d(a, f^k(x)) + s \cdot d(a, f^{k-1}(x)) \xrightarrow{k \to \infty} 0,
\end{aligned}
\tag{2.1.8}
$$

it follows that $d(a, f(a)) = 0$, i.e., $f(a) = a$.

Finally, the uniqueness of the fixed point needs to be proven. Suppose there is another fixed point $b \in X$, $f(b) = b$. Then

$$
d(a, b) = d(f(a), f(b)) \le s \cdot d(a, b), \tag{2.1.9}
$$

holds only when $d(a,b) = 0$, i.e., $a = b$. This concludes the proof. ♦

The contractive mapping theorem is the key theorem in fractal imaging. In this and the following chapters it will evolve into theorems that assure the existence of fractals in our imaging model.

2.1.3 METRICS IN THE n-DIMENSIONAL VECTOR SPACE $\Re^n$

In the n-dimensional vector space $\Re^n$ of real numbers (n will be 2 or 3 in most of our applications), the most frequently used metrics are *L^p-metrics*, which are defined by the formula

$$
d(x, y) = \left(\sum_{i=1}^{n} |x_i - y_i|^p \right)^{\frac{1}{p}}, \tag{2.1.10}
$$

for all $x = (x_1, x_2, \cdots, x_n)$, $y = (y_1, y_2, \cdots, y_n) \in \Re^n$ and $p \ge 1$. In the limiting case, when p tends to infinity, the metric becomes the *supremum metric*, i.e., the *L^∞-metric*, which is given by the formula

$$d(x, y) = \sup_{1 \le i \le n} |x_i - y_i| . \tag{2.1.11}$$

The case $p = 2$ is the most frequently used metric. It is also called the *Euclidean metric* or the *root mean square error metric (rmse)*.

A metric d is said to be *translative* if the following extra property holds:

4. (*translativity*) $d(x+z, y+z) = d(x, y)$ for all $x, y, z \in \Re^n$,

i.e., the metric is invariant under the translation operator. In consequence, all L^p-metrics, $p \ge 1$, are *translative*.

For any translative metric $d : \Re^n \times \Re^n \longrightarrow [0, \infty)$, a function, called the *norm* of the metric d, $\| \cdot \| : \Re^n \longrightarrow [0, \infty)$, can be defined by

$$\| x \| = d(x, O) \tag{2.1.12}$$

for all $x \in \Re^n$. Reciprocally, any function $\| \cdot \| : \Re^n \longrightarrow [0, \infty)$ is said to be a *norm* in the real vector space $\Re^n$ if its associate function $d : \Re^n \times \Re^n \longrightarrow [0, \infty)$ defined by

$$d(x, y) = \| x - y \| \tag{2.1.13}$$

for all $x, y \in \Re^n$ is a metric.

The norm of an L^p-metric, $p \ge 1$, is called an L^p*-norm*, denoted by $\| \cdot \|_p$.

For a metric space $(\Re^n, d)$, *the unit sphere* is defined to be the set

$$\left\{ x \in \Re^n \mid d(x, O) = 1 \right\}, \tag{2.1.14}$$

where $O = (0, 0, \cdots, 0) \in \Re^n$ is the *origin* of the vector space.

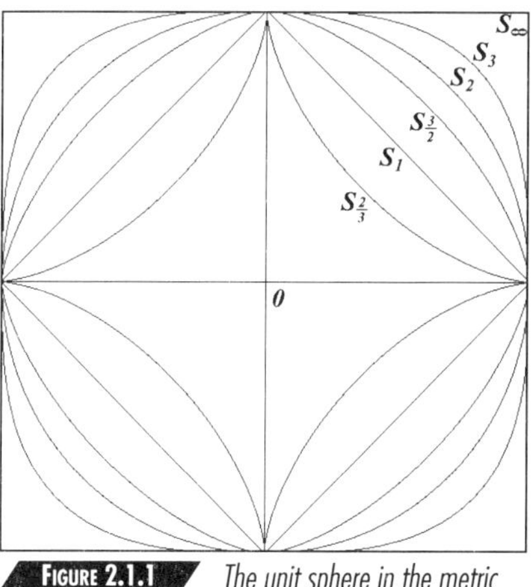

FIGURE 2.1.1 The unit sphere in the metric space $(\Re^2, L^p)$

For the case $n = 2$, the planar case, the unit sphere S_p of an L^p-metric, $p \geq 1$, is drawn in Figure 2.1.1. As p moves from ∞ to 1, the unit sphere S_p shrinks from a square ($p = \infty$), to a circle ($p = 2$), to a diamond shape ($p = 1$).

2.1.4 THE SPACE OF COMPACT SUBSETS IN $\Re^n$

Given a metric d on the n-dimensional space $\Re^n$, an associated metric, called the *Hausdorff metric* of d and using the same notation d, is defined in a space of subsets of $\Re^n$ by the following formula:

$$d(A, B) = \max\left\{ \sup_{x \in A}\left\{ \inf_{y \in B} d(x, y) \right\}, \sup_{y \in B}\left\{ \inf_{x \in A} d(x, y) \right\} \right\}, \quad (2.1.15)$$

for any A, B subsets of $\Re^n$.

The Hausdorff metric is the most natural metric in comparing objects in an ideal geometric space, like the fractal objects illustrated in Chapter 1. To make the above definition explicit, the definition of the metric is decomposed into several trivial steps.

Step 1: *The distance from a point* $x \in \Re^n$ *to a subset* $B \subset \Re^n$ is naturally defined as the distance between the point x and a point y in B that is one of the closest points in B to x:

$$d_x(B) = \inf_{y \in B} d(x, y). \quad (2.1.16)$$

As shown in Figure 2.1.2, $d_x(B) = 0$ when $x \in B$.

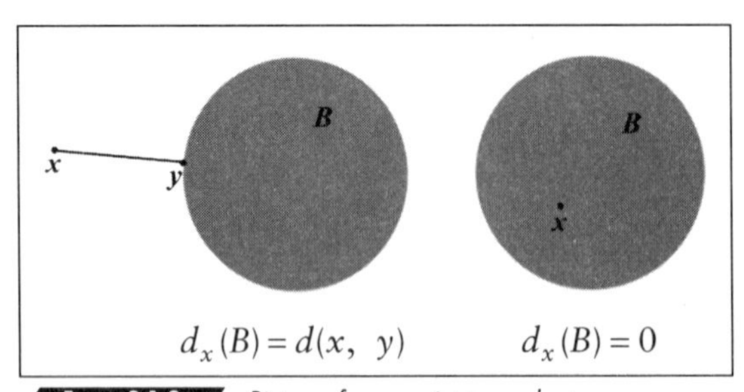

FIGURE 2.1.2 *Distance from a point to a subset*

Step 2: *The distance from a subset* $A \subset \Re^n$ *to a subset* $B \subset \Re^n$ is defined as the distance from one of the furthest points x in A to the subset B:

$$d_A(B) = \sup_{x \in A} d_x(B). \tag{2.1.17}$$

As shown in Figure 2.1.3, $d_A(B) = 0$ when $A \subset B$.

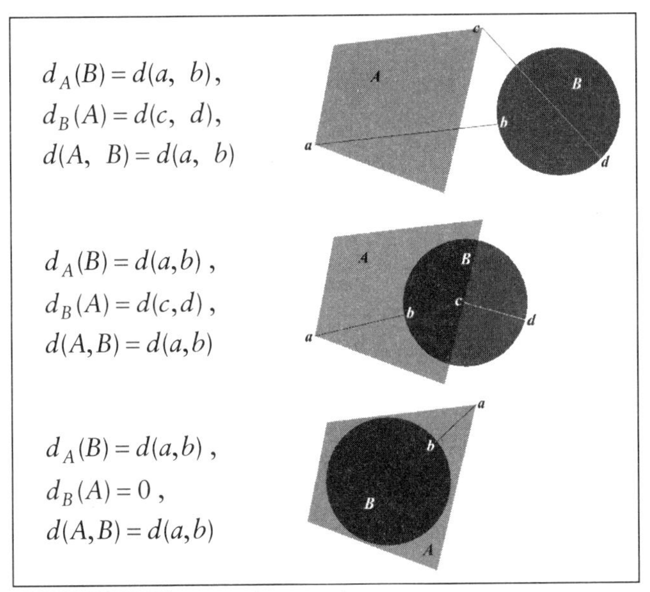

FIGURE 2.1.3 *Distance between subset*

Step 3: Extending the previous concept to a metric that satisfies our definition and from the identity and symmetry formulas, it is easy to define the *distance between a subset* $A \subset \Re^n$ *and a subset* $B \subset \Re^n$ to be

$$d(A,\ B) = \max\left\{ d_A(B),\ d_B(A) \right\}, \tag{2.1.18}$$

as shown in Figure 2.1.3.

Is the function d well-defined on the space of all subsets of $\Re^n$? Is it a metric? Unfortunately, the answer is "no" to both questions. However, the situation can be fixed by imposing some conditions on the subsets that are studied. Instead of the space of all subsets, only the *compact* subsets will be considered.

A subset A in $\Re^n$ is *compact* if and only if

1. It is *bounded,* i.e., there is a constant δ such that $d(x,\ y) \le \delta$ for any two points $x, y \in A$. It is equivalent to say that the set A has a finite *diameter:*

$$\text{diameter of } A \stackrel{\text{def}}{=\!=} \sup_{x,y\in A} d(x, y) < \infty; \tag{2.1.19}$$

2. It is *closed,* i.e., for any sequence $\{x_i\}_{i=1,2,\cdots}$ in A that converges in $\Re^n$ to some point $x \in \Re^n$, the limit point x is in the set A as well.

The *boundedness* condition insures that the distance function is well-defined, because no supremum could go to infinity. The *closedness* condition guarantees that any two subsets of null distance are identical. The space of all compact subsets of $\Re^n$ is denoted by $\mathcal{H}(\Re^n)$.

As an exercise, one can prove that *the Hausdorff distance defined on the space of compact subsets $\mathcal{H}(\Re^n)$ is a metric.*

2.2 IMAGE MODELS

A black and white image can be viewed as a compact subset in the plane $\Re^2$. Yet grayscale and color images are more interesting. Color images are typical extensions of the grayscale representation of images, since a color image can be viewed as several grayscale images, e.g., as a decomposition of *red, green,* and *blue* channels (cf. Chapter 11). Therefore, only grayscale images will be discussed except where color decomposition is an issue.

In most engineering applications an image is usually modeled as a piecewise continuous function on a rectangular support. Barnsley saw a problem in that model. When looking at a specific section of the night sky using a telescope, instead of seeing bigger stars, one should actually see more stars. What does this tell us? The implication is that there is something incomplete in the function model, since zooming inward at any differentiable point, the function is increasingly flat.

To solve this dilemma, Barnsley gave a new definition of image as a measure on a rectangular support ([BH]). Here, this concept is extended further by saying that mathematically *a real-world image is a distribution.* This model not only explains the night sky example nicely, but it also fits much better the physical world of optics, vision, and imaging as understood in physics.

Indeed, not only is the function space a subset of the distribution space, but it is also dense. Therefore, the distribution models of images can always be approximated by the function models. Furthermore, as will be explained, the *digital pixelation* (i.e., analog image signal numerical sampling) is exactly an image distribution acted on by a specifically chosen set of testing functions.

2.2.1 DISTRIBUTIONS

Quite often, physicists and engineers avoid the notation of distributions by forcing themselves to use some singular indefinable functions to describe their theories. These singular functions appear often when some energy sources, such as light and heat sources, are involved. For instance, in our previous discussion of the night sky example, the *Dirac function* δ, which will be used as the subsampling filter (Figure 2.3.3*a*), can be also used to model a star. This function can be defined verbally as follows:

> *It is equal to 0 everywhere except at one point, e.g., $x = 0$, where it is equal to infinity and possesses a total mass unit, i.e., integration at this point, or over any neighborhood containing this point, is equal to 1.*

In fact, for any function φ, the integral

$$< \delta, \varphi > \overset{\text{define}}{=} \int_{+\infty}^{-\infty} \delta(x)\varphi(x)\,dx = \varphi(0). \qquad (2.2.1)$$

The Dirac function δ is one of those "singular functions" that have meaning only when there is another function that interacts with it. In mathematics, this is not a *function*, it is a *functional*, for it does not have values itself, but rather it generates testing values only when it is tested by other functions.

Keeping the discussion in the real vector space $\Re^n$, what kind of testing function set will enable us to test a functional completely? It is necessary and sufficient to have a set of functions that is able to test the functional at any small neighborhood of any given point, for example, $\{\varphi_{\rho,\xi}\}_{\rho>0,\xi\in\Re^n}$ the set of functions, where $\varphi_{\rho,\xi}$, is defined by

$$\varphi_{\rho,\xi}(x) = \varphi(\rho(x-\xi)), \text{ where } \varphi(x) = \begin{cases} e^{-\frac{1}{1-|x|^2}} & \text{for } |x| < 1; \\ 0 & \text{otherwise,} \end{cases} \qquad (2.2.2)$$

for all $\rho > 0$ and for all ξ and $x \in \Re^n$. Figure 2.2.1 shows the function $\varphi_{0,0}(x) = \varphi(x)$. Certainly, any function set that contains this set forms a testing set.

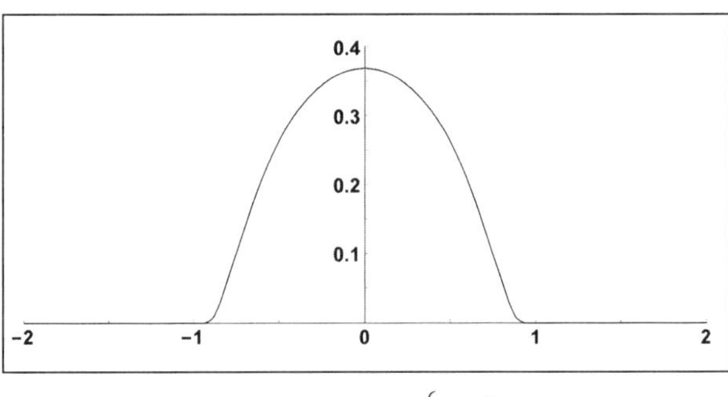

FIGURE 2.2.1 The function $\varphi(x) = \begin{cases} e^{-\frac{1}{1-|x|^2}} & \text{for } |x| < 1; \\ 0 & \text{otherwise.} \end{cases}$

Given a function φ, the support of φ is defined to be the set where φ is not zero: $\operatorname{supp}\varphi = \left\{ x \in \Re^n \,\middle|\, \varphi(x) \neq 0 \right\}$. The function φ is said to have *compact support* if the set supp φ is compact.

In general, the set $\mathcal{K} = \mathcal{K}(\Re^n)$ of all differentiable real functions with compact support is used as the set of testing functions. These functions are also often called *window functions*, and their supports are also called *windows*.

A *distribution* f is defined to be a finite linear continuous functional, that is, a map $f : \mathcal{K} = \mathcal{K}(\Re^n) \longrightarrow \Re$ satisfying the following conditions:

Linearity: $f(\alpha \cdot \varphi + \beta \cdot \psi) = \alpha \cdot f(\varphi) + \beta \cdot f(\psi)$ for any α, $\beta \in \Re$ and φ, $\psi \in \mathcal{K}$.

Continuity: for any testing function sequence $\{\varphi_n\}_n \subset \mathcal{K}$ that has limit zero, i.e., $\varphi_n \xrightarrow{n\to\infty} 0$, the sequence of numbers $\{f(\varphi_n)\}_n$ goes to zero: $f(\varphi_n) \xrightarrow{n\to\infty} 0$.

Conventionally, the following equivalent notations are more commonly used:

$$f(\varphi) = (f, \varphi) = \int_{\Re^n} f(x)\,\varphi(x)\,dx\,. \tag{2.2.3}$$

The space of distributions will be denoted by $\mathcal{D} = \mathcal{D}(\Re^n)$.

For any subset $\Delta \subset \Re^n$, a distribution f is said to have *support* in Δ if

$$(f, \varphi) = 0 \text{ for any } \varphi \in \mathcal{K} \text{ with } (\operatorname{supp}\varphi) \cap \Delta = \varnothing\,. \tag{2.2.4}$$

The space of distributions having support Δ is denoted by $\mathcal{D}(\Delta)$.

Let $\mathcal{F} = \mathcal{F}(\Re^n)$ denote the local integral functions in $\Re^n$, and let $\mathcal{F}(\Delta)$ denote the local integral functions whose supports are in Δ, for some set $\Delta \subset \Re^n$. The inclusion map $\mathcal{F}(\Re^n) \subset \mathcal{D}(\Re^n)$ is exactly given by the notation, i.e.,

$$f(\varphi) = (f, \varphi) = \int_{\Re^n} f(x)\,\varphi(x)\,dx \text{ for all } f \in \mathcal{F}\,. \tag{2.2.5}$$

Exercise: Prove that $\mathcal{F}(\Delta) \subset \mathcal{D}(\Delta)$ for any set $\Delta \subset \Re^n$.

2.2.2 IMAGE AS PLANAR DISTRIBUTION

An image is a distribution on a rectangular support. For a planar rectangle $\Delta = [\,0, W\,) \times [\,0, H\,) \subset \Re^2$, an *image* is an element of $\mathcal{D}(\Delta)$. In this model, the philosophic difference between an image "out there" and a picture people subjectively see is inherently given. When people see an image, a collection of testing functions is involved. The tested values are transmitted to our brain, then shuffled, reorganized, prioritized, and analyzed in our brain to create a subjective picture.

For example, it has been verified that the human eye measures the light flux through an area, not at a point, with a weight filter that looks like a windowed Gaussian, also known as the Gabor transform, similar to the testing function drawn in Figure 2.2.1.

When a scanner or digital camera captures an image, a built-in finite set of testing functions are activated. Each testing function produces a value, a so-called *picture element*, or *pixel*.

Distribution theory is one of the most beautiful pieces of classical mathematics. To know more about it you may decide to take a *Functional Analysis* course or read one of the many books on this topic, such as Rudin's *Functional Analysis* [R] or Gelfand and Shilov's *Generalized Functions* ([GS]). One step further, a good *differential equations* course or book will show you how the concept of distributions as solutions to heat and wave equations fits in our physical reality. This topic will be revisited in Chapter 9, where the Fourier transform will be discussed.

2.2.3 PICTURE AS PERCEIVED IMAGE

In fact, the words *image* and *picture* do not have quite the same meaning. An image is a distribution as discussed in the previous paragraph, while a picture can be defined as a perceived version of an image by a specific receiver, e.g., a human being.

In fact, to us, a monochrome picture P of some image f seems to be a function

$$P : \Delta = [\,0, W\,] \times [\,0, H\,] \longrightarrow [\,0, 1\,], \tag{2.2.6}$$

where 0 is the darkest black and 1 is the brightest white people can sense. However, this is not another model, but a result of the model previously presented.

In reality, an object is seen by light going through human eyes. An *eye filter* is inevitably presented in this process. The eye filter is not a part of the image, but a part of our human vision system, which varies from human to animal, even from human to human, and changes with different distances between the eye and the image and with different eye conditions at different viewing times. This eye filter can conceivably be modeled as a local testing function

$$\omega : (-\varepsilon, \varepsilon\,) \times (-\varepsilon, \varepsilon\,) \longrightarrow [\,0, 1\,] \tag{2.2.7}$$

centered at the focal point to generate a stimulation signal. Thus, the value of the picture P at the point (x, y), for any $(x, y) \in \Delta$, has the formula

$$P(x, y) = \iint_{\Delta} f(\xi, \eta)\omega(\xi - x, \eta - y)d(\xi, \eta), \tag{2.2.8}$$

which will be called the *convolution* of the image f and the eye filter ω. It is equivalent to say that the image f is tested by the set of testing functions $\{\omega_{(\xi,\eta)}\}_{(\xi,\eta)\in\Delta}$, where $\omega_{(\xi,\eta)}(x, y) = \omega(x-\xi, y-\eta)$ for all (ξ, η) and $(x, y) \in \Delta$. Thus the picture P is the collection of testing values.

In practice, nothing about the image is known. Only a few characteristic properties can be known about the filter; the picture is mainly what people receive. Therefore, people often have to live in the world they can see, even though what they see is a very limited illusion. To most people, there is no difference between the illusive picture that they see and the image of reality that they can comprehend. It is within this ambiguity that this book stays most of the time.

In fact, any picture can be viewed as an image because of the following inclusion:

$$\mathcal{F}(\Delta) \subset \mathcal{D}(\Delta). \tag{2.2.9}$$

So, a picture to us will be as the same as an image. When it is described as a function, it can be displayed with a special coordinate system. As shown in Figure 2.2.2, its rectangular support is placed flat on the page, and its function value is displayed in different brightness intensities.

FIGURE 2.2.2 Resolution-independent system

This coordinate system is called an *image resolution-independent coordinate system* and is used in fractal image representations. The brightness of the image at a point (x, y) is denoted by $P(x, y)$. The graph of the image,

$$\text{Graph}\,(P)=\left\{(x,y,P(x,y))\in\Re^3 \;\middle|\; (x,y)\in\Re^2\right\}, \tag{2.2.10}$$

is a subset of the 3-dimensional vector space, though images could be viewed as 3-dimensional sets and the Hausdorff metric used as the metric. But in practice this metric is not trivial to calculate. Instead of it, the *L^p-metric*, for any $p \geq 1$, is widely accepted, especially when $p = 1$ or 2. Its definition is similar to those defined for vectors in $\Re^n$. For any images P and Q,

$$d(P,Q)=\|P-Q\|_p=\left(\int_\Delta |P(x,y)-Q(x,y)|^p\,dx\,dy\right)^{1/p}. \tag{2.2.11}$$

In this case, the image space will be chosen as the *space of L^p-integrable functions*:

$$\mathcal{P}(\Delta)=\mathbf{L}^p(\Delta)\subset\mathcal{F}(\Delta)\subset\mathcal{D}(\Delta). \tag{2.2.12}$$

As p goes to infinity, the limiting metric is

$$d(P,Q)=\|P-Q\|_\infty=\sup_{(x,y)\in\Delta}|P(x,y)-Q(x,y)|. \tag{2.2.13}$$

When $p = 2$, an *inner product* can be defined by

$$\langle P,Q\rangle=\int_\Delta P(x,y)\cdot Q(x,y)\,dx\,dy, \tag{2.2.14}$$

which determines the norm on $\mathbf{L}^2(\Delta)$. This inner product can be extended to the distribution space.

In the case $p = 2$, the *L^2-metric* is also called the *root of mean square error (rmse) metric.*

The *peak signal-to-noise ratio* (*PSNR*), used to measure the difference between two images, is defined as

$$PSNR=20\cdot\log_{10}\left(\frac{b}{\|P-Q\|_2}\right), \tag{2.2.15}$$

where b is the largest possible value of the signal, which is 1 here, and 255 in our typical digital case.

2.2.4 IMAGE IN DISCRETE PIXEL VALUES

In computer imaging, images are captured by using digital cameras or scanners. These images are then stored as finite sets of equally distributed sample pixels. In Figure 2.2.3, A picture P of the same image f is described in a sample grid of w columns and h rows, which is a *pixel representation* of the image f in an *image resolution-dependent coordinate system,*

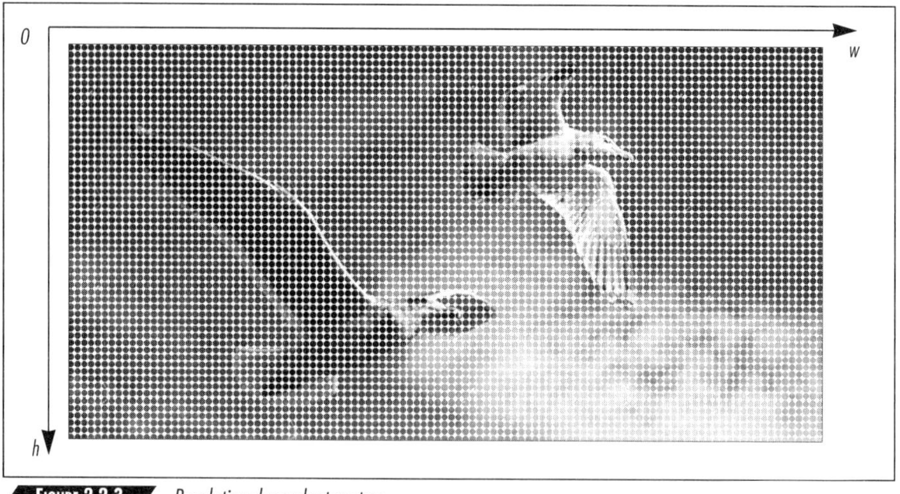

FIGURE 2.2.3 Resolution-dependent system

$$P:\{1,2,\cdots,w\}\times\{1,2,\cdots,h\} \longrightarrow \{0,\ 1,\cdots,\ 255\}^C, \qquad (2.2.16)$$

designated by $P=\{P[i,j]\}_{1\le i\le w,1\le j\le h}$, $w\times h$ integers, where 0 is the darkest black, 255 is the brightest white, and C is the number of color channels.

As mentioned earlier, the digitization and pixelation at each image location, that is, at (i, j), on each channel $0\le c<C$, is really the image distribution f applied to a testing function, which in this case is also called a *sampling filter* $\phi^c_{(i,j)}$, that is,

$$P^c[i,j] \quad = \quad \iint_{\Re^2} f(x,y)\,\phi^c_{(i,j)}(x,y)\,dx\,dy\,. \qquad (2.2.17)$$

The collection of the sample values $P[i,j]=\{P^c[i,j]\}$ at the single location (i, j) is called a *pixel*. Clearly, the sampling filter set $\phi_{(i,j)}=\{\phi^c_{(i,j)}\}$ is completely characterized by the capture device. For simplicity, the monochrome case $C = 1$ will be used most of the time.

It is natural to model the sampling filter as pixel location independent, because it is characterized by an image capturing device or eyes, not by images. Thus, writing $\phi=\phi_{(0,0)}$,

$$\phi_{(i,j)}(x,y)=\phi(x-i,y-j), \text{ for all } i,j,x,\ y\in\Re^2. \qquad (2.2.18)$$

As a result, the digitization formula becomes exactly the convolution between the image distribution f and the sampling function ϕ.

Often, the sampling function is chosen to be a symmetric function, i.e.,

$$\phi(x,y)=\phi(-x,y)=\phi(x,-y)=\phi(-x,-y), \text{ for all } (x,y)\in\Re^2, \qquad (2.2.19)$$

since a picture should not change even it it has been turned upside down.

To fit the convoluted values into the given range, a normalization factor can be assigned to yield

$$\iint_{\Re^2} \phi(x,y)\, dx\, dy = 1 \ (\text{or } 255). \tag{2.2.20}$$

In conclusion, the *sampling filter* characterizes completely the digitization of a camera or scanner.

In the extreme, constant, case, the *mean function* is given by

$$\phi(x,y) = \begin{cases} \dfrac{1}{\varepsilon^2} & \text{for all } -\dfrac{\varepsilon}{2} < x, y < \dfrac{\varepsilon}{2}; \\ 0 & \text{otherwise.} \end{cases} \tag{2.2.21}$$

This filter is called a *square averaging filter* of width ε. The limit of this averaging filter as the width ε tends to be zero is the *subsampling filter*, i.e., the *Dirac function*. One may choose a more complicated, and perceptually more accurate, filter, such as a windowed *Gaussian*: $\phi(x, y) = \alpha e^{-\beta(x^2+y^2)}$, for some constants α and β. All of these are illustrated in Figure 2.2.4.

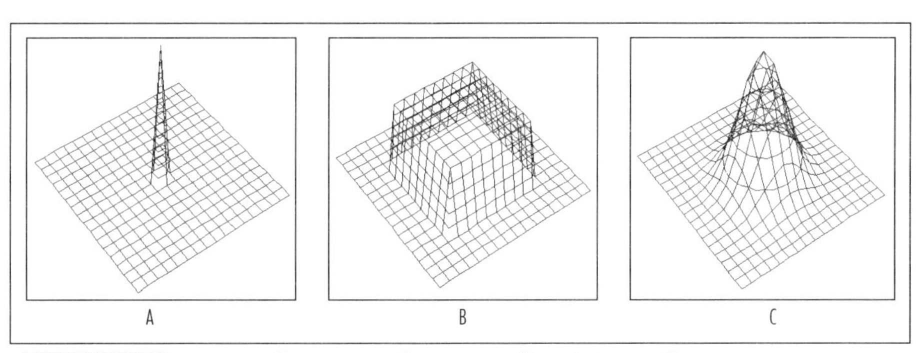

FIGURE 2.2.4 *Digitization filters: (A) Dirac filter; (B) Mean filter; (C) Gaussian filter*

In a resolution-dependent pixel system, the *L^p-metric*, for any $p \geq 1$, has the formula

$$d(P,Q) = \|P - Q\|_p = \left(\sum_{j=1}^{h} \sum_{i=1}^{w} |P[i,j] - Q[i,j]|^p \right)^{1/p}, \tag{2.2.22}$$

for any images P and Q.

Again, the *peak signal-to-noise ratio* (*PSNR*), used to measure the difference between two images, is defined as

$$PSNR = 20 \cdot \log_{10}\left(\frac{255}{\| P - Q \|_2}\right). \tag{2.2.23}$$

From now on, only the averaging and the subsampling down-sampling cases will be studied unless otherwise mentioned.

When a digital picture P is given, the corresponding default image f in the distribution space will be considered as an image distribution

$$f = \sum_{j=1}^{h} \sum_{i=1}^{w} P[i,j] \cdot \delta_{(i,j)}. \tag{2.2.24}$$

Now the image model is complete. From this point forward there is no need to distinguish between a picture and its image. Viewing an image as a distribution will be implicitly assumed throughout the whole book.

Interestingly enough, independent from image modeling, Forte and Vrscay have found inherent advantages in using distribution theory to study fractal object [FV], as demonstrated in their latest research work in inverse problems of IFS theory.

2.3 AFFINE TRANSFORMATIONS

In mathematics, a *transformation* f in the n-dimensional real space $\Re^n$ is just a function from $\Re^n$ to itself. Generically, a transformation is expected to be bijective.

The most interesting and useful transformations of real space $\Re^n$ are the *affine transformations* studied in *linear algebra* and the *analytic transformations* studied in *complex analysis*.

This book will focus only on affine transformations, and most of the work presented here can be generalized as analytic transformations. Some of the study in that direction has been presented in Barnsley's book [B].

An *affine transformation* f: $\Re^n \longrightarrow \Re^n$ is a transformation that can be written as

$$f(x) = Ax + b = \begin{pmatrix} a_{11} & a_{12} & \cdots & a_{1n} \\ a_{21} & a_{22} & \cdots & a_{2n} \\ \vdots & \vdots & \ddots & \vdots \\ a_{n1} & a_{n2} & \cdots & a_{nn} \end{pmatrix} \cdot \begin{pmatrix} x_1 \\ x_2 \\ \vdots \\ x_n \end{pmatrix} + \begin{pmatrix} b_1 \\ b_2 \\ \vdots \\ b_n \end{pmatrix}, \tag{2.3.1}$$

where $A = (a_{ij})_{i,j=1}^{n}$ is an $n \times n$ matrix in $\Re^{n \times n}$, called the *deformation matrix* of f, and $b = (b_j)_{j=1}^{n}$ is a vector in $\Re^n$, called the *translation vector* of f.

Given a norm $\|\cdot\|$ on the vector space $\Re^n$, the *norm* of an affine transformation *f*, or a deformation matrix *A*, is defined by the following formula:

$$\|f\| = \|A\| = \sup_{x \in \Re^n} \frac{\|A(x)\|}{\|x\|} = \sup_{x \in \Re^n, \|x\|=1} \|A(x)\|. \qquad (2.3.2)$$

An affine transform *f*, or a deformation matrix *A*, is said to be *contractive* if $\|A\| < 1$. Furthermore, it is said to be *contractive within the factor s*, for some $0 \le s \le 1$, if $\|A\| < s$.

Using the above definition, it is not difficult to deduce that for the L^p-metrics, $p=1, 2, \infty$, their corresponding norms for the deformation matrices have the following intuitive formulas:

$$\|A\|_1 = \max_{1 \le j \le n} \sum_{i=1}^{n} |a_{ij}|,$$

$$\|A\|_2 = \text{square root of the largest eigenvalue of } A^T A,$$

$$\|A\|_\infty = \max_{1 \le i \le n} \sum_{j=1}^{n} |a_{ij}|,$$

for any $n \times n$ matrix $A = (a_{ij})_{i,j=1}^{n}$ in $\Re^{n \times n}$.

In the 2-dimensional case, $n=2$, a deformation matrix can be decomposed into the following four steps: *scaling*, *stretching*, *skewing*, and *rotating*.

Figure 2.3.1 illustrates this decomposition.

1. *Scale:* $A_s = \begin{pmatrix} s & 0 \\ 0 & s \end{pmatrix}, \quad s \ge 0;$

2. *Stretch:* $A_t = \begin{pmatrix} 1 & 0 \\ 0 & t \end{pmatrix}, \quad A_t A_s = \begin{pmatrix} s & 0 \\ 0 & st \end{pmatrix};$

3. *Skew:* $A_u = \begin{pmatrix} 1 & u \\ 0 & 1 \end{pmatrix}, \quad A_u A_t A_s = \begin{pmatrix} s & stu \\ 0 & st \end{pmatrix};$

4. *Rotate:* $A_\theta = \begin{pmatrix} \cos\theta & -\sin\theta \\ \sin\theta & \cos\theta \end{pmatrix}, \quad 0 \le \theta < 2\pi;$

For convenience of discussion, a matrix $A = \begin{pmatrix} a_{11} & a_{12} \\ a_{21} & a_{22} \end{pmatrix}$ and its decomposition parameters $\{s, t, u, \theta\}$ are considered to be the same.

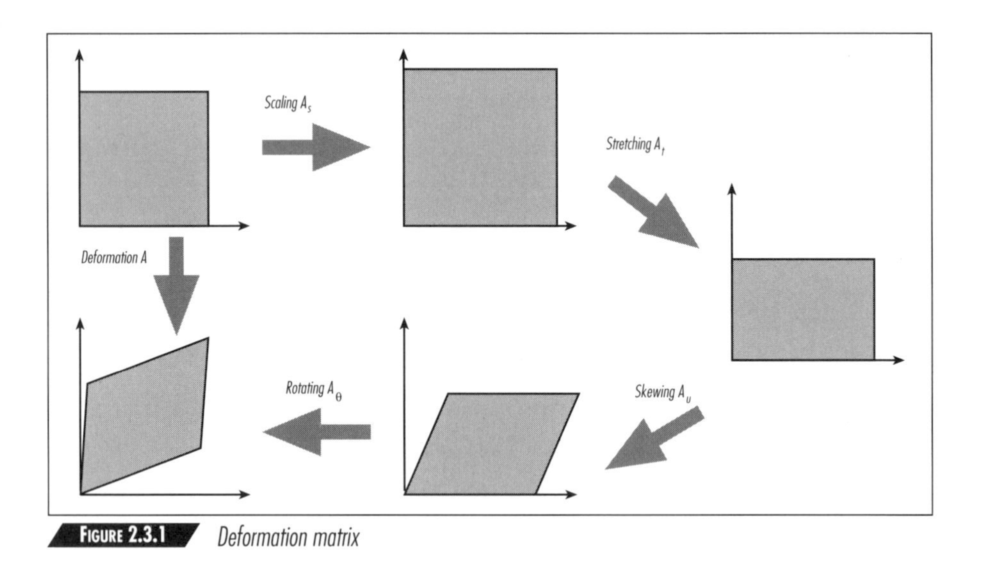

FIGURE 2.3.1 Deformation matrix

For a given translative metric *d*, i.e., a metric defined by a norm, an affine transformation *f*, or a deformation matrix *A*, is called an *isometry* if it keeps the distance function invariant, i.e., for any $x, y \in \Re^n$,

$$d(x, y) = d(f(x), f(y)) = d(Ax, Ay). \tag{2.3.3}$$

In the L^2-metric there is no nontrivial isometric scaling and no nontrivial isometric skewing. The only nontrivial isometric stretching is the flip $\begin{pmatrix} 1 & 0 \\ 0 & -1 \end{pmatrix}$, and all rotations are isometric. *The only isometric deformations are the rotations, possibly composed with the flip.*

TABLE 2.3.1 Index of spatial forms

index	isometry	matrix	index	isometry	matrix
0	*identity*	$\begin{pmatrix} 1 & 0 \\ 0 & 1 \end{pmatrix}$	4	*(x = y) flip*	$\begin{pmatrix} 0 & 1 \\ 1 & 0 \end{pmatrix}$
1	*x flip*	$\begin{pmatrix} -1 & 0 \\ 0 & 1 \end{pmatrix}$	5	*90° rotation*	$\begin{pmatrix} 0 & -1 \\ 1 & 0 \end{pmatrix}$
2	*y flip*	$\begin{pmatrix} 1 & 0 \\ 0 & -1 \end{pmatrix}$	6	*270° rotation*	$\begin{pmatrix} 0 & 1 \\ -1 & 0 \end{pmatrix}$
3	*180° rotation*	$\begin{pmatrix} -1 & 0 \\ 0 & -1 \end{pmatrix}$	7	*(x + y = 0) flip*	$\begin{pmatrix} 0 & -1 \\ -1 & 0 \end{pmatrix}$

Among all rotations, four preserves the orientation of a square, namely, the identity, the 90° rotation, the 180° rotation, and the 270° rotation. Composing these rotations with the flip, the following eight deformation matrices are obtained. They will be indexed and called the *standard indexed spatial matrices*, as shown in Table 2.3.1.

Algebraically, it seems more natural to place the four orientation-preserving rotations first, then the orientation-reversing flips. However, geometrically, the identity, the horizontal and vertical flips, and the double-flip which is the 180°-rotation are the most frequently used ones. Thus, this is the reason in ordering the spatial form indices.

2.4 ITERATED FUNCTION SYSTEMS

Now is the time to show the theory and algorithms for the fractals presented in Chapter 1. The sets of affine transformations used there are *iterated function systems*.

Consider the vector space $\Re^n$ with a complete translative metric d. An *iterated function system (IFS)* in this space is a finite set of *contractive (affine) transformations* $W = \{w_1, w_2, \cdots, w_m\}$. The *contractivity factor* s is defined to be the maximum of the contractivity factors of the transformations: $s = \max\{\|w_1\|, \|w_2\|, \cdots, \|w_m\|\}$.

Given an IFS $W = \{w_1, w_2, \cdots, w_m\}$, define its *associated transform* on the space of compact subsets, $\mathcal{H}(\Re^n)$, by

$$W(B) = w_1(B) \cup w_2(B) \cup \cdots \cup w_m(B), \tag{2.4.1}$$

for each $B \in \mathcal{H}(\Re^n)$.

2.4.1 IFS DECODING THEOREM

IFS DECODING THEOREM:

Let $W = \{w_1, w_2, \cdots, w_m\}$ be an iterated function system with contractivity factor s. Then its associated transform $W : \mathcal{H}(\Re^n) \longrightarrow \mathcal{H}(\Re^n)$ is a contractive mapping in the space $\mathcal{H}(\Re^n)$ with the corresponding Hausdorff metric d with the same contractivity factor s. That is,

$$d(W(B), W(C)) \leq s \cdot d(B, C), \tag{2.4.2}$$

for all $B, C \in \mathcal{H}(\Re^n)$.

Consequently, W has a unique fixed point $A \in \mathcal{H}(\Re^n)$, i.e., a compact subset A in $\Re^n$ that obeys

$$A = W(A) = \bigcup_{k=1}^{m} w_k(A). \tag{2.4.3}$$

The compact subset $A \in \mathcal{H}(\Re^n)$ described in the theorem is called the *attractor*, or *deterministic fractal*, of the IFS $W = \{w_1, w_2, \cdots, w_m\}$.

Moreover, for any compact subset $B \in \mathcal{H}(\Re^n)$, the sequence

$$B, W(B), W^2(B) = W(W(B)), \cdots, W^k(B) = W(W^{k-1}(B)), \cdots \tag{2.4.4}$$

converges to the limit $A \in \mathcal{H}(\Re^n)$.

The theorem is a direct consequence of the *contractive mapping theorem*, as can be shown by verifying that the associated transformation W is contractive. It provides a precise method for the construction of the fixed point $A \in \mathcal{H}(\Re^n)$ from the IFS W. Hence, the above theorem is also called the *IFS decoding theorem*. The *deterministic algorithm* is literally deduced from it.

2.4.2 DETERMINISTIC ITERATION ALGORITHM

C language code will now be used to generate some amazing objects.

Before introducing the algorithm, let us set a few coding conventions that will be used throughout the book.

In accordance with an unwritten rule that most of us follow, constants and labels are named in all capital letters, functions and global variables start with a capital letter, and local variables use only lowercase letters. The data types are defined as follows:

```
#define     CHAR     signed char
#define     BYTE     unsigned char
#define     SHORT    short
#define     WORD     unsigned short
#define     INT      long
#define     LONG     long
#define     DWORD    unsigned long
```

The error messages are limited to OK and FAIL.

```
#define     OK       0
#define     FAIL     -1
```

And memory allocation and deallocation use two functions that can be defined to whatever fits one's system:

```
INT GetMemory( (void *)pointer, LONG size_in_byte );

INT FreeMemory( (void *)pointer );
```

The algorithm directly following the IFS decoding theorem is called the *deterministic algorithm*. Here is the C program:

```
1     #define N   3        // vector space dimension, N = 2,3,4,...

2     void IFS_Deterministic_Iteration(

      //  parameter:
3         LONG     iteration,  // number of iterations

      //  input:   The IFS
4         SHORT    num,        // number of transf. in an IFS
5         SHORT    *a[N][N],   // deformation matrices in 1/1024
6         SHORT    *b[N],      // translation vectors

      //  output:  The set of attractor points
7         LONG     *count,     // number of points in attractor
8         float    *attractor[N]
9     ){

10        LONG     i,j,k,n,m;  // working variables
11        float    z;          // working variables

      //  Start from one point, e.g. attractor[.][0] = {0,0,...,0}
12        *count = 1;
13        for(i=0;i<N;i++) attractor[i][0] = 0;

          // Iterate the attractor:
14        while((iteration--)>0)
15        {
          //  Set m to the end of the attractor buffer:
16            m = *count;

          //  Loop all attractor points:
17            for(n=0;n<*count;n++)
18            {
          //  Loop all transformation
19                for(k=0;k<num;k++)
20                {

          //  Map attractor[.][n] to attractor[.][m]
          //  using k-th transform
21                    for(j=1;j<N;j++)
22                    {
23                        z = b[j][k]*1024;
24                        for(i=0;i<N;i++)
25                            z += a[j][i][k]*attractor[i][n];
26                        attractor[j][m] = (z/1024);
27                    }
28                    m++:
29                }
          //  Overwrite attractor[.][n] with the latest one
30                m--;
31                for(i=0;i<N;i++)
32                    attractor[i][n] = attractor[i][m];
33            }

          // The attractor points increased by a factor of "num"
```

```
34          *count *= num;
        }
35      return;
36  }

//  Line 5: multiply the decimal number by 1024 to simplify
//          the calculation, (cf. lines 25 & 26).
```

If $n > 2$, the high-dimension attractor is projected onto the display plane; it could be any function $\pi : \Re^n \longrightarrow \Re^2$.

```
1   #define W    1024    // width of the plot resolution
2   #define H    1024    // height of the plot resolution

3   void Display_Attractor(

    //  input:  set of attractor points
4       LONG    count,          // number of points in attractor
5       float   *attractor[N]

    //  output: A black(0)-white(255) image of dimension WxH
6       BYTE    fractal[H][W]
7   ){

8       SHORT   i;        // working variables
9       LONG    n;        // working variables
10      SHORT   x,y;      // plotting coordinates in R(2)
11      float   v[N];     // active point of the fractal in R(N)

        // Set the image to white:
12      memset(&fractal[0][0],255,W*H);

        // Project every point in the attractor to the plane
13      for(n=0;n<count;n++)
14      {
15          for(i=0;i<N;i++) v[i] = attractor[i][n];
16          Projection(v,&x,&y);
17          fractal[x][y] = 0;
18      }
19      return;
20  }

//  Line 16:Projection(v,&x,&y) maps vector v in R(N) to (x,y) in R(2)
```

In most cases, the projection can be assumed to be an affine transformation, which is exactly an $n \times 2$ matrix,

$$\pi = \begin{pmatrix} \pi_{11} & \pi_{12} & \cdots & \pi_{1n} \\ \pi_{21} & \pi_{22} & \cdots & \pi_{2n} \end{pmatrix}. \tag{2.4.5}$$

As a C routine, this projection is implemented as follows:

```
1   SHORT   pi[2][N]={...};  // the predefined projection matrix

2   void Projection(

    //  input:  a vector in N-dimensional vector space R(N)
```

```
3        float   v[N],

     //  output: projected coordinate in R(2)
4        SHORT   *x, *y
5    ){

6        SHORT   i;        // working variables
7        LONG    xx,yy;    // working variables

8        xx = pi[0][0]*v[0];
9        yy = pi[1][0]*v[0];
10       for(i=1;i<N;i++)
11       {
12           xx += pi[0][i]*v[i];
13           yy += pi[1][i]*v[i];
14       }
15       *x = (xx/1024);
16       *y = (yy/1024);

17       return;
18   }
```

The best way to illustrate this algorithm is through visual means. Figure 2.4.1 displays the following four affine transformations used to generate the *maple leaf*.

$$w_1\begin{pmatrix}x\\y\end{pmatrix}=\begin{pmatrix}0.49&0.01\\0&0.62\end{pmatrix}\begin{pmatrix}x\\y\end{pmatrix}+\begin{pmatrix}25\\-2\end{pmatrix},\quad w_2\begin{pmatrix}x\\y\end{pmatrix}=\begin{pmatrix}0.27&0.52\\-0.40&0.36\end{pmatrix}\begin{pmatrix}x\\y\end{pmatrix}+\begin{pmatrix}0\\56\end{pmatrix},$$

$$w_3\begin{pmatrix}x\\y\end{pmatrix}=\begin{pmatrix}0.18&-0.73\\0.50&0.26\end{pmatrix}\begin{pmatrix}x\\y\end{pmatrix}+\begin{pmatrix}88\\8\end{pmatrix},\quad w_4\begin{pmatrix}x\\y\end{pmatrix}=\begin{pmatrix}0.04&-0.01\\0.50&0\end{pmatrix}\begin{pmatrix}x\\y\end{pmatrix}+\begin{pmatrix}52\\32\end{pmatrix}. \quad (2.4.6)$$

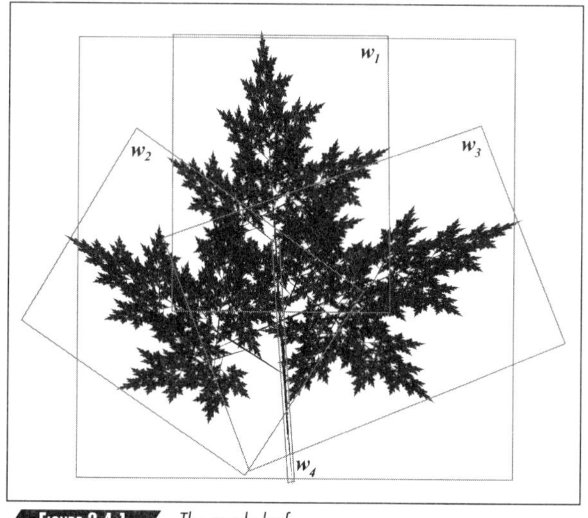

FIGURE 2.4.1 *The maple leaf*

The process of generating the fractal *maple leaf* using the deterministic algorithm is displayed in Figure 2.4.2 step by step showing several iterations. The initial image is shown as iteration zero.

The main shortfall of the above algorithm is that it may take a great deal of computer memory. Actually, the attractor set's buffer takes

$$\mathtt{N \cdot num^{iteration} \cdot sizeof(SHORT)}$$

bytes. If there are four transformations, the practical limit of the iteration number is about 10 or 11. However, one always can eliminate the duplicate points,

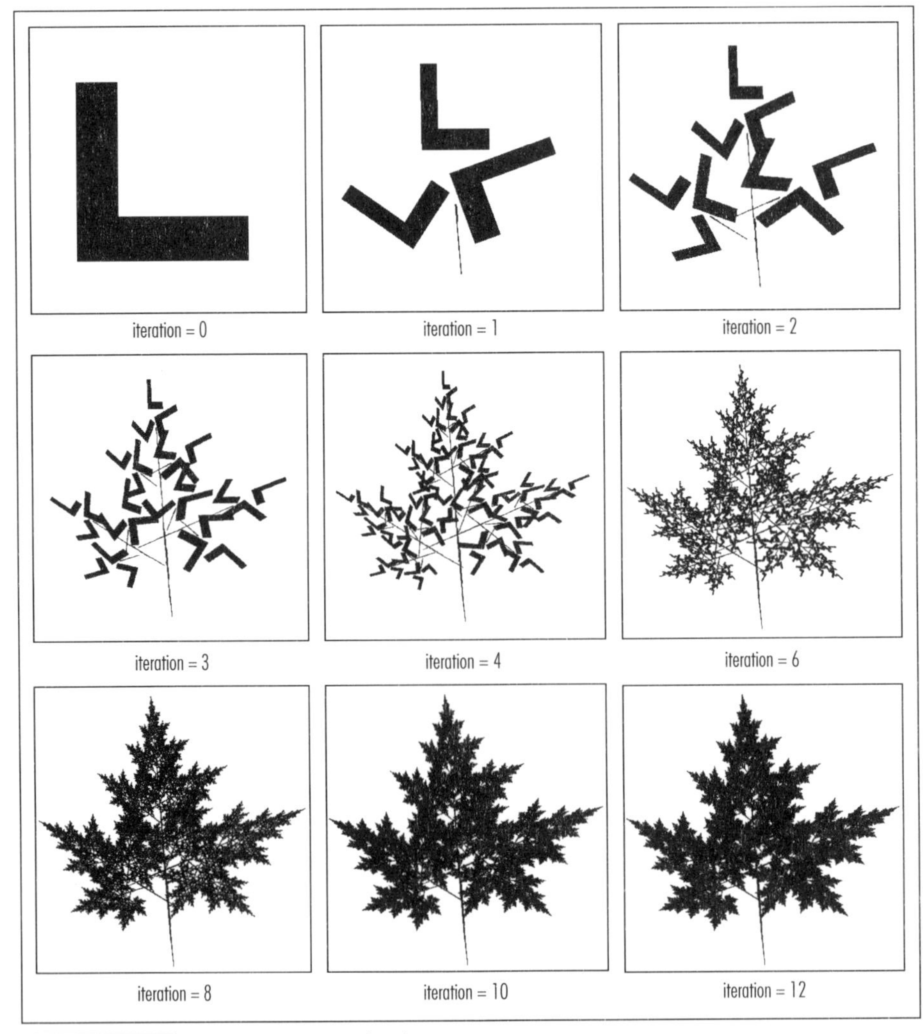

FIGURE 2.4.2 *Deterministic iteration algorithm*

after some precision rounding, to insure that the attractor set remains in a maintainable size, though this may reduce image resolution.

In the $n = 2$ planar case, this elimination can be done exactly, as the display projection resolution rounding.

Is there a better way to solve this memory problem? In the next section a beautiful answer to this question will be given.

2.4.3 RANDOM ITERATION ALGORITHM

Returning to our fractals in Chapter 1, a given set of affine transformations $W = \{w_1, w_2, \cdots, w_m\}$ is an IFS, the amazing fractal is its attractor $A \in \mathcal{H}(\Re^n)$, and the magic performance is exactly the construction of the *random iteration algorithm.*

What will it happen if we keep choosing one transformation much more frequently than the others in the random transformation selecting procedure? To answer this question a new concept needs to be introduced first: *IFS with probabilities.*

An *IFS* $W = \{w_1, w_2, \cdots, w_m\}$ *with probabilities* $P = \{p_1, p_2, \cdots, p_m\}$ is an IFS with a positive number associated to each transformation; the total sum of the probabilities is 1. That is,

$$p_i > 0 \text{ for all } i = 1, 2, \cdots, m, \text{ and } \sum_{i=1}^{m} p_i = 1. \tag{2.4.7}$$

The fractals constructed in the introduction use equal probabilities. However, to be efficient and to cover the space as quickly and evenly as possible, the best procedure is to set the value of each probability to be proportional to the volume of its corresponding transformation. The volume of a transformation is defined as the volume of the transformed unit cube, which is exactly the *absolute value of the determinant* of the deformation matrices. In conclusion, the *default probabilities* should be set to

$$p_i = \frac{|\det A_i|}{\sum_{j=1}^{m} |\det A_j|}, \tag{2.4.8}$$

where we assume that $W_i(x) = A_i x + b_i$ for all $i = 1, 2, \cdots, m$.

Here is the C program:

```
1   #define N    3       // vector space dimension, N = 2,3,4,...
2   #define W    1024    // width of the plot resolution
3   #define H    1024    // height of the plot resolution

4   void IFS_Random_Iteration(
```

```
   // parameter:
5      LONG    iteration,  // number of iterations

   // input: IFS with probabilities
6      SHORT   num,          // number of transforms in an IFS
7      SHORT   *a[N][N],     // deformation matrices in 1/1024
8      SHORT   *b[N],        // translation vectors
9      SHORT   *p,           // integers proportional to probabilities

   // output: A black(0)-white(255) image of dimension WxH
10     BYTE    fractal[H][W]
   ){
11     SHORT   i,j,k;        // working variables
12     SHORT   x,y;          // plotting coordinates in R(2)
13     float   u[N];         // working point in R(N)
14     float   v[N];         // active point in fractal R(N)

       // Calculate accumulate probabilities:
15     for(k=1;k<num;k++) p[k] += p[k-1];

       // Set the image to white:
16     memset(&fractal[0][0],255,W*H);

       // Set the initial point, e.g. v=(0,0,...,0):
17     for(i=0;i<N;i++) v[i] = 0;

       // Iterate the random plotting:
18     while((iteration--)>0)
       {
       //  plot the point v:
19         Projection(v,&x,&y);
20         fractal[x][y] = 0;

       //  Determine the transformation through random process;
21         x = Randomize(p[num-1]);
22         for(k=0;k<num;k++)
           {
23             if(x<p[k]) break;
           }

24     //  Get new point u = av + b:
25         for(j=0;j<N;j++)
           {
26             u[j] = b[j][k]*1024;
27             for(i=0;i<N;i++) u[j] += a[j][i][k]*v[i];
28             u[j] /= 1024;
           }

       //  Copy u to v:
29           for(i=0;i<N;i++) v[i] = u[i];
       }   // end of iteration

30     return;
   }

// Line 7: multiply the decimal number by 1024 to simplify
//         the calculation, (cf. lines 26 & 28).
// Line 9: same trick for integer calculation
// Line 19:Projection (v,&x,&y) maps vector v in R(N) to (x,y) in R(2)
```

```
//  Line 21: Randomize(K) = random number among 0,1,...,K-1

// Remark: To avoid plotting points at start which may not be in the attractor, you
//      may modify the program such that the first several, e.g. 100, points will
//      skip from plotting
```

Back to the same example: the *maple leaf*, Figure 2.4.3, displays this fractal in various stages, for different iteration numbers, by applying the random iteration algorithm with the default probabilities. The *random iteration algorithm* is much faster than the deterministic one, since the latter plots only one pixel in each iteration.

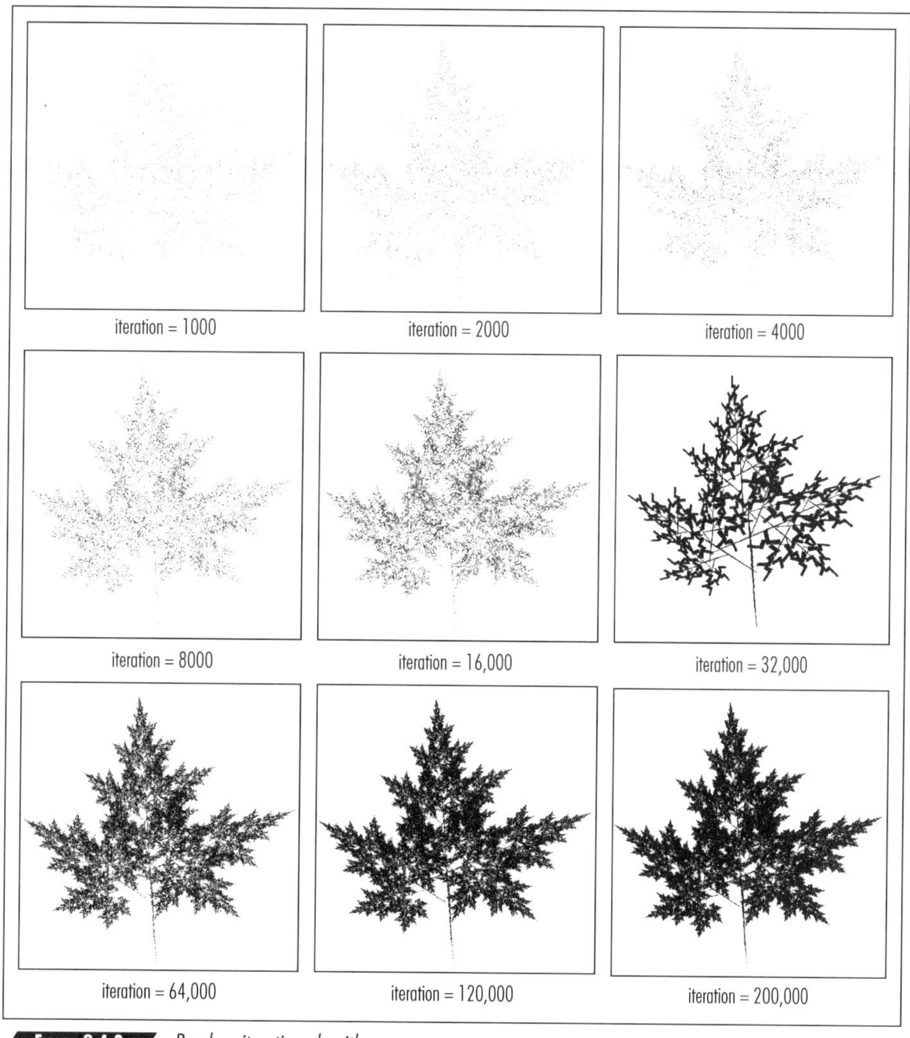

FIGURE 2.4.3 *Random iteration algorithm*

2.4.4 IFS ENCODING THEOREM

Having outlined the construction of the image from an IFS, what about the inverse problem? Given an arbitrary compact subset in $\Re^n$, can an IFS be found to generate it? Intensive research has been done in this area [VR], and fractal image compression is actually the fruit of the research in these directions. This section will introduce only a very small portion of the results that impact real-world imagery.

Instead of answering the inverse problem directly, the question can be simplified through different approaches:

What kinds of compact subsets are generated by IFSs?

What are the commonly shared characteristics of those compact subsets generated by IFSs?

For a given compact subset and an IFS, is the subset the attractor of the IFS?

Michael Barnsley's *collage theorem* [B], also called the *IFS encoding theorem*, gives some insight on the last question. As a matter of fact, it provides a tool to estimate the distance between a given compact subset and the attractor of a given IFS.

THE COLLAGE THEOREM:

Let $W=\{w_1,w_2,\cdots,w_m\}$ be an iterated function system with contractivity factor s, $0\le s<1$. Let $W:\mathcal{H}(\Re^n)\longrightarrow\mathcal{H}(\Re^n)$ denote its associated transform, and let $A\in\mathcal{H}(\Re^n)$ denote the attractor of W. Then

$$d(L,A) \le \frac{d(L,W(L))}{1-s} \tag{2.4.9}$$

for any $L\in\mathcal{H}(\Re^n)$.

What this theorem implies is that *for a given compact subset* $L\in\mathcal{H}(\Re^n)$ *and for a given IFS* $W=\{w_1,w_2,\cdots,w_m\}$ *such that* $d(L, W(L))$ *is very small, the attractor of* W, $A\in\mathcal{H}(\Re^n)$, *is very close to* L.

This theorem and its variations become the foundation of fractal image compression. Most digital image compression methods required in real world applications are *lossy* compression methods, i.e., the decompressed images must be visually similar, not necessarily identical, to the original images. So there is always a tradeoff between the compression ratio and the reconstruction error. In fractal image compression the task is to find an IFS whose attractor is very close to a given image.

2.5 RECURRENT ITERATED FUNCTION SYSTEMS

Now that black and white binary images have been discussed, this section will show how to create grayscale images using *IFS with probabilities* and will expand the current IFS system to a more generalized *recurrent IFS*.

2.5.1 IFS WITH PROBABILITIES

In the random iteration algorithm, the fractal construction uses a set of probabilities, one positive number for each transformation. What will the fractal look like if it is plotted with different probabilities other than the default one?

According to the decoding theorem, it will converge to the same image—the attractor of the IFS. The images will appear identical after enough iterations. However, the procedure itself is much more interesting than one might have imagined.

Figure 2.5.1 shows four leaves generated using four different sets of probabilities stopped at iteration 200,000. It is remarkable to see that some of them have sharp stems and some of them have fuzzy edges. Nevertheless, none of them converges to the attractor faster than the IFS using the default probabilities.

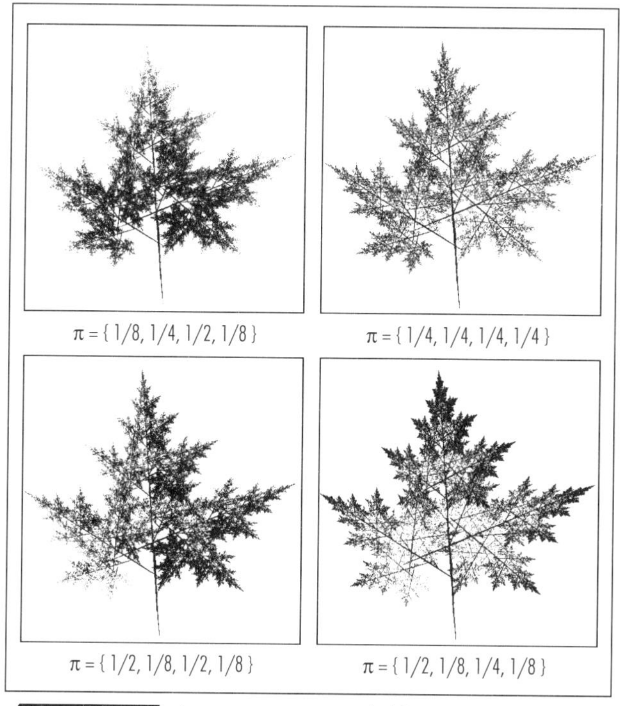

FIGURE 2.5.1 *Iteration using various probability sets*

2.5.2 GRAYSCALE PHOTOCOPY MACHINE

Our first idea for turning the black and white image into a grayscale one is Michael Barnsley's *Photocopy Machine* ([B], [BH]). He expanded the iterated function systems theory to the space of images $\mathcal{P}(\Re^n)$.

Given an IFS $W = \{w_1, w_2, \cdots, w_m\}$ with probabilities $\{p_1, p_2, \cdots, p_m\}$, an associated contractive map $W : \mathcal{P}(\Re^n) \longrightarrow \mathcal{P}(\Re^n)$ is given by the formula

$$W(f) = \sum_{i=1}^{m} p_i \cdot f \circ w_i^{-1} , \tag{2.5.1}$$

for all image functions $f \in \mathcal{P}(\Re^n)$. The process of getting $W(f)$ from f is called a *Markov process*, or an *iteration*. The corresponding decoding and encoding theorems in this space have been proven by John Elton and others [BEH], [B]. Intuitively, this construction can be explained in the following two ways.

In the deterministic way, the description follows the definition directly—given a constant lump sum of light energy that is the total brightness of the image, the IFS with probabilities is really a set of transformation lenses corresponding to the

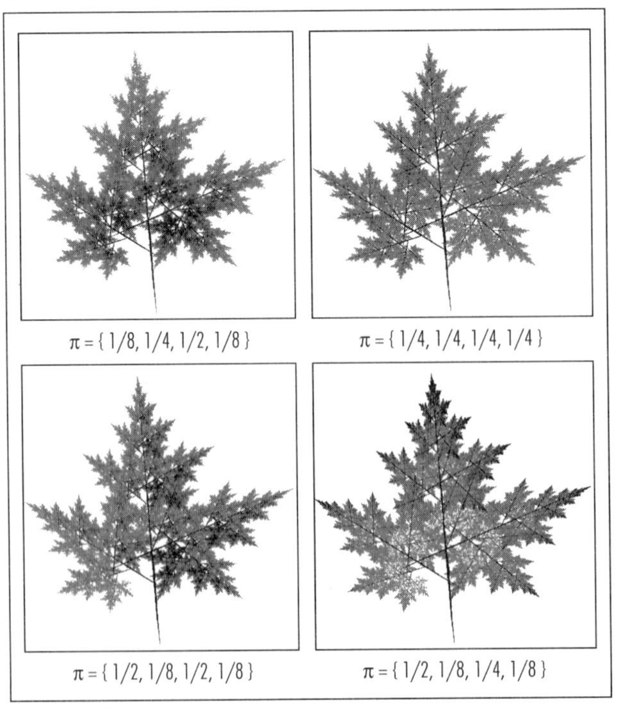

FIGURE 2.5.2 *Grayscale images generated by IFS with probabilities*

set of transformations redistributing this light energy. The reflection of each lens is exactly described by the corresponding affine transformation, and the reflection parameter is exactly the corresponding probability. Each reflection cycle is the Markov process formulated previously. In repeating this process, gradually a well-converged image will appear as the attractor.

In the random case, the negative image can be explained in the following manner. In a dark room there is a photon that contains all the light energy of the image. It bounces randomly, following the rule given in the random iteration algorithm. In each location of its bouncing trajectory, it leaves some constant amount of light energy behind until nothing is left.

Figure 2.5.2 shows the grayscale images generated using the same IFS with the same probabilities as in Figure 2.5.1. By tuning the probability set of a given IFS, new and different pictures are created from the same IFS (Figure 2.5.3).

In the black and white case the location (where some point hits) is all that is needed, but in the grayscale case the numbers of hits at every location are important as well. As a consequence, for obtaining a decent picture, more iteration points—considered as *bouncing photons*—become necessary. Each picture in Figures 2.5.2 and 2.5.3 has been iterated over 5,000,000 "*photons.*"

FIGURE 2.5.3 *Maple leaf with* $\pi = \{5/16, 5/16, 5/16, 1/16\}$

2.5.3 RECURRENT IFS

After seeing how IFSs with probabilities generate grayscale images, Barnsley, Elton, and Hardin [BEH] generalized the concept to that of *recurrent iterated function system* (RIFS), which is an IFS with a set of probability sets—one probability set attached to each transformation.

More precisely, let $W=\{w_1,w_2,\cdots,w_m\}$ be an IFS. A *recurrent structure* is a matrix $P=\{p_{ij}\}_{i,j=1,2,\cdots,m}$ of probabilities with the following properties:

1. (*probability*) $\sum_{j=1}^{m} p_{ij}=1$, for all $i=1,\ 2,\cdots,\ m$;
2. (*irreducibility*) for any $i,j=1,\ 2,\cdots,\ m$, there is some finite sequence

$$s_0,s_1,\cdots,s_k$$

such that $s_0=i$, $s_k=j$, and $\prod_{h=1}^{k} p_{s_{h-1}s_h}>0$.

The procedure of generating an image using RIFS is the same as using IFS with probabilities. Instead of following the same probability set all the time, an RIFS follows a probability set determined by the transformation in each iteration. That is, after applying the *i*th transformation, the probability set will be $\{p_{i1},p_{i2},\cdots,p_{im}\}$ for all $i=1,\ 2,\cdots,\ m$.

The generalization does increase our power to create new images. Let us look at the simplest case: an IFS of four transformations. Each transformation maps the unit square into one of the four different corners, without any rotation or flip, with the contraction factor 2, defined as follows:

$$w_1\begin{pmatrix}x\\y\end{pmatrix}=\begin{pmatrix}0.5&0\\0&0.5\end{pmatrix}\begin{pmatrix}x\\y\end{pmatrix}+\begin{pmatrix}0\\0\end{pmatrix},\quad w_2\begin{pmatrix}x\\y\end{pmatrix}=\begin{pmatrix}0.5&0\\0&0.5\end{pmatrix}\begin{pmatrix}x\\y\end{pmatrix}+\begin{pmatrix}0\\50\end{pmatrix},$$

$$w_3\begin{pmatrix}x\\y\end{pmatrix}=\begin{pmatrix}0.5&0\\0&0.5\end{pmatrix}\begin{pmatrix}x\\y\end{pmatrix}+\begin{pmatrix}50\\0\end{pmatrix},\quad w_4\begin{pmatrix}x\\y\end{pmatrix}=\begin{pmatrix}0.5&0\\0&0.5\end{pmatrix}\begin{pmatrix}x\\y\end{pmatrix}+\begin{pmatrix}50\\50\end{pmatrix}.$$

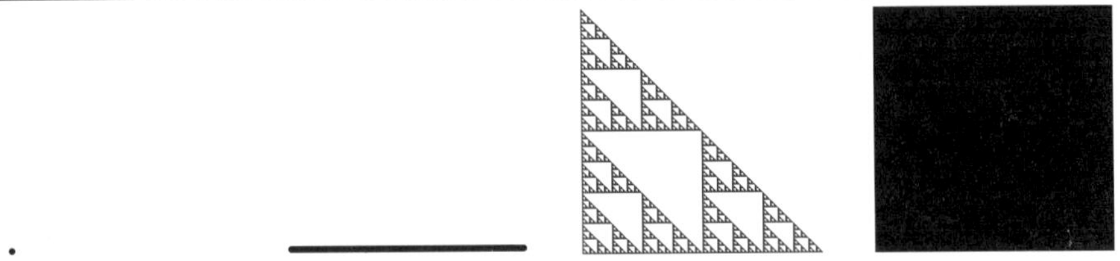

$$\pi=\begin{pmatrix}1&0&0&0\\1&0&0&0\\1&0&0&0\\1&0&0&0\end{pmatrix}\quad \pi=\frac{1}{2}\begin{pmatrix}1&1&0&0\\1&1&0&0\\1&1&0&0\\1&1&0&0\end{pmatrix}\quad \pi=\frac{1}{3}\begin{pmatrix}1&1&1&0\\1&1&1&0\\1&1&1&0\\1&1&1&0\end{pmatrix}\quad \pi=\frac{1}{4}\begin{pmatrix}1&1&1&1\\1&1&1&1\\1&1&1&1\\1&1&1&1\end{pmatrix}$$

FIGURE 2.5.4 One, two, three, and four transformations

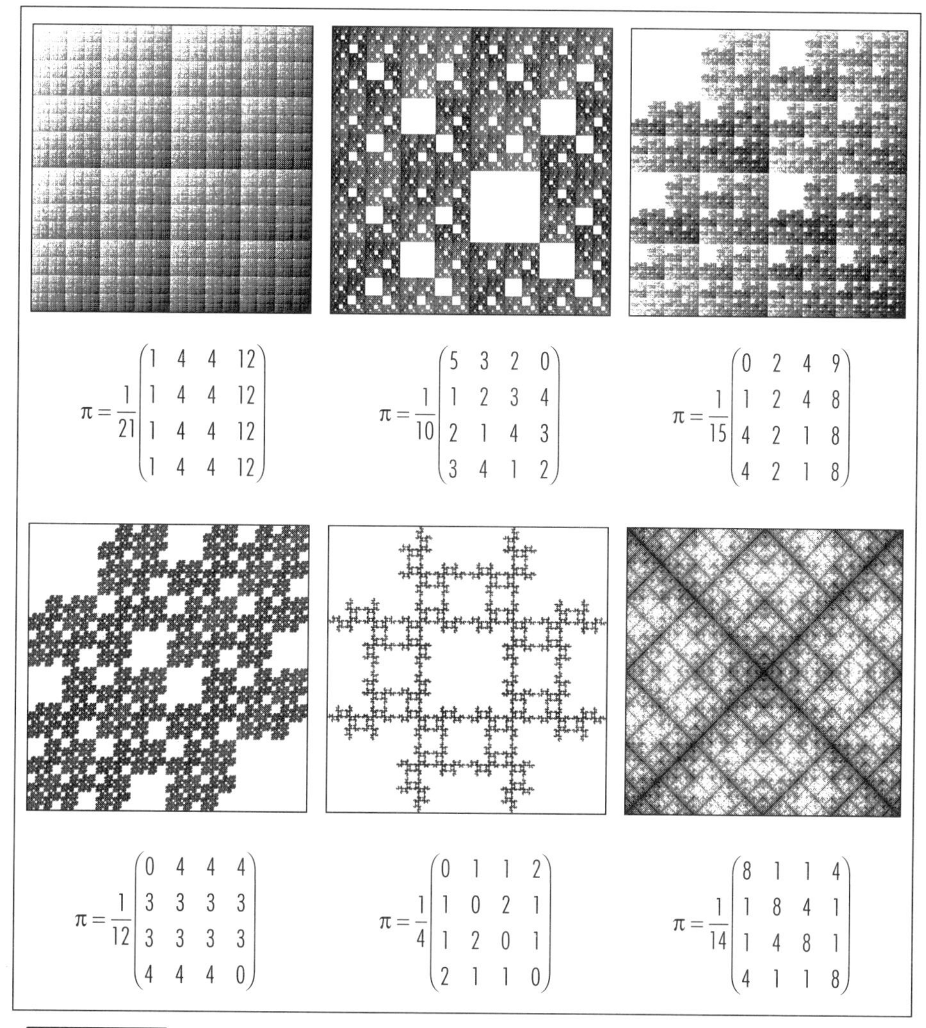

FIGURE 2.5.5 *More probability matrices*

In the most trivial case and the reducible cases, the attractors of these four transforms are a point, a segment, a Sierpinski triangle, and a square tile, as the attractors of the IFSs $\{w_1\}$, $\{w_1, w_2\}$, $\{w_1, w_2, w_3\}$, and $\{w_1, w_2, w_3, w_4\}$, respectively (Figure 2.5.4).

Figure 2.5.5 shows the same IFS with several different probability matrices.

2.5.4 RECURRENT IFS FOR VECTOR OF IMAGES

All mathematical proofs of the existence and the uniqueness of fractal attractors and RIFSs are omitted. More emphasis is given to the algorithmic construction of fractals. The theoretic proofs can be found in [B].

A further generalization of IFS is implemented in Barnsley's VRIFS system. Instead of working on the image space $\mathcal{P}(\Re^n)$, he introduced RIFSs in the image vector space $\left(\mathcal{P}(\Re^n)\right)^k$, for $k > 0$. This allows one to work on k image plates at the same time.

By adding additional image plates, and by allowing transformations across image plates, Barnsley created enough room and right kind of tool to construct more sophisticated images. One of those plates realizes the final output image. Figure 2.5.6, *Birds and Trees,* is a nice picture purely generated by using the system of Barnsley and Crawford.

FIGURE 2.5.6 *Birds and Trees using VRIFS system (from [B])*

The above algorithms not only shaped the fractal real-world image compression technology that will be introduced here, but they also formed the basis for many fractal computer graphic design systems [DHN]. Incorporating these algorithms into other computer graphic techniques, extending to 3-D or 4-D, you can have your own art studio.

2.5.5 FRACTAL LOCAL ZOOM

A nice feature of the random iteration algorithm is local zoom. A small piece of a fractal example can be plotted only when a pixel in some iteration falls within

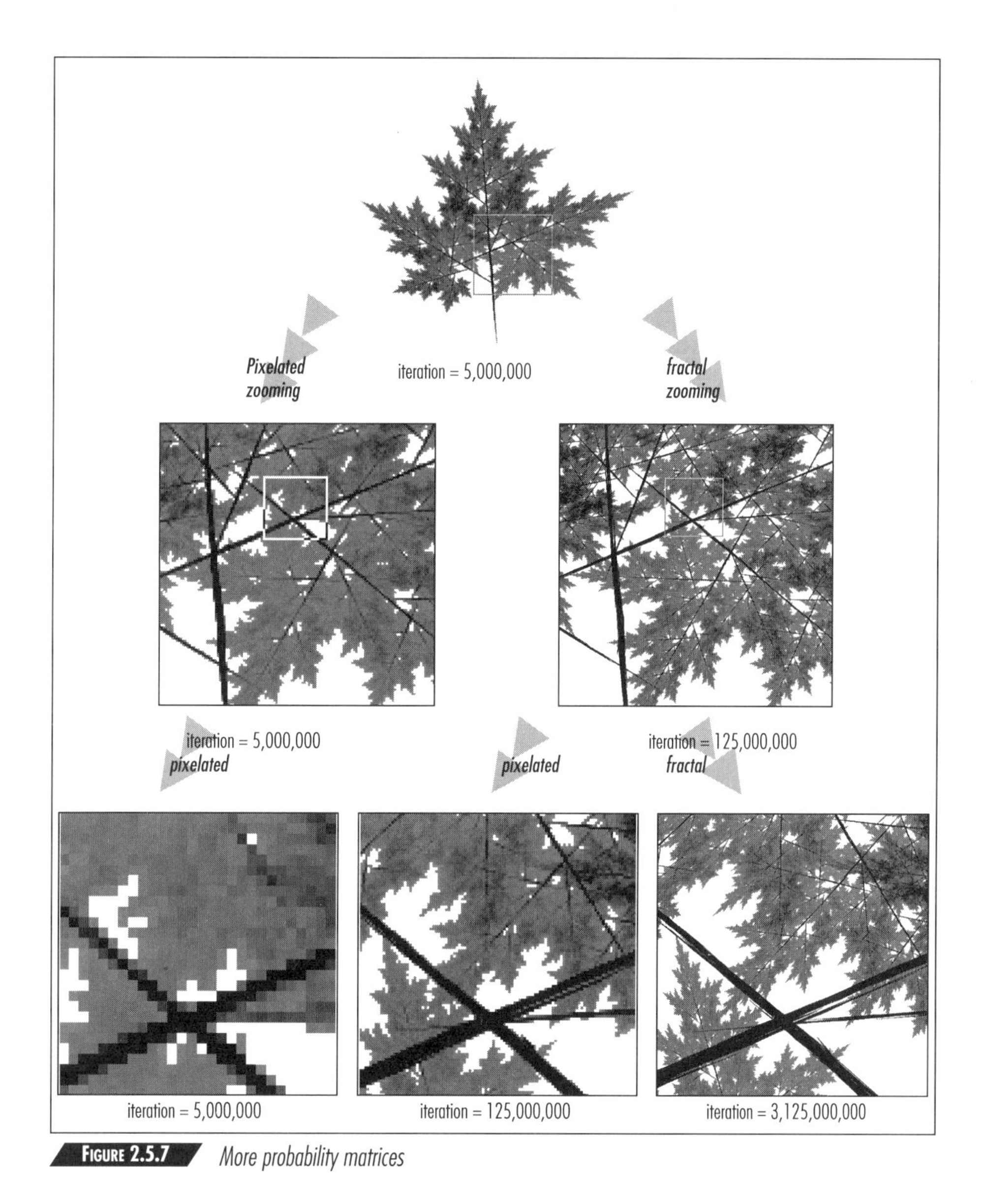

FIGURE 2.5.7 *More probability matrices*

that piece. By restricting attention to a small region, one can display a portion of a fractal image without having to use the amount of computer memory that could hold the full image at this display resolution. The bottom right piece of Figure 2.5.7, a 512×512 image section, is just one of the $25 \times 25 = 225$ pieces of the maple leaf picture. The full image has a resolution of $12{,}800 \times 12{,}800$, about 150 gigabytes, which is beyond today's computer capacities.

This feature represents the unique property of fractals: *resolution-independence*. That is, the same fractal representation can be *decoded* to various output devices in the best resolution for each of them. This future-proof character enables us to store an image at the level of quality beyond today's hardware limit. In Chapter 8, this exciting property will be explored further to real-world image fractal representations.

Having arrived at this point in the book, the reader will have acquired many techniques that will make possible the creation of some truly amazing images. Using creativity and imagination, talents and skills, the reader will now be able to illustrate in a way heretofore unimagined.

Fantasy, or Imagination, are the names given to the faculty of reproducing copies of originals once felt. The imagination is called "reproductive" when the copies are literal; "productive" when elements from different originals are recombined so as to make new wholes.

—Imagination

William James

3 Fractal Imaging Model

Not only does fractal mathematics produce objects that are fascinating to view. It is also a rather powerful tool and a new way of thinking. It is so different from the engineering models we used to work with. As a matter of fact, until today, almost all mathematical models provided in the engineering world have encouraged everything to be smooth, continuous, and differentiable. We approximated everything under the assumption of *rectifiability*, i.e., any part of a model is assumed to be flat and smooth under magnification—the simplest assumption one can make theoretically. And such an assumption generates enough solutions for most practical applications.

Magnifying the elliptic Earth, we see mountains and valleys, rivers and seas, in all different shapes and sizes. Magnifying a green mountain, we see trees and flowers, birds and beasts, in all colors. But it doesn't matter how true it is, when we blow up a picture of the earth, we blur the unknown with flat regions.

Fractals challenged the assumption of rectifiability with the hypothesis of *self-similarity*. It declares that *instead of assuming zero complexity, it is much better to approach the magnified world with the same complexity as the current world we have known*. This is where fractal, as a new philosophy and methodology, differs from the other, conventional, technologies.

Based upon this methodology, to use fractals in photographic imaging is really to seek self-similarity in images and to use that self-similarity to predict, to clone, and to magnify unknown details.

3.1 FRACTAL ELEMENTS

Look at a real image in Figure 3.1.1. We see a spatial similarity: two sailing boats. What do we need to connect them? We identify the region of the first boat as the *reference region*, or simply *R-region*; the region of the second boat as the *destination region*, or simply *D-region*; and a *transformation* between them.

Note that in other literature the destination region has been called *domain region*, *range region*, *target region*, *decoded region*, etc., and the reference region has been called *range region*, *domain region*, *source region*, *codebook region*, etc.

To describe a destination region using its corresponding reference region, a transformation mapping has to be globally defined. We always can get the destination region by applying the transformation to the reference region, and vice versa. Moreover, a destination region can be masked by a global function that sets the active region to 1 and inactive region to 0. Hence, we define a *fractal element*, $\Phi = (\mu, T)$, to be a pair consisting of a *destination region mask function* μ and a *global transformation* T *in the image space*.

Let $\Delta = [0, w] \times [0, h] \subset \Re^2$ be the image's rectangular support. Let $\mathcal{P}(\Delta)$ be the space of images having the support Δ. Then a destination region mask func-

FIGURE 3.1.1 Fractal Element: *Reference region, destination region, transformation*

tion μ is a function $\mu : \Delta \longrightarrow [0, 1]$, and a global transformation T is just a function $T : \mathcal{P}(\Delta) \longrightarrow \mathcal{P}(\Delta)$.

Given two compact image support sets $\Delta \subset \tilde{\Delta} \subset \Re^2$, we have a natural *inclusion map* $\mathcal{P}(\Delta) \xrightarrow{\subset} \mathcal{P}(\tilde{\Delta})$, which identifies an image $f \in \mathcal{P}(\Delta)$ to its trivial extension $\tilde{f} \in \mathcal{P}(\tilde{\Delta})$, i.e., $\tilde{f}\big|_{\Delta} = f$ and $\tilde{f}\big|_{\tilde{\Delta} \setminus \Delta} = 0$. We also have a natural *restriction map* $\mathcal{P}(\tilde{\Delta}) \xrightarrow{|_{\Delta}} \mathcal{P}(\Delta)$, obtained from the restriction that maps $\tilde{f} \in \mathcal{P}(\tilde{\Delta})$ to $f = \tilde{f}\big|_{\Delta} \in \mathcal{P}(\Delta)$.

Let us define the *image space* to be the union of all of them:

$$\mathcal{P}(\Re^2) \stackrel{\text{def}}{=\!=\!=} \bigcup_{\Delta \in \mathcal{H}(\Re^2)} \mathcal{P}(\Delta) = \bigcup_{\Delta \text{ rectangular}} \mathcal{P}(\Delta), \tag{3.1.1}$$

where the last equality is obtained from the fact that any rectangular support is a compact set, and any compact set can be contained in a rectangular region. As a consequence, we naturally have the *restriction* $\mathcal{P}(\Re^2)\big|_{\Delta} = \mathcal{P}(\Delta)$.

Thus, in general, a fractal element will be defined as a pair of functions $\Phi = (\mu, T)$, where the mask function $\mu : \Re^2 \longrightarrow [0, 1]$ has compact support, i.e., the closure of the set $\mu^{-1}\big((0,1]\big)$, $\overline{\mu^{-1}\big((0,1]\big)}$, is a compact subset of $\Re^2$, and the transformation $T : \mathcal{P}(\Re^2) \longrightarrow \mathcal{P}(\Re^2)$ is a global transformation in the image space.

In principle, in a fractal element, it is natural to let the region with more information (in general, the larger one) be the reference region, since it will be used to represent the destination region.

3.2 FRACTAL IMAGING MODEL

A *fractal imaging model* is defined to be a set of fractal elements $W = \{\Phi_i = (\mu_i, T_i)\}_{i=1,2,\cdots,m}$. We define its associated *Markov process* to be the map $W : \mathcal{P}(\Re^2) \longrightarrow \mathcal{P}(\Re^2)$ defined by the formula

$$W(f)(x,y) = \sum_{i=1}^{m} \mu_i(x,y) \cdot T_i(f)(x,y), \tag{3.2.1}$$

for any $(x,y) \in \Re^2$ and $f \in \mathcal{P}(\Re^2)$.

A *fractal imaging model with seed*, (W, s), is a fractal imaging model $W = \{\Phi_i = (\mu_i, T_i)\}_{i=1,2,\cdots,m}$ with an initial image $s \in \mathcal{P}(\Re^2)$ attached to it.

In some literature, a seed image is also called a *condensation set*, or *fractal transform template* (FTT). When a complete fractal description of the seed image is given as well, the seed image is also called an *invisible image*, or *ghost image*. Later, ghost images will be used in video compression for rate control.

A system (W, s) is said to be a *fractal model of an image* $a \in \mathcal{P}(\Re^2)$ if the image $a \in \mathcal{P}(\Re^2)$ is the *attractor* of the fractal imaging model with seed (W, s), i.e., the image $a \in \mathcal{P}(\Re^2)$ is the limit of the following sequence of Markov processes in the image space $\mathcal{P}(\Re^2)$:

$$s, W(s), W^2(s) = W(W(s)), \cdots, W^k(s) = W(W^{k-1}(s)), \cdots. \tag{3.2.2}$$

In general, a fractal imaging model $W = \{\Phi_i = (\mu_i, T_i)\}_{i=1,2,\cdots,m}$ is said to be a *fractal model of an image* $a \in \mathcal{P}(\Re^2)$ if for any seed image $s \in \mathcal{P}(\Re^2)$ the fractal model with seed (W, s) is a fractal model of the image $a \in \mathcal{P}(\Re^2)$.

Similarly, a fractal imaging model with seed (W, s) is said to be a *fractal model of an image* $a \in \mathcal{P}(\Delta)$ if the image $a \in \mathcal{P}(\Delta)$ is the limit of the following sequence of Markov processes in the image space $\mathcal{P}(\Delta)$:

$$s|_{\Delta}, W(s)|_{\Delta}, W^2(s)|_{\Delta}, \cdots, W^k(s)|_{\Delta}, \cdots. \tag{3.2.3}$$

In general, a fractal imaging model $W = \{\Phi_i = (\mu_i, T_i)\}_{i=1,2,\cdots,m}$ is said to be a *fractal model of an image* $a \in \mathcal{P}(\Delta)$ if for any seed image $s \in \mathcal{P}(\Re^2)$ the fractal model with seed (W, s) is a fractal model of the image $a \in \mathcal{P}(\Delta)$.

FIGURE 3.2.1 Maple leaf *in a fractal model with seed: (A) The seed image* $s \in \mathcal{P}(\Re^2)$; *(B) The attractor of (W, s) in* $\mathcal{P}(\Re^2)$; *(C) The attractor image* $a \in \mathcal{P}([0, 2] \times [0, 2])$.

3.2.1 AN EXAMPLE

The fractal model (W, s) defined below represents the image $a \in \mathcal{P}(\Delta)$, where $\Delta = [0,2] \times [0,2]$, as shown in Figure 3.2.1*c*.

The seed image $s \in \mathcal{P}(\Re^2)$ is pictured in Figure 3.2.1*a*. The fractal model contains two fractal elements: $W = \{(\mu_1, T_1), (\mu_2, T_2)\}$, defined by

$$\mu_1(x,y)=\begin{cases}1-((x-1)^2+(y-1)^2), & \text{if } (x-1)^2+(y-1)^2\le 1,\\ 0, & \text{otherwise,}\end{cases} \tag{3.2.4}$$

for all $(x, y)\in\Re^2$, which is a fuzzy mask of a round region;

$$T_1(f)(x,y)=255-f(x+2,\ y+\sin 4\pi x), \tag{3.2.5}$$

for any $f\in\mathcal{P}(\Re^2)$ and all $(x, y)\in\Re^2$, that transforms the seed into the image support;

$$\mu_2(x,y)=\begin{cases}1, & \text{if } x\in[2,4] \text{ and } y\in[0,2],\\ 0, & \text{otherwise,}\end{cases} \tag{3.2.6}$$

for all $(x, y)\in\Re^2$, which preserves the seed area; and

$$T_2(f)(x,y)=f(x,y)\ , \tag{3.2.7}$$

for any $f\in\mathcal{P}(\Re^2)$ and all $(x, y)\in\Re^2$, which is the identity transform.

3.2.2 DESTINATION REGION MASK FUNCTION

Given a fractal model $W=\{\Phi_i=(\mu_i,T_i)\}_{i=1,2,\cdots,m}$, the *image support* is defined to be the compact set

$$\Delta=\bigcup_{i=1}^{m} D_i=\bigcup_{i=1}^{m}\overline{\mu_i^{-1}((0,1])}\,. \tag{3.2.8}$$

Consequently, all destination regions $\left\{D_i=\overline{\mu_i^{-1}((0,1])}\right\}_{i=1,2,\cdots,m}$ form a *covering* of the image support.

A destination region image covering is said to be a *partition* if

$$D_i\cap D_j=\varnothing \tag{3.2.9}$$

for any $1\le i<j\le m$.

A destination region image covering is said to be *binary* if all masking functions are mapped to either 0 or 1, i.e.,

$$\mu_i\ :\Re^2\longrightarrow\{\,0,1\,\}, \tag{3.2.10}$$

for any $1\le i\le m$.

The simplest image masking on the rectangular support $\Delta=[0,w]\times[0,h]$ is the *unit square binary partition* $\{\mu_k\}_{k=1,2,\cdots,w\cdot h}$, defined as

$$\mu_k(x,y)=\begin{cases}1, & \text{if } (i-1)<x<i \text{ and } (j-1)<y<j,\\ 0, & \text{otherwise,}\end{cases} \tag{3.2.11}$$

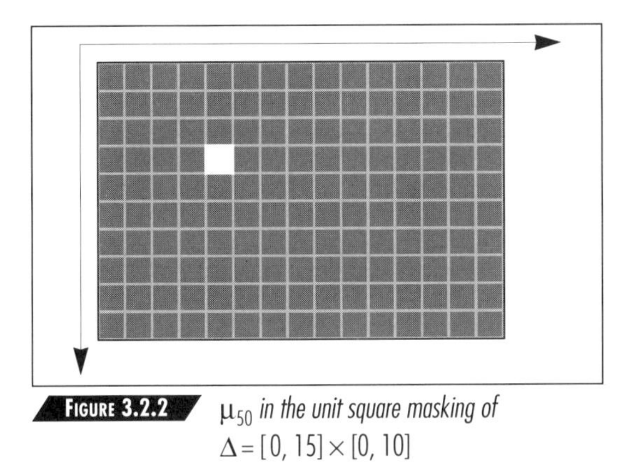

FIGURE 3.2.2 μ_{50} *in the unit square masking of* $\Delta = [0, 15] \times [0, 10]$

for any $1 \leq k = i + j \cdot w \leq w \cdot h$ and all $(x, y) \in \Re^2$ (Figure 3.2.2).

By letting a mask function $\mu : \Re^2 \longrightarrow [0,1]$ taking values between 0 and 1, we actually allow destination regions, $D = \mu^{-1}\left((0,1] \right)$, with fussy boundaries.

To use masking functions in imaging is equivalent to handling the image in manageable pieces; it is a generalization of the concepts of *image partition*, *image segmentation*, and *image localization*. They have been named differently in different places, such as *weighted function*, *probability, function with measure*, and *filter parameter*.

3.2.3 IMAGE TRANSFORMATION

Given a fractal model $W = \{\Phi_i = (\mu_i, T_i)\}_{i=1,2,\cdots,m}$, another important consideration is the set of *image transformations* $\{T_i\}_{i=1,2,\cdots,m}$. In general, we assume that such a transformation T can be decomposed into two components: the *spatial component* σ and the *intensity component* λ.

The *spatial component* σ identifies the active reference region R with the destination region D, which is assumed to be independent of color intensity in all our approaches. Quite often, we also set local coordinate systems for both reference region R and destination region D. Let (x_D, y_D) denote the *local origin* of D, and (x_R, y_R) the *local origin* of R. The *spatial component* X can be assumed to have the following form:

$$\chi(x, y) = (x_R, y_R) + \sigma(x - x_D, y - y_D), \tag{3.2.12}$$

for all $(x, y) \in \Re^2$.

In real, practical applications, we will limit the spatial component to be affine; in other words, the function S is exactly a deformation matrix in $\Re^{2\times 2}$.

The *intensity component* λ is actually the light-adjustment function after the image piece has been mapped from reference region to destination region. It is assumed, naturally, to be dependent only on the relative location and the local intensity brightness. Thus, it can be modeled as a function

$$\begin{aligned} \lambda: \quad \Re^2 \times \Re & \longrightarrow \quad \Re, \\ (x, y, z) & \mapsto \quad \lambda(x, y, z), \end{aligned} \tag{3.2.13}$$

for all $(x, y) \in \Re^2$ and $z \in \Re$, where $z \in \Re$ represents the intensity axis.

Combining all these functions, the transformation T has the following formula:

$$T(f)(x + x_D, y + y_D) = \lambda(x, y, f((x_R, y_R) + \sigma(x, y))), \tag{3.2.14}$$

for all $(x, y) \in \Re^2$.

In most applications the intensity adjustment λ can be further decomposed into two parts: the *contrast adjustment* γ and the *brightness adjustment* β. Both depend only on the spatial location, i.e., by moving and mapping an object from the reference region to the destination region, the color intensity adjustment depends only on the light condition of their relative locations and has nothing to do with the object itself. In this case, the intensity component λ can be written as

$$\lambda(x, y, z) = z \cdot \gamma(x, y) + \beta(x, y), \tag{3.2.15}$$

and the image transformation T can be written as

$$T(f)(x + x_D, y + y_D) = \gamma(x, y) \cdot f((x_R, y_R) + \sigma(x, y)) + \beta(x, y), \tag{3.2.16}$$

for all $(x, y) \in \Re^2$ and $z \in \Re$.

All image transformations used in the image models in this book have the above form, including the ones presented in previous chapters. Before we end this section let us complete the fractal model of the maple leaf.

3.2.4 THE MAPLE LEAF

As we have shown in Chapter 2, a maple leaf is generated by the iterated function system $W = \{w_1, w_2, w_3, w_4\}$, where

$$w_1\begin{pmatrix} x \\ y \end{pmatrix} = \begin{pmatrix} 0.49 & 0.01 \\ 0 & 0.62 \end{pmatrix}\begin{pmatrix} x \\ y \end{pmatrix} + \begin{pmatrix} 25 \\ -2 \end{pmatrix}, \quad w_2\begin{pmatrix} x \\ y \end{pmatrix} = \begin{pmatrix} 0.27 & 0.52 \\ -0.40 & 0.36 \end{pmatrix}\begin{pmatrix} x \\ y \end{pmatrix} + \begin{pmatrix} 0 \\ 56 \end{pmatrix},$$

$$w_3\begin{pmatrix} x \\ y \end{pmatrix} = \begin{pmatrix} 0.18 & -0.73 \\ 0.50 & 0.26 \end{pmatrix}\begin{pmatrix} x \\ y \end{pmatrix} + \begin{pmatrix} 88 \\ 8 \end{pmatrix}, \quad w_4\begin{pmatrix} x \\ y \end{pmatrix} = \begin{pmatrix} 0.04 & -0.01 \\ 0.50 & 0 \end{pmatrix}\begin{pmatrix} x \\ y \end{pmatrix} + \begin{pmatrix} 52 \\ 32 \end{pmatrix},$$

with the probabilities $\{p_1 = 0.31, p_2 = 0.31, p_3 = 0.31, p_4 = 0.07\}$. How does this fit in our model? It is quite simple. We have exactly four fractal elements, and each affine transformation in the IFS W corresponds to one of them.

All four fractal elements have the identical masking function, which masks a large enough area to accommodate the whole leaf, e.g.,

$$\mu_1(x,y)=\mu_2(x,y)=\mu_3(x,y)=\mu_4(x,y)=\begin{cases}1, & \text{if } (x,y)\in[-100,100]\times[-100,100],\\ 0, & \text{otherwise,}\end{cases}$$

for all $(x, y)\in\Re^2$. For each $i=1, 2, 3, 4$, the image transformation T_i is constructed in the way described in the previous paragraph, that is, from a spatial component, which is exactly the iterated function $\chi_i=w_i$; and an intensity component, which is defined by the probability p_i:

$$\lambda_i(x,y,z)=K\cdot p_i\cdot z\,, \tag{3.2.17}$$

for any $(x, y)\in\Re^2$and $z\in\Re$, where

$$K=\frac{1}{\sum_{i=1}^{4}p_i\cdot\left|\det w_i\right|} \tag{3.2.18}$$

is the positive number that preserves the total brightness constant in the Markov process. If we want the represented leaf to have a total brightness B, then the seed image can be any image with a total brightness B. The image generated by this model is exactly the image shown in Figure 3.2.1*a*, the seed image for the example in Section 3.2.1, by a coordinate system shifting and rescaling. So the four fractal elements of the seed image could be added to this model. Consequently, the image shown in Figure 3.2.1*c* can be represented by a fractal model of six fractal elements with an arbitrary seed image. The total brightness of the seed image in the rectangular area $[2, 4]\times[0, 2]$ is required to be a predetermined constant.

3.3 PARTITIONED ITERATED FUNCTION SYSTEMS

The simplest general approach to image representation using the fractal model is the *partitioned IFS theory*, which was introduced by Barnsley, Sloan, and Jacquin in 1988 ([BS], [J2]) for image compression. Here we introduce their theory using the language of fractal elements presented in the last section.

3.3.1 FRACTAL PARAMETERS

Given a fractal model $(W=\{\Phi_i=(\mu_i,T_i)\}_{i=1,2,\cdots,m},\ s)$ on a rectangular image support $\Delta=[0,w]\times[0,\ h]\subset\Re^2$, it is called a *partitioned iterated function system* if

1. The masking functions form a binary partition on the image support Δ, i.e., $\mu_i : \Re^2 \longrightarrow \{0,1\}$ for any $1 \le i \le m$, and $\mu_i^{-1}(1) \cap \mu_j^{-1}(1) = \varnothing$ for any $1 \le i < j \le m$.

2. The spatial component of any image transformation is an affine transformation, i.e., the spatial component χ_ι of T_i, for any $1 \le i \le m$, has the form

$$\chi_i \begin{pmatrix} x + x_D^i \\ y + y_D^i \end{pmatrix} = \sigma_i \begin{pmatrix} x \\ y \end{pmatrix} + \begin{pmatrix} x_R^i \\ y_R^i \end{pmatrix} = \begin{pmatrix} s_{00}^i & s_{01}^i \\ s_{10}^i & s_{11}^i \end{pmatrix} \begin{pmatrix} x \\ y \end{pmatrix} + \begin{pmatrix} x_R^i \\ y_R^i \end{pmatrix}, \qquad (3.3.1)$$

for any $(x, y) \in \Re^2$, where $(x_D^i, y_D^i) \in \Re^2$ is a chosen destination region origin and $(x_R^i, y_R^i) \in \Re^2$ is a chosen reference region origin.

3. The intensity component of any image is generated from a constant brightness adjustment β_i, a local brightness tilting $\tau_i = (t_0^i, t_1^i)$, and a constant contrast adjustment γ_i. Moreover, the contrast adjustment is contractive, i.e., the intensity component λ_i of T_i, for any $1 \le i \le m$, has the form

$$\lambda_i(x, y, z) = \gamma_i \cdot z + t_0^i \cdot x + t_1^i \cdot y + \beta_i, \qquad (3.3.2)$$

for any $(x, y) \in \Re^2$ and $z \in \Re$, where β_i is a constant, called *brightness shift*, or β-*value*, that adjusts the image brightness, and where γ_i is a constant, called *contrast scaling*, or γ-*value*, that adjusts the image contrast with a factor $|\gamma_i| < 1$.

Having the spatial and intensity components unified, the image transformation T_i is defined by

$$T_i \begin{pmatrix} x + x_D^i \\ y + y_D^i \\ z \end{pmatrix} = \begin{pmatrix} s_{00}^i & s_{01}^i & 0 \\ s_{10}^i & s_{11}^i & 0 \\ t_0^i & t_1^i & \gamma_i \end{pmatrix} \begin{pmatrix} x \\ y \\ z \end{pmatrix} + \begin{pmatrix} x_R^i \\ y_R^i \\ \beta_i \end{pmatrix}, \qquad (3.3.3)$$

for any $1 \le i \le m$. Thus in this model, for any $1 \le i \le m$, the ith fractal element is completely characterized by these parameters:

$$\left((x_D^i, y_D^i), (x_R^i, y_R^i), \sigma_i = \begin{pmatrix} s_{00}^i & s_{01}^i \\ s_{10}^i & s_{11}^i \end{pmatrix}, \tau_i = (t_0^i, t_1^i), \gamma_i, \beta_i \right), \qquad (3.3.4)$$

which is called a *fractal code* of the fractal element. Furthermore,

(x_D^i, y_D^i)	the destination region origin, is called *the D-origin*,
(x_R^i, y_R^i)	the reference region origin, is called *the R-origin*,
σ_i	the spatial transformation matrix, is called *the s-form*,
τ_i	the intensity tilting vector, is called *the t-form*,
γ_i	the contrast adjustment, is called *the γ-value*, and
β_i	the brightness adjustment, is called *the β-value*.

Become familiar with these six parameters, for we will use them very often from now on.

3.3.2 THE COLLAGE THEOREM

Before we state the theorem, let us revisit the Markov process associated to a fractal model and describe it in a partitioned IFS.

Given a partitioned IFS fractal model $W = \{\Phi_i = (\mu_i, T_i)\}_{i=1,2,\cdots,m}$, where T_i has the affine parameters we have just defined,

$$T_i = \left((x_D^i, y_D^i), (x_R^i, y_R^i), \sigma_i = \begin{pmatrix} s_{00}^i & s_{01}^i \\ s_{10}^i & s_{11}^i \end{pmatrix}, \tau_i = (t_0^i, t_1^i), \gamma_i, \beta_i \right), \tag{3.3.5}$$

the associated Markov process is defined by

$$W(f)(x, y) = \begin{cases} \gamma_i \cdot f\left(\sigma_i^{-1} \begin{pmatrix} x - x_D^i \\ y - y_D^i \end{pmatrix} + \begin{pmatrix} x_R^i \\ y_R^i \end{pmatrix} \right) + \tau \begin{pmatrix} x - x_D^i \\ y - y_D^i \end{pmatrix} + \beta_i, \\ \qquad \text{if } (x, y) \in \mu_i^{-1}(1) \subset \Re^2, \text{ for some } 1 \le i \le m; \\ 0, \qquad \text{otherwise}; \end{cases} \tag{3.3.6}$$

for any $(x, y) \in \Re^2$ and $f \in \mathcal{P}(\Re^2)$.

Let the *maximal γ-value* be denoted by

$$p = \max_{1 \le i \le m} |\gamma_i|, \tag{3.3.7}$$

which is less than 1 by assumption. The *collage theorem for partitioned IFS* in the image space using the *supremum* metric, i.e., the L^∞-metric, is the following:

THE COLLAGE THEOREM:

Given a partitioned IFS $W = \{\Phi_i = (\mu_i, T_i)\}_{i=1,2,\cdots,m}$ with the maximal γ-value p, $p < 1$, then there is a unique image, $a \in \mathcal{P}(\Re^2)$, in the image space that is represented by this fractal model. Moreover,

1. For any image $f \in \mathcal{P}(\Re^2)$, if

$$d(f,\ W(f)) \leq \varepsilon \tag{3.3.8}$$

for some $\varepsilon > 0$, then

$$d(f,\ a) \leq \frac{\varepsilon}{1-p}. \tag{3.3.9}$$

2. For any image $f \in \mathcal{P}(\Re^2)$, we have

$$d(W^n(f),\ a) \leq p^n, \tag{3.3.10}$$

for all integers n.

The proof is straightforward and computational. We will leave it as an exercise.

This theorem is the foundation of the fractal compression algorithm. That is, given an *original image* $o \in \mathcal{P}(\Re^2)$, an *encoder* will find a partitioned IFS fractal model $W = \{\Phi_i = (\mu_i, T_i)\}_{i=1,2,\cdots,m}$ with the maximal γ-value p such that $d(o,\ W(o))$is very small. In this case, this fractal model represents an image $a \in \mathcal{P}(\Re^2)$ whose difference from the original image is bounded by

$$d(o,\ a) \leq \frac{d(o,\ W(o))}{1-p}. \tag{3.3.11}$$

The difference $d(o,\ W(o))$ is called the *collage error*, and the difference $d(o,\ a)$ is really the *decoded error.*

The above theorem does not give a useful error boundary, but it provides motivation with a methodical idea that works with few exceptions.

One of the extreme exceptions is shown in Figure 3.3.1*a*, an image formed by black (1) and white (255) vertical stripes of 1-pixel width. Let us consider the following fractal model parameterized by a number p, $p < 1$. The masking functions divide the image into the 4×4 square partition. For each 4×4 square destination region, the image transformation is constructed from the spatial map, which is a simple translation mapping the very first top-left 4×4 square to the destination region, and the intensity map, which is given by the formula

$$z \mapsto p \cdot z + 127\,(1-p), \tag{3.3.12}$$

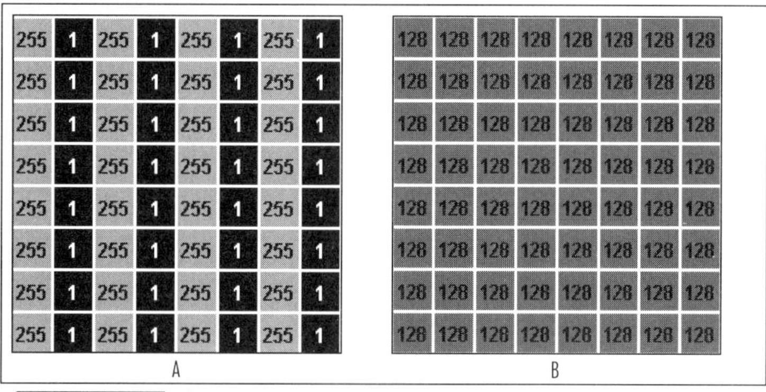

FIGURE 3.3.1 An extreme case

for any $z \in \Re$. Following the definitions step by step, it is easy to verify that the image represented by this model is the flat image of the constant intensity value 128 (Figure 3.3.1*b*). Thus, the decoded error is 127, the worst possible error one can get. But the collage error is $127(1-p)$, which can be made as small as we wish by tuning the value p.

In fact, in this extreme case, the first fractal element can either be replaced with a piece of seed image, or split into a few more fractal elements. Actually, the flat approximation of the striped image is not a bad representation of visual reality. It is exactly what we see at a distance.

Note in the collage theorem that the first part of it shows that by using a partitioned IFS fractal model no seed image is needed. The second part of the theorem gives an iteration estimation. Indeed, the iteration process can be well controlled by the error upper bound p^n. For example, for $p=3/4$, $n=3$ gives the first bit correctly, $n=5$ gives two correct bits, $n=8$ gives three correct bits, $n=10$, 13, 15, 17, and 20 give 4, 5, 6, 7, and 8 correct bits, respectively, since $\left(\frac{3}{4}\right)^3 < \frac{1}{2}$, $\left(\frac{3}{4}\right)^5 < \left(\frac{1}{2}\right)^2$, $\left(\frac{3}{4}\right)^8 < \left(\frac{1}{2}\right)^3$, and so on. Thus, the theorem provides a guideline for fractal decompression algorithms.

3.4 SQUARE MASKING FRACTAL REPRESENTATION

There has been enough theory for this chapter. It's time for applying the ideas to photographic digital imagery. As with learning anything else, we always start from the simplest case.

First, put the following list of restrictions on the first fractal modeling system:

1. The *masking functions* form a square partition. That is, an image is cut into same-sized square pieces, i.e., every square piece has $n \times n$ pixels, for some integer n. Each piece is a destination region, having exactly one fractal code attached to it.

2. The *six parameters* of each image transformation have the following conditions:

   a. The *D-origin* is always pointed to the top-left corner of the square destination region.

   b. The *R-origin* is always pointed to the even pixel grid of the digital image.

   c. The *s-form* is $1/2\sigma$, where σ is one of the eight isometries listed in Section 2.3.1.

   d. The *t-form* is null, i.e., $\tau = (0\ 0)$.

   e. The *γ-value* is the constant $\gamma = 3/4$.

   f. The *β-value* is an integer between –127 to 127, assuming that the maximal digital intensity value is 255.

There is no reason to limit fractal codes to such a small set of choices in the vast number of possible fractal models. However, even for such a small collection, many problems still need to be solved. These choices will be gradually expanded from this set in different directions for a better-performing, advanced compression system.

3.4.1 ENCODING AND DECODING ALGORITHMS

The procedure for finding a fractal model for a given image is called *encoding*, *compression*, or searching for a *fractal image representation*. The reverse process of generating an image from a fractal model is called *decoding*, *decompression*, or displaying a fractal format image.

Start from the compression algorithm (see Figure 3.4.1), which consists of the following major steps:

For the moment, a 4×4 square partition will be used, and only 8-bit monochrome digital images with a dimension exactly fitting the partition are considered, i.e., the image width and height are multiples of 4. The discussion can be easily generalized to any $n \times n$ square partition. An image of arbitrary dimension can always be padded to be multiple of 4. And Chapter 11 will deal with color images in detail.

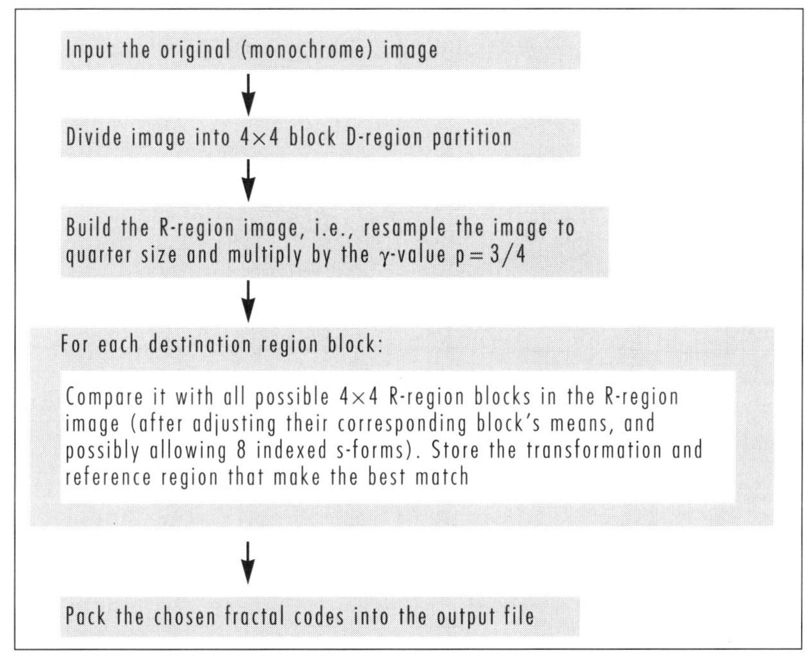

FIGURE 3.4.1 *Fractal image encoder*

1. *Input the original image.* Read the original image into the designated image buffer.

2. *Create reference region list.* Prepare the list of all possible reference regions for a destination region to match. In our case, we resample the original image using an averaging method to get a shrinkage in the ratio 2:1, since all s-forms are 2:1, and then multiply the image by a factor of 3/4.

3. *Initialize destination region list.* Allocate the space for all possible fractal elements and set the default values.

4. *Search for fractal match.* Given a destination region, loop over all possible reference regions to find the best match using a given metric. This is the computationally most involved step of the whole algorithm.

5. *Select fractal elements.* We will select all fractal elements in this case, since there are no overlapped codes.

6. *Pack the fractal codes.* Store the fractal parameters into a file by further lossless packing using some entropy coding methods.

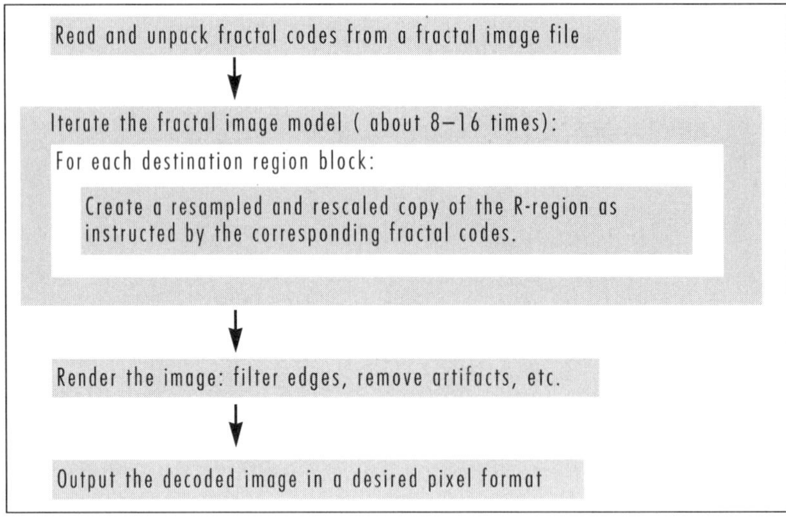

FIGURE 3.4.2 *Fractal image decoder*

Decompression is exactly the reverse procedure, as shown in Figure 3.4.2.

1. *Input the fractal codes.* Read and unpack the code file of the fractal model.
2. *Do decompression iteration* about 8 to 16 times. In each iteration, we process one destination region at a time, by mapping the resampled and rescaled reference region, as instructed by the corresponding fractal code, to the destination region.
3. *Render the decoded image* (optional). Filter the block edges to reduce the blockiness, and detect and remove certain types of artifacts.
4. *Output the decoded image* in some desired pixel format, such as BMP, TIFF, Targa, Raster, or GIF.

When an image is described in a fractal model, it is purely in mathematical transforms. In other words, the meaning of *pixel* used in the compression process is no longer relevant here. Even though an image is encoded in the grid of 4×4 for each destination region, it can still be decoded in any resolution as long as the mathematical precision is accurate. This is a unique property inherent in fractal image modeling and is called *resolution independence.* It enables us to display the same fractal image in various resolutions that fit different output devices.

3.4.2 FRACTAL ENCODING CLASS

Here is the prototype of our first fractal compression codec, which will be called *Encoder—version 1.00.* The program is written in the C language for monochrome images having dimension a multiple of 4 not exceeding 512×512. The restriction is purely for simplifying our discussion.

Therefore, we may assume that an image has the following structure:

```
1   typedef struct monochrome_image
2   {
3       SHORT    w;      // image width
4       SHORT    h;      // image height
5       BYTE     *px;    // image pixel buffer
6   }GIMAGE;
```

And a fractal code has the following structure:

```
1   typedef struct fractal_codes_100
2   {
3       WORD     dx;     // D-region origin x-component
4       WORD     dy;     // D-region origin y-component
5       WORD     rx;     // R-region origin x-component
6       WORD     ry;     // R-region origin y-component
7       SHORT    q;      // intensity beta-value
8       WORD     s;      // spatial matrix index
9   }FRACODE;
```

The compressor converts an input image, which is assumed in `GIMAGE` structure, to a fractal model, which is an array of fractal codes in `FRACODE` structure:

```
1   INT Fractal_Compressor( );
```

It will be built as the major component of the class `Encoder_100`:

```
1   #define N    4      // D-region resolution in pixel
2   #define NXN  4*4    // number of pixels in a D-region

3   class Encoder_100 {

        public :
4               Encoder_100();
5               ~Encoder_100();

6       INT     Fractal_Compressor( );
7       INT     Fractal_Packer( );

8       GIMAGE  image;          // original image

9       LONG    fcode_num;      // number of fractal codes
10      FRACODE *fcode;         // array of fractal codes

11      LONG    fdata_len;      // length of fractal data in byte
12      BYTE    *fdata;         // buffer of fractal data

        private:

13      LONG    *Dist,*DTbl[512];    // metric table
14      BYTE    PTbl[1024];          // gamma-value table

15      LONG    STbl[8][NXN];   // s-form table
16      GIMAGE  R_image;        // R-region image
```

```
17      BYTE    *R_mean;         // R-region means
18      INT     Initialize_R_Region( );

19      BYTE    Dblk[NXN];         // D-region
20      INT     Global_Searching(FRACODE *fc);
21  };
```

Now we have our main compression routine implemented exactly as the algorithm shown in the skeleton of the previous paragraph.

```
1   INT Encoder_100::Fractal_Compressor( )
2   {
3       LONG    k;           // long variables
4       BYTE    *px0,*px1;   // byte pointer variables
5       SHORT   x,y;         // D-region coordinates
6       FRACODE *fc;         // active fractal code pointer

        // Check image dimension:
7       x = image.w/N;
8       y = image.h/N;
9       if((x*N != image.w)||(y*N != image.h)) return(FAIL);
10      fcode_num = x*y;

        // Initialize R-region for matching
11      k = ((image.w*image.h)>>2);
12      GetMemory(R_image.px, k);
13      GetMemory(R_mean, k);
14      Initialize_R_Region( );

        // Loop D-regions in scan line order
15      fc = fcode;
16      for(y=0;y<image.h;y+=N){
17      for(x=0;x<image.w;x+=N){

            // Read in one D-region
18          memcpy((px1=Dblk),(px0=(image.px+x+y*image.w)),N);
19          for(k=1;k<N;k++) memcpy((px1+=N),(px0+=image.w)),N);

            // Set fractal code to default:
20          fc->dx = fc->rx = x;
21          fc->dy = fc->ry = y;
22          fc->s = fc->q = 0;

            // Call the R-region matching routine:
23          Global_Searching(fc);

            // Go to next code
24          fc++;
25      }}
26      FreeMemory(R_mean);
27      FreeMemory(R_image.px);

28      return(OK);
29  }
```

The reference region initialization routine first creates the reference region image, which is obtained by averaging the original image to its half dimension, i.e., quarter size, and multiplying a γ-value factor, which is 3/4 in our case here. Then this routine generates eight s-form maps, one for each indexed s-form.

Finally, it calculates the reference region intensity means, for each possible reference region with the origin at the pointed location. The reason for having intensity means is simply for reduced computation since they are repeatedly used in the searching routine.

```
1   INT Encoder_100::Initialize_R_Region( )
2   {
3       LONG    x,y;
4       LONG    z;
5       BYTE    *px0,*px1,*px2;

        // Get R-region image dimension:
6       R_image.w = (image.w>>1);
7       R_image.h = (image.h>>1);

        // Get R-region image by averaging:
8       px0 = R_image.px;
9       px2 = (px1 = image.px) + image.w;
10      for(y=0;y<R_image.h;y++){
11          for(x=0;x<R_image.w;x++){
12              *(px0++) = PTbl[px1[0]+px1[1]+px2[0]+px2[1]];
13              px1+=2; px2+=2;
14          }
15          px2 = (px1 += image.w) + image.w;
16      }

        // Get R-region indexed s-form map:
17      for(y=0;y<N;y++){
18      for(x=0;x<N;x++){
19          STbl[0][x+y*N] = x + y*R_image.w;
20          STbl[1][x+y*N] = (N-1-x) + y*R_image.w;
21          STbl[2][x+y*N] = x + (N-1-y)*R_image.w;
22          STbl[3][x+y*N] = (N-1-x) + (N-1-y)*R_image.w;
23          STbl[4][x+y*N] = x*R_image.w + y;
24          STbl[5][x+y*N] = x*R_image.w + (N-1-y);
25          STbl[6][x+y*N] = (N-1-x)*R_image.w + y;
26          STbl[7][x+y*N] = (N-1-x)*R_image.w + (N-1-y);
27      }}

        // Get R-region NxN block mean screen:
28      px0 = R_mean;
29      px1 = R_image.px;
30      for(x=(R_image.h-N+1)*R_image.w-N+1);x>0;x--){
31          z = (NXN>>1);
32          for(y=0;y<NXN;y++) z+=px1[STbl[0][y]];
33          *(px0++) = (z/NXN);
34          px1++;
35      }

36      return(OK);
37  }
```

The searching routine is the most time-consuming routine in the whole encoding process. It will be improved gradually. Here the simplest exhaustive version is given.

```
1   INT Encoder_100::Global_Searching(
2       FRACODE *fc
3   ){

4       LONG    k,z,m,q,best;   // long variables
5       BYTE    *px;            // r-region origin pointer
6       LONG    x,y;            // r-region origin coordinates
7       LONG    *sss;           // active s-form pointer
8       LONG    *dif;           // active metric pointer

        // Calculate D-region mean:
9       m = (NXN>>1);
10      for(y=0;y<NXN;y++) m+=Dblk[y];
11      m /= NXN;

        // Loop R-regions:
12      best = 0x2BAD2BAD;
13      for(y=(R_image.h-4)*Image.w;y>=0;y-=Image.w){
14      for(x=R_image.w-4;x>=0;x--){

            // Get R-region origin:
15          px = R_image.px + (z=x+y);

            // Offset the metric table by the beta-value:
16          dif = Dist + (q = m - R_mean[z]);

            // Check q is between -128 to 127
17          if(!((q+128)&0xFF00)){

            // Try all 8 indexed s-forms:
18          for(s=0; s<8; s++){
19              sss = STbl[s];
20              z = 0;
21              for(k=0;k<NXN;k++) z+=dif[px[sss[k]]-Dblk[k]];

                // Take it if it is better:
22              if(z<best){
23                  fc->rx  = x;
24                  fc->ry  = y;
25                  fc->s   = s;
26                  fc->q   = q;
27                  best = z;
28              }
29          }}
30      }}
31      return(OK);
32  }
```

The metric table and the γ-value table are generated in the class constructor.

```
1   void Encoder_100::Encoder_100( )
2   {

3       LONG    k;

    //  Initialize metric table:
4       Dist = DTbl + 256;
        // use L(2):
5       for(k=0;k<256;k++) Dist[k] = Dist[-k] = k*k;
        // use L(1): (replace line above with line below)
```

```
        // for(k=0;k<256;k++) Dist[k] = Dist[-k] = k;

    //  Initialize gamma table: p = 3/4, PTbl[k] = (k*4)*(3/4)
6       for(k=0;k<1024;k++)  PTbl[k] = ((k*3+8)>>4);

7       return;
8   }
```

The fractal code structure has a size of 12 bytes. There is no reason to store all 12 bytes, since some bits are always null and some of the entries can be deduced. So when compression is an issue, the fractal codes can be packed with many fewer bytes. Chapter 12 will focus on this in depth. For right now, the straightforward bit packer will be given.

What are the unnecessary bits in the fractal code structure? If fractal codes are placed in a specific order, such as the scan line order, the entire destination region origin coordinates become redundant. The reference region origin can be limited to a number between 0 and `(R_image.w*R__image.h-1)`. The s-form is one of the eight choices, i.e., 3 bits. And the β-value has been fit to exactly 8 bits.

For example, for a 512×512 image, exactly 27 bits are needed to each fractal code. If an $n \times n$ square partition is coded, for $n = 3, 4, 5, 6, 7, 8, \ldots$, the compression ratios will be 2.6, 4.7, 7.4, 10.7, 14.5, 19.0, ..., respectively.

Hence, the input of the packer will be the fractal codes, and the output will be a string of bits, which will be padded to bytes.

```
1   INT Encoder_100::Fractal_Packer( )
2   {
3       LONG    k,z;
4       BYTE    r;          // R-region origin bit number
5       BYTE    *ptr;       // packed data active pointer
6       BYTE    b;          // cursor of bit position

        // Get the number of bits for R-region origin:
7       k = R_image.w*R_image.h - 1;
8       r = 0;
9       while((k>>r)) r++;

        // Calculate the packed size, and
        // set the data buffer to NULL:
10      fdata_len = ((fcode_num*(r+11)+7)>>3);
11      memset(fdata,0,fdata_len);

        // Loop the fractal codes:
12      ptr = fdata;
13      b = 0;
14      for(k=0;k<fcode_num;k++){

            // R-region origin: (r bits)
15          z = fcode[k].rx + fcode[k].ry*R_image.w;
16          *ptr |= ((z<<=b)&255);
17          b += r;
18          while(b>7){
19              b-=8;
20              *(++ptr) = ((z>>=8)&255);
21          }

            // s-form index: (3 bits)
```

```
22          z = fcode[k].s;
23          *ptr |= ((z<<=b)&255);
24          if((b+=3)>7){
25              b-=8;
26              *(++ptr) = ((z>>=8)&255);
27          }

            // beta-value: (8 bits)
28          z = fcode[k].q + 128;
29          *ptr |= ((z<<=b)&255);
30          *(++ptr) = ((z>>=8)&255);
31      }

32      return(OK);
33  }
```

By now, a complete class of compression modules has been presented that can be adapted and compiled on any computer platform. In the next paragraph, the corresponding decompression class will be explained.

3.4.3 FRACTAL DECODING CLASS

As the inverse of the last section, we have most functions and variables well paired.

The decompressor converts a fractal model, i.e., an array of fractal codes in FRACODE structure, to a digital image in the form of GIMAGE structure:

```
1   INT Fractal_Decompressor( );
```

It is the major component of the class Decoder_100:

```
1   #define N     4      // D-region resolution in pixel
2   #define NXN   4*4    // number of pixels in a D-region

3   class Decoder_100 {

        public :
4                   Decoder_100();
5                   ~Decoder_100();

6       INT      Fractal_Decompressor( WORD iteration );
7       INT      Fractal_Unpacker( );
8       GIMAGE   image;           // decoded image
9       LONG     fcode_num;       // number of fractal codes
10      FRACODE  *fcode;          // array of fractal codes
11      LONG     fdata_len;       // length of fractal data in byte
12      BYTE     *fdata;          // buffer of fractal data

        private:

13      BYTE     *Cut,CTbl[768];      // truncate to [0,255
14      BYTE     PTbl[1024];          // gamma-value table

15      LONG     STbl[8][NXN];        // s-form table
16      GIMAGE   R_image;             // R-region image
17      INT      Initialize_S_Form( );

18      INT      Decoding_One_Code(FRACODE *fc);
19  };
```

What follows is the main decompression routine implemented for *Decoder—version 1.00*.

```
1   INT Decoder_100::Fractal_Decompression(
2       WORD    iteration
3   ){
4       LONG    i,k,z;
5       FRACODE *fc;

        //  Create a temporary working image for iteration:
6       GetMemory(R_image.px,(z=image.w*image.h));

        //  Set decoded screen to grey 128:
7       Initialize_S_Form( );

        //  Set decoded screen to grey 128:
8       memset(image.px,128,z);

        //  Iteration Loop:
9       for(i = iteration; i>0; i--){

        //  Copy the decoded image to the working image:
10      memcpy(image.px,R_image.px,z);

        //  Decode one fractal code at a time
11          fc = fcode;
12          for(k=fcode_num;k>0;k--){
13              Decoding_One_Code(fc);
14              fc ++;
15          }
16      }
17      return(0);
18  }
```

Initialize the s-form map in the full dimension (in comparison with the half dimension case in *Encoder*).

```
1   INT Decoder_100::Initialize_S_Form( )
2   {

3       LONG    x,y;

        // Get R-region image dimension:
4       R_image.w = image.w;
5       R_image.h = image.h;

        // Get R-region indexed s-form map:
6       for(y=0;y<N;y++){
7       for(x=0;x<N;x++){
8           STbl[0][x+y*N]=(( x + y*R_image.w )<<1);
9           STbl[1][x+y*N]=(( (N-1-x) + y*R_image.w )<<1);
10          STbl[2][x+y*N]=(( x + (N-1-y)*R_image.w )<<1);
11          STbl[3][x+y*N]=(( (N-1-x) + (N-1-y)*R_image.w )<<1);
12          STbl[4][x+y*N]=(( x*R_image.w + y )<<1);
13          STbl[5][x+y*N]=(( x*R_image.w + (N-1-y) )<<1);
14          STbl[6][x+y*N]=(( (N-1-x)*R_image.w + y )<<1);
15          STbl[7][x+y*N]=(( (N-1-x)*R_image.w + (N-1-y) )<<1);
16      }}

17      return(OK);
18  }
```

Corresponding to the searching routine, the decoding routine handles one fractal code at a time.

```
1   INT Decoder_100::Decoding_One_Code(
2       FRACODE *fc
3   ){

4       LONG    x,y;
5       BYTE    *dpx;         // D-region pointer
6       BYTE    *rpx;         // R-region origin
7       BYTE    *px1,*px2;    // R-region pointers
8       BYTE    *ccc;         // trunc. pointer after q-shifting
9       LONG    *sss;         // s-form pointer

        // Set D-origin, R-origin, s-form map, q-adjustment:
10      dpx = image.px + fc->dx + fc->dy*image.w;
11      rpx = R_image.px + fc->rx + fc->ry*R_image.w;
12      sss = STbl[fc->s];
13      ccc = Cut + fc->q;

        // Bring R-region into D-region:
14      for(y=0;y<NXN;y+=N){
15          for(x=0;x<N;x++){
16              px2 = (px1 = rpx + sss[y+x]) + R_image.w;
17              dpx[x] = ccc[PTbl[px1[0]+px1[1]+px2[0]+px2[1]]];
18          }
19          dpx += image.w;
20      }

21      return(OK);
22  }
```

Analogously, some shared tables, such as the truncation table and the γ-value table, are generated in the class constructor.

```
1   void Decoder_100::Decoder_100( )
2   {

3       LONG    k;

        // Initialize p-table: PTbl[k] = (k/4) * (3/4):
4       for(k=0;k<1024;k++) PTbl[k] = ((k*3+8)>>4);

        // Initialize byte cut table:
5       Cut = CTbl + 256;
6       memset(Cut-256,0,256);
7       memset(Cut+256,255,256);
8       for(k=0;k<256;k++) Cut[k] = k;

9       return;
10  }
```

In contradistinction to the packer, the unpacker will alter the packed fractal data into the fractal codes in FRACODE structure.

```
1   INT Decoder_100::Fractal_Unpacker( )
2   {
3       LONG    x,y,z;
```

```
4       LONG    w_half;      // half of the image width
5       LONG    r_mask;      // R-origin data mask
6       BYTE    r;           // R-region origin bit number
7       BYTE    *ptr;        // packed data active pointer
8       BYTE    b;           // cursor of bit position
9       FRACODE *fc;         // active fractal code

        // Check image dimension:
10      x = image.w/N;
11      y = image.h/N;
12      if((x*N != image.w)||(y*N != image.h)) return(FAIL);
13      fcode_num = x*y;
14      fdata_len = ((fcode_num*(r+11)+7)>>3);

        // Get the number of bits for R-region origin:
15      z = (w_half = (image.w>>1))*(image.h>>1) - 1;
16      r = 0;
17      while((z>>r)) r++;
18      r_mask = (1<<r) - 1;      // mask a number to <r bits

        // Loop the fractal codes:
19      ptr = fdata;
20      b = 0;
21      fc = fcode;
22      for(y=0;y<image.h;y+=N){
23      for(x=0;x<image.w;x+=N){

            // D-region origin:
24          fc->dx = x;
25          fc->dy = y;

            // R-region origin in full dimension: (r bits)
26          z = ptr[0]|(ptr[1]<<8)|(ptr[2]<<8)|(ptr[3]<<8);
27          z = ((z>>b)&rrr);
28          fc->rx = ((z%w_half)<<1));
29          fc->ry = ((z/w_half)<<1));

30          b += r;
31          while(b>7){
32              b-=8; ptr++;
33          }

            // s-form index: (3 bits)
34          z = ptr[0]|(ptr[1]<<8);
35          fc->s = ((z>>b)&7);
36          if((b+=3)>7){
37              b-=8; ptr++;
38          }

            // beta-value: (8 bits)
39          z = ptr[0]|(ptr[1]<<8);
40          fc->q = ((z>>b)&255)-128;
41          ptr++;

42          fc++;
43      }}

44      return(OK);
45  }
```

The first fractal compression system is complete. An exercise for the reader is to make it run on a computer.

Keep in mind that with slight modification in the unpacker, a fractal image can be decoded in a different resolution from the resolution in which the image was generated. Thus,

Exercise: Code an image in 4×4 destination blocks and decode its fractal representation in 8×8 and 6×6 destination blocks.

3.5 SAMPLE IMAGES

Since this is a technical book, the first set of sample images are the ones generally used in digital image compression and processing research: *Lena, Plava Laguna, Golden Hill,* and *Mandrill* as shown in Color Plates 2(a), 3(a), 3(b), and 2(b) respectively. These form the collection of *standard sample images.* Except for *Lena,* these images have been used as the standard JPEG testing images and they were digitized for JPEG by the *Independent Broadcasting Authority* (IBA). For scientists using common testing images, the advantage is clear: If you are familiar with these images you are able to judge more intuitively research results illustrated in these images and better comprehend the statistics gathered from these images.

It can be redundant to use the same image over and over again. Thus, where the statistical comparison is not important, other interesting images are also used as sample illustrations.

3.5.1 STANDARD SAMPLE IMAGES

What do you need to know about these sample images? In this section, we explain why these images are widely used and how to apply them as tools to judge your and others' work. Once you understand this, they no longer serve as testing images for gathering performance statistics, but become testing machines themselves.

The most commonly used testing image in digital image compression is *Lena,* a picture of Swedish born *Lenna Sjooblom,* that first appeared in *Playboy,* November 1972 [T]. This legendary face with a Mona Lisa smile can be found in almost every digital imaging journal, particularly the past decade. It is one of the earliest digitized images. Technically, the image is not a difficult image to be compressed or processed, yet it has some problematic areas. The striped texture on her hat and the irregular texture of her hair can cause problems in many compression techniques. The most important part of the image is her eyes, including the pupil, rim, and eyelash, which are surrounded by the smooth regions of her face. Scientists choose this image because it is easy but not trivial. Furthermore, for a researcher to spend long hours in front of a testing image, this picture is a pleasant one.

The image *Golden Hill* is an intricate and complicated image—full of subtleties. Eight roofs are covered by eight different textures; brick chimneys stand out as the most geometric objects; windows and doors are filled with fine lines and squares; and the nearby road and fence are in contrast to the faraway mountain and trees. There are also many small but recognizable objects: a man, a bicycle, a gasoline barrel with a rope around it, a drainage ditch on the roadside, a packed car, red flowers at the doorstep, and white sheep in the distant field. It is a detailed photograph!

The Hotel Lotus-Lila in *Plava Laguna* is a natural image with many man-made objects. It is a good image for testing the intricacies of geometric patterns. The hotel building, windows, balcony, stairs, brick wall, light poles, and trees provide a good combination of geometry and, equally important, texture. The man with a suitcase adds some motion to the image. The colorful title is helpful for determining how the text in an image is handled. The red chairs on the balcony are difficult in both color and geometry. Further subtleties that should be considered are the window curtains, a TV antenna, and a crack on the building's roof.

Mandrill, also called *Baboon*, is an image with extremely divergent colors and varied intensities. The smooth nose and ring-shaped eyes on the hairy face of the baboon with the contrasting mustache provides a perfect picture for analyzing and inspecting how the smooth region and the noise region is relatively balanced—how the recognizable geometric shapes and unrecognizable noises are distinguished. For this reason, it has been a favorite image for visual quality research. It is also widely used for color study, because of the richness of its reds, blues, yellows, and greens.

There are some other widely used testing images, e.g., *Boats*, *Barb*, *Peppers*, *Balloons*, etc.

The measurements that will be used in assessing the image quality are mainly the *root mean square metric*, i.e., L^2*-distance*, or its equivalent *peak signal-to-noise ratio*, and the *average of absolute differences*, i.e., L^1*-distance*, for the visual quality is very important. The following quality assessment standard that gives any reconstructed image a 0 to 10 *quality scale* is recommended. It has the following criteria:

10 *lossless:* numerically lossless, i.e., every bit is identical;

9 (*visually*) *perfect:* they are indistinguishable side by side in a 20" 1024×768 monitor actual display;

8 (*visually*) *transparent:* the difference cannot be identified by displaying one at a time in a 20" 1024×768 monitor actual display with a one-minute delay;

7 (*visually*) *appealing:* without looking at the original image, no fraud, other than the defects from the original one, can be found;

6 *agreeable:* it is good enough to toss the original one away;

5 *acceptable:* it is a good image; no complaints come from those who have never seen the original one;

4 *usable:* it is an OK image of average quality;
3 *cryable:* it has some obvious problems;
2 *laughable:* it still can be used to identify the original image;
1 *useless:* the original image cannot be identified.

TABLE 3.5.1 *Subjective quality assessment*

Quality Scale		Goodness Scale		Impairment Scale	
10	lossless	5	Excellent	1	Not noticeable
9	perfect	5	Excellent	1	Not noticeable
8	transparent	5	Excellent	2	Just noticeable
7	appealing	4	Good	2	Just noticeable
6	agreeable	4	Good	3	Definitely noticeable
5	acceptable	3	Fair	3	Definitely noticeable
4	usable	3	Fair	4	Impairment
3	cryable	2	Poor	5	Somewhat objectionable
2	laughable	1	Unsatisfactory	6	Definitely objectionable
1	useless	1	Unsatisfactory	7	Extremely objectionable

There are a few other well-known subjective scales: the *goodness scale* and the *impairment scale*. Their approximate relations are illustrated in Table 3.5.1. The rating is done by a *mean opinion score (MOS)*, which is the mean of all scores given by participants.

More discussion on image visual quality assessment is given in Chapter 13.

3.5.2 CHOOSING FRACTAL CODES

Vector quantization (VQ) is a well-established compression technique [GG]. When the fractal algorithm was first presented, some people concluded that the fractal method was VQ with the codebook generated from the image itself. Then they asked, "How can the image itself carry a better codebook, since the universal one can be trained from all possible images?".

After having understood the fractal modeling presented in this chapter, you should be able to see its fundamental difference from a VQ scheme. However, purely looking at the simple coding system previously presented, it is true that it can be translated as a VQ with a self-generating codebook. Is the self-generated "codebook" better than a universal one? So far, we don't know the exact answer. Maybe, someday, there will be a perfect universal codebook. It might address images as the Mandelbrot set represents Julia sets. One can imagine that for symbolizing all images in all resolutions, this *codebook* may well be a perfect fractal itself.

Until we find the ultimate universal super-image, fractal elements, as self-generated codes, are very efficient choices. In this paragraph, some evidence to support this argument will be presented. The experiment is not completely conclusive due to the incomplete technology, but it does provide the intuition and a sense of confidence in the argument.

The standard images will be coded from various other images, i.e., by taking reference regions from some other images, comparing the collage errors of the different combinations, and seeing whether an image is more accurate when coded from itself than from some other images. The result presented in Table 3.5.2 really speaks for itself and confirms our belief.

TABLE 3.5.2 *Collage error comparison.* Diagonal sum = 21.38896

D \ R	Lena	Plava Laguna	Golden Hill	Mandrill
Lena	**3.16244**	3.30780	3.68997	3.83686
Plava Laguna	5.01200	**4.47773**	6.08250	6.04348
Golden Hill	4.04245	3.88428	**3.90527**	4.07135
Mandrill	10.36655	9.80930	10.96436	**9.84352**
SUM	22.58344	21.47911	24.64210	23.79521

In the table, each row shows the collage errors of the given image coded from the specified images, and each column shows the collage errors of the specified images coded from the given one. Thus, the bold numbers on the diagonal line are the self-referenced ones, which are consistently the smallest or very close to the smallest number in each row.

Overall, the total error on the diagonal line gives the best total matching, i.e., the smallest error, in comparison with the sum of each column. Interestingly enough, the image *Plava Laguna* could serve as a good reference image, for it gave three best matches out of four. As discussed before, this image has a variety of different objects in different scaling, including the building geometry, colorful text, and brick textures.

3.5.3 FRACTAL CODE TILING SIZES

In the square partition method, an image is represented in square blocks of any given size. What is the right block size? That really depends on what it will be used for. Here, we present only a comparison. Table 3.5.3 provides statistics for some standard images represented in 3×3, 4×4, 6×6, 8×8, 12×12, and 16×16, and Figure 3.5.1 and Color Plate 4 display some of them. Notice that the difference between the decoded error from the collage error increases as the error increases.

The next chapter will study how to obtain an optimal mixture of those blocks for every given image. More evidence will confirm the fact that the right tile size always depends on the viewing distance. There is not a simple solution.

FIGURE 3.5.1 Plava Laguna *in various square partitions*

Table 3.5.3 *Various square partitions*

L^2	Lena		Golden Hill		Plava Laguna	
N	*collage*	*decoded*	*collage*	*decoded*	*collage*	*decoded*
3	1.8226	1.9985	2.2442	2.5214	2.6099	2.8596
4	3.1624	3.3704	3.8911	4.1520	4.6964	5.0193
6	5.1928	5.4832	6.3425	6.6842	8.4617	9.1457
8	6.6277	6.9640	7.8332	8.2982	11.9872	13.7905
12	9.2456	9.8277	10.5412	11.0294	16.5069	17.9802
16	10.9586	11.6998	12.3663	12.9910	20.3227	21.5922

3.5.4 Iteration numbers

In the second section (Section 3.3.2) the number of iterations needed to get all of the information of the fractal codes has been theoretically estimated to be a number less than 20 for $p=3/4$. Now let us check it out and see how close to this boundary we get in practical cases. Table 3.5.4 illustrates the L^2-differences of the decoded images in each iteration i in comparison with the original image, the decoded attractor, and the image from the previous iteration.

The interesting conclusion from Table 3.5.4 is that both images fully converge at iteration 23! Fate or coincidence?

Exercise: Find the smallest iteration number that all decoded images numerically stabilize.

Color Plate 5 shows an image decoded step-by-step in various iteration numbers.

3.5.5 Fractal code annealing

Having seen the difference between the decoded error and collage error, it is reasonable to ask whether the best set of codes is found within the given constraints. If not, can a better set of codes be obtained?

Let us analyze the algorithm first. When an image is compressed, a destination region has been compared with all possible reference regions in the collage errors, and the one with the least error chosen. For $p=3/4$, the chosen sample case, the decoded error could be as high as four times larger than the collage one. The least collage error reference region has no meaning as the least decoded error reference region. It gives only an error lower bound. Thus, the reference region with the least collage error is not necessarily the best choice, but an adequate recommendation.

Are there ways to improve fractal codes for smaller decoded errors? A good method of improving fractal image representation has been proven to be *compression with annealing*. Given an image, first compress this image in a normal way to get the first fractal representation. Then decompress the fractal representation and compress the original image again by taking the reference

TABLE 3.5.4 *Decompression, iteration, and convergence*

L^2	Lena			Golden Hill		
i	*original*	*attractor*	*previous*	*original*	*attractor*	*previous*
0	48.016402	47.803839		49.891805	49.595791	
1	33.164421	32.869468	50.160133	31.820888	31.390948	46.880547
2	17.771090	17.348753	31.199474	17.035684	16.403380	27.298880
3	7.801571	6.959135	14.851330	8.794332	7.672172	12.666736
4	4.604374	3.074657	5.086840	5.859413	4.059589	4.579958
5	3.792194	1.708593	1.685629	4.797178	2.340094	1.911426
6	3.528083	1.050378	0.860085	4.398248	1.401360	1.071217
7	3.432070	0.735213	0.613695	4.250532	0.894098	0.736818
8	3.395904	0.564582	0.463661	4.197342	0.652312	0.553793
9	3.381499	0.436850	0.357641	4.177849	0.499110	0.418407
10	3.375080	0.338697	0.275903	4.170598	0.382284	0.320890
11	3.372592	0.263274	0.213079	4.167727	0.293353	0.245123
12	3.371729	0.207675	0.161815	4.166111	0.225981	0.187052
13	3.371436	0.164863	0.126290	4.164938	0.177777	0.139508
14	3.371220	0.133857	0.096240	4.164147	0.138837	0.111036
15	3.371217	0.109148	0.077488	4.164061	0.109532	0.085314
16	3.371045	0.089525	0.062439	4.163969	0.090394	0.061856
17	3.370769	0.071443	0.053950	4.164169	0.074552	0.051118
18	3.370539	0.054966	0.045638	4.164149	0.058659	0.046013
19	3.370425	0.042791	0.034499	4.164252	0.042702	0.040217
20	3.370346	0.031433	0.029035	4.164183	0.031128	0.029232
21	3.370448	0.015747	0.027204	4.164124	0.021036	0.022944
22	3.370388	0.005167	0.014875	4.164116	0.011220	0.017794
23	3.370394	0.000000	0.005167	4.164091	0.000000	0.011220
24	3.370394	0.000000	0.000000	4.164091	0.000000	0.000000

regions (not from the original image but) from the decoded image while the destination regions are still using the original image for matching. And then keep repeating this process until there is no further improvement in quality. As shown in Table 3.5.5, by using the decoded image to predict the decoded errors, the decoded image qualities have been definitely improved in the first few annealing steps, despite of the fact that the collage errors were getting worse. The results illustrated in Table 3.5.5 were obtained from 6×6 square block representations.

Though this annealing method is effective, it requires a lot of compression time, since the compression procedure has to be repeated many times. Nonetheless, the compression speed increases by keeping the fractal codes whose decoded errors are not too far from their collage errors and modifying

TABLE 3.5.5 *Compression with annealing*

L^2	Lena		Golden Hill		Plava Laguna	
anneal	*collage*	*decoded*	*collage*	*decoded*	*collage*	*decoded*
0	5.1928	5.4832	6.3425	6.6842	8.4617	9.1457
1	5.4982	5.4139	6.7369	6.6577	9.0709	8.9467
2	5.4599	5.3764	6.7120	6.6288	9.0345	8.8562
3	5.4786	5.3852	6.7062	6.6119	9.0420	8.8014
4	5.4678	5.3796	6.6907	6.5928	9.0184	8.8417
5	5.4744	5.3780	6.7054	6.5874	9.0324	8.7919

only a small percentage of the worst ones. For example, if the annealing process is restricted and applied only to the worst 10% of the blocks of all codes, which are the blocks that have the highest ratio between their decoded errors and corresponding collage errors, the quality improves as fast as the full annealing shown in Table 3.5.5.

What is the best percentage of codes that should be compressed again? The question will be left to the reader as an exercise. Notice the difference between the decoded error and the collage error. There is a question left open in fractal image representation:

Problem: Could the decoded errors be approached directly in a compression system instead of using the collage errors?

There may not exist a complete solution. However, some partial results could be a new basis for better algorithms.

There are images I need to
complete my own reality

— The American Night

Jim Morrison

IMAGE PARTITION 4

The *fractal imaging model* has been established. In addition, one of the simplest forms, *fractal model using uniform square partition and some simple affine transformations*, has been shown and applied to real-world images. Now, in the next few chapters, details will be filled into each component of the image model as we extend this simple case to some more general models.

This chapter will focus on the masking functions. As one may notice, the true reason for the masking functions is to separate a complicated image into manageable pieces so that each piece can be handled with ease. Logically, it makes sense that the image is split based on the image's physical content. However, it is quite difficult in practice because the research on digital image segmentation presents many stumbling blocks. Also, there is not a simple and efficient way to store the image segmentation information. Furthermore, an image-independent partition could give a quite good image representation, such as the uniform square image partition we have used in the previous chapter. Realistically, the practical approaches fall somewhere between the image-independent partition and the image-dependent segmentation, and there are more likely to be some partitions of destination regions that have some fixed geometric shapes but various image-dependent scales.

As with dividing land into countries and states, the goal here in image partition is the same: for managing, controlling, and handling. It involves an art of balance—balancing between idealism and realism, between theoretical clarity and practical naturality. When a straight line was drawn between the United States and Canada, we didn't know how to avoid splitting someone's house between two separate countries. When the border between China and Russia was established along the Amur river, it sparked an inevitable century of conflict. Similarly, real-world imaging allows us to segregate a picture either into regular geometric pieces, with no concern for the image content, or into some irregular pieces that are defined ambiguously by some image content segmentation algorithm.

What is the best way to split a picture for image compression? There are many factors that need to be considered. The *mathematical simplicity,* the *engineering manageability*, and the *performance quality* are the main elements in choosing a scheme. As a common principle, all methods are intended to cut the complicated and detailed regions into smaller pieces and to group the simple and predictable ones together.

This chapter will start from a general discussion about image partitioning and covering, then confine it to partitioning by showing several well-known methods. The *quadtree partition method*, which is the most simple, effective, and commonly used method, will be discussed in detail. And in the end, a few more choices of image partitioning and masking are discussed briefly.

4.1 IMAGE DISTORTION RATE

Before demonstrating numerous fractal models, a comparison standard needs to be set. That is, for any two fractal models of the same real-world image, what is the criterion for us to say one is *better* than the other?

Mathematically, a commonly used model is based on the L^p-metric of vector spaces. In practice, it has been proven to be computationally effective but visually senseless. Actually, later in this book, in Chapter 13, the fact that there even *does not exist* such a universal criterion will be more generally convincing, for different models may suit different viewing conditions, and different eyes may have different view flavors.

In scientific research, when there is a missing piece in a common approach, we substitute something for this missing piece and continue by hoping one day the missing piece or some better substitutes are found. That is the approach we take here in compensating for the lack of a universal comparison criterion.

Let us assume there is an "acceptable" *image distortion measurement*, theoretically, which is exactly a metric in the image space,

$$\delta : \mathcal{P}(\Re^2) \times \mathcal{P}(\Re^2) \longrightarrow [0,\infty), \tag{4.1.1}$$

e.g., L^p-metric, for some $p \geq 1$. In fact, in our current implementation throughout the book, except in Chapter 13 (where a human visual perceptive metric is introduced), we use L^2-metric, i.e., the *square root of the mean square error (rmse)*. As a homework project, you are encouraged to repeat the whole story with your own favorite metric. For the discussion of this section, the distance function δ will remain abstract.

Given two fractal models P and Q for the same image O, the one having *higher quality* is defined to be the one that has smaller error based on the image distortion measurement function δ. In real applications, quality is not the only issue. Quite often a trade-off has to be made among all the factors, such as *image quality*, *compression ratio*, *compression speed*, and *decompression speed*.

The *image distortion rate*, λ, is defined to be a positive parameter that gives the ratio of importance between the image quality and the compressed data size.

THE COSTING CRITERION:

Given a fixed image distortion rate λ, the image model Q is said to be more efficient than the model P with respect to the distortion rate λ if and only if the errors with cost penalties satisfy the formula

$$\text{error}(Q) + \text{size}(Q) * \lambda < \text{error}(P) + \text{size}(P) * \lambda. \tag{4.1.2}$$

The formula can be rewritten:

$$\lambda < (\text{error}(P) - \text{error}(Q)) / (\text{size}(Q) - \text{size}(P)). \tag{4.1.3}$$

This formula is called the *costing criterion*. It implies that the *image distortion rate* λ is really the boundary line of the ratio between the error improvement and additional file size cost. On the *error-size graph* in Figure 4.1.1, the distortion rate is exactly the slope of the curve.

The *distortion rate* λ will be used as the compression parameter that tunes the image distortion and file size. When the distortion rate $\lambda = 0$, the best available fractal elements should always be chosen; these elements give the best fractal

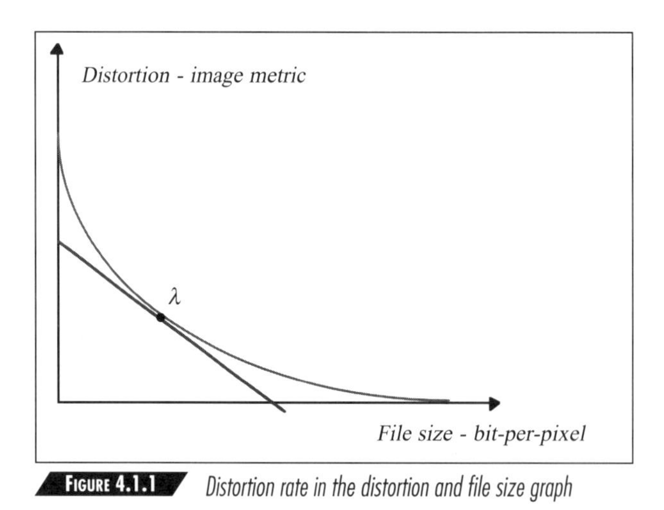

FIGURE 4.1.1 Distortion rate in the distortion and file size graph

representation codes in code data that may give no compression. On the other extreme, when the distortion rate is large enough, the cheapest fractal elements become the only best choices that give the smallest representation.

4.2 IMAGE MASKING AND COVERING

To represent the whole image, the union of the destination regions of all fractal elements must cover the entire image. It is easy to prove that any covering can be reduced to a partitioning, i.e., a covering that has no two pieces of destination regions overlapping. Wherever two destination regions may overlap, the fractal representations in the intersection can be tested to determine which is inferior to the other. The destination region with the lesser result can then be reduced so that the destination regions do not overlap. Possibly, then, a better reference region match can be found for the remaining area of the reduced destination region. Indeed, there is no reason to represent the same part of the image twice. Therefore, image partitioning has been adopted and assumed in most image compression and representation systems.

4.2.1 OVERLAPPED IMAGE COVERING

In this book as well, we will assume image partitioning in our implementation. Nonetheless, we would like to share an interesting example of overlapping destination regions, discovered recently. The point is that in scientific research, directions are chosen based on judgments and criteria. When "no" to a certain approach is concluded, it is better to know the exact meaning and implications of such a decision. Otherwise, some good alternative choices could be missed.

Let us start from an 8×8 square partition, plus another 8×8 square partition that is shifted 4 pixels right and 4 pixels down. Joining them together, an

FIGURE 4.2.1 *Overlapped 8× 8 covering*

0	1/8	1/4	1/2	1/2	1/4	1/8	0
1/8	1/4	1/2	3/4	3/4	1/2	1/4	1/8
1/4	1/2	3/4	7/8	7/8	3/4	1/2	1/4
1/2	3/4	7/8	1	1	7/8	3/4	1/2
1/2	3/4	7/8	1	1	7/8	3/4	1/2
1/4	1/2	3/4	7/8	7/8	3/4	1/2	1/4
1/8	1/4	1/2	3/4	3/4	1/2	1/4	1/8
0	1/8	1/4	1/2	1/2	1/4	1/8	0

FIGURE 4.2.2 *8× 8 block digital masking function*

overlapped covering (Figure 4.2.1) is given. Most points in the image are in exactly two destination regions. Which representation do we want to choose?

We will choose both, after applying a fixed masking function. The result will be the sum of both masked representations, as explained in the definition of a fractal model.

The masking function is given digitally in Figure 4.2.2, which is similar to a low-pass filter. The advantage of this method is that there are no blocky artifacts, i.e., the recognizable edges between the partition region edges, which can become noticeable when the destination regions become larger, as demonstrated in the last chapter.

When we store the fractal codes of the image using this method, we are doubling the number of codes that are needed to represent the destination regions. However, we don't need to store the intensity shifting or mean value in the second partition, which saves about one-third of each fractal code. So, if the fractal elements of an 8×8 partition require β *bpp*, then the overlapped case will require $(5/3)\beta$ *bpp*, a size that is comparable to the 6×6 square partition, which requires $(64/36)\beta$ *bpp*.

Now let us compare the 8×8 overlapped covering with the 6×6 square partitioning. Figure 4.2.3 shows that the 6×6 case produces an image with sharp detail, which looks better than the 8×8 one. However, in a close-up view, as shown in Figure 4.2.4, the 8×8 one is obviously more pleasant, without the blockiness that the 6×6 one has.

This discussion actually provides a practical example of a well-known fact: *the quality of a visual image depends upon the viewing distance and the output device* (which is a technical way of saying, Beauty is in the eye of the beholder).

FIGURE 4.2.3 Lena *represented using different domain coverings*

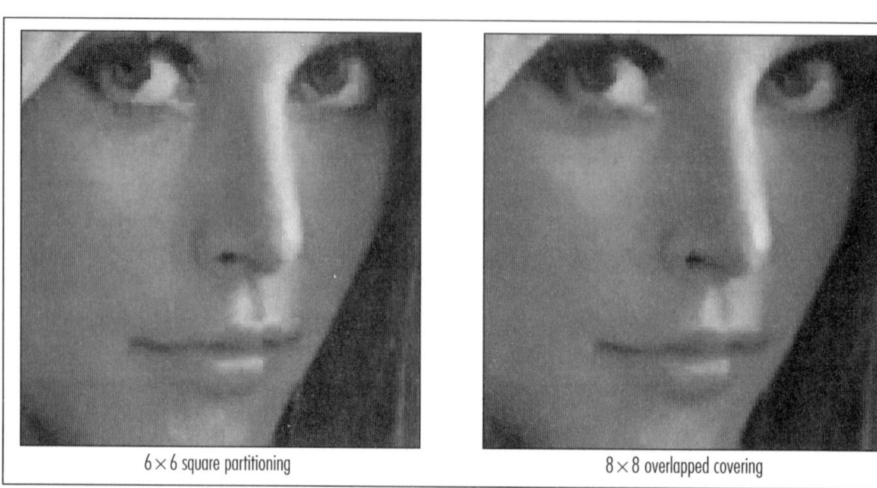

FIGURE 4.2.4 *Close look at* Lena *in Figure 4.2.3.*

4.2.2 IMAGE PARTITION

Methods of image partitioning can be classified as *image-independent tiling* and *image-dependent segmentation.*

Image-independent tiling is any partitioning scheme that does not take the structure of the image into account, but instead uses simple geometric shapes to tile the image. This method has the advantage of simplicity: the shapes of the domain regions can be simple (Figure 4.2.5), and the data can be structured simply, so that the decompressor can easily decode the domain tiles. The disadvantage is that this method cannot take into account regional difficulties and the natural connections between areas: exactly the advantage of image-dependent segmentation (Figure 4.2.6). Image-dependent segmentation will be the way of

FIGURE 4.2.5 *Simple geometric tiling*

FIGURE 4.2.6 *Image segmentation of* Lena

FIGURE 4.2.7 *Tiling from* The World of M.C. Escher *[E]: (A) Birds and Fish; (B) Reptiles; (C) Mosaic II (Reprinted from* The World of M.C. Escher, *courtesy of Cordon Art © 1997, Baarn, Holland. All rights reserved.)*

the future. So far, however, there is for our purpose no satisfactory algorithm to segment an image.

Planar tiling is an interesting mathematical field in its own right. In Figure 4.2.7, some master pieces created by Ecsher are illustrated [E]. Though they are not the practical choices, it is nice to know that they exist.

As mentioned in the beginning of this chapter, the current reality is a compromise between both methods. All of them are based on an image-independent tiling strategy, with an image-dependent splitting criterion. The most popular one is the *quadtree partition*.

4.3 QUADTREE PARTITION

Given two digital images on the same rectangular support in an L^p-metric, $p \geq 1$, the distance, i.e., the error, between these two images is given by the formula

$$d(P,Q) = \left(\sum_{i,j} \left| P[i,j] - Q[i,j] \right|^p \right)^{1/p} . \qquad (4.3.1)$$

Thus, a compression algorithm that uses a costing criterion to minimize the error is equivalent to one that minimizes the sum of total pixel distortions to some power:

$$\sum_{i,j} \left| P[i,j] - Q[i,j] \right|^p . \qquad (4.3.2)$$

Therefore, the minimization procedure in this case can be localized. In fact, we always improve the region that gives the biggest error improvement per increasing bit, which is not necessarily the region with the worst error. As a consequence, one may notice that this metric may not provide the best method to ensure maximum visual quality, but it does prevent the compressor from wasting its time on a region that might not be improvable.

4.3.1 FRACTAL CODE HIERARCHY

Let us assume an image representation of five layers of fractal codes (in most applications three layers are more than enough). That is, the combination of fractal codes whose destination regions are squares of the dimensions $16N \times 16N$, $8N \times 8N$, $4N \times 4N$, $2N \times 2N$, and $N \times N$ will be used, where the number N is called the *base tile width*. This number will be set to 1 in this chapter to make the discussion simpler. In this construction, for each fractal code that is not on the bottom layer, there are exactly four fractal codes that can be found in the next lower layer, which will be called its *children*, and this code is also called the *parent* of these four children. This parent–child relation between adjacent layers gives the *pyramid hierarchy tree* structure of the quadtree fractal codes.

As a matter of fact, in the chosen case, the fractal representation could cover from 100:1 — very high compression — to numerically lossless.

For a given destination region area, which block size should be used? As studied in the first section of this chapter, it is totally determined by the distortion rate equation we introduced earlier. That is, for a given distortion rate λ, the *adjusted error* of a fractal element of a destination region B is defined to be

$$(\text{error of the block } B) + (\text{code size in bits of } B) \cdot \lambda .$$

We will choose one of the possible mixtures of these fractal elements that produces the minimal total adjusted error.

In the quadtree image partition case, the exact procedure can be done locally as explained in the following example. Let us consider one of the largest possible destination regions, in this case a 16×16 block area, of a given image. Assume that the adjusted errors of all possible fractal codes on its sub-blocks are shown in Figure 4.3.1*a*. That is, the adjusted error of using one 16×16 fractal code is 707; the adjusted errors of four 8×8 fractal codes are 175, 176, 178, and 177, and so on. Figure 4.3.1*b* shows how such a choice is made by an ascending procedure from the bottom layer up to the top one by comparing and updating two adjacent layers at a time.

In precision, start from the smallest block size, called the *bottom layer block*, and set the partition to be the uniform square partition in the smallest block size. Move one layer up, and for each fractal code in that layer, compare between the adjusted error of this code and the sum of the adjusted errors of its four child-blocks that cover exactly the same area in the up-to-date partition. If the adjusted error of this code gives a smaller error, modify the partition by replacing all of its child-block codes with this code. Alternatively, keep the partition unchanged but update the adjusted error with the adjusted error sum of its four children. Then repeat this process by moving a layer up until the top layer is done. The result is the desired quadtree image partition.

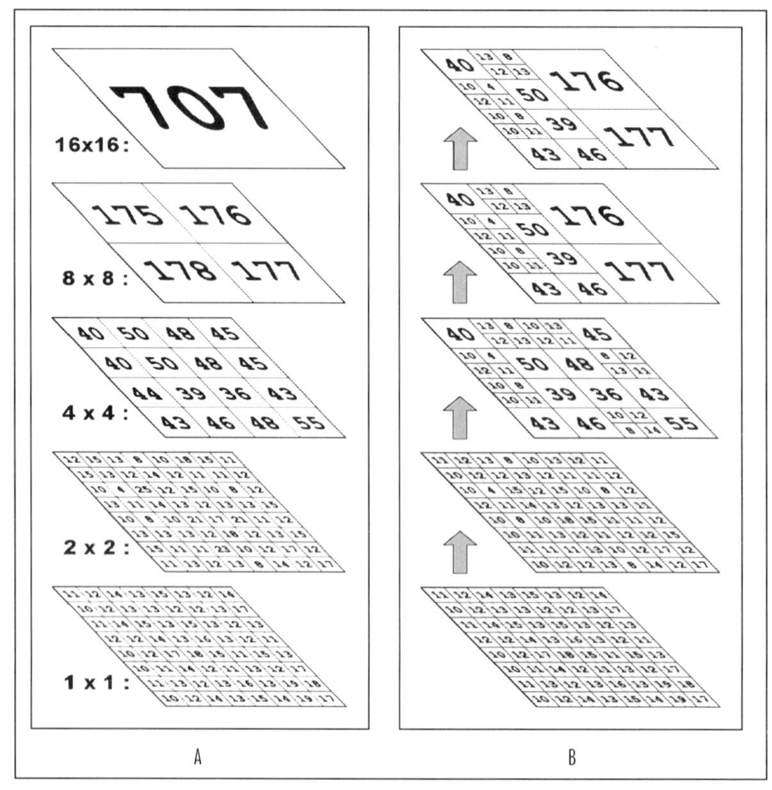

FIGURE 4.3.1 (A) Sample adjusted errors; (B) Quadtree selection

4.3.2 QUADTREE CODE SELECTION ALGORITHM

The best way to show the idea of an algorithm is to draw a skeleton, and the best way to show the detail of an implementation is to give the exact programming codes. The algorithm for selecting fractal codes will be presented in these languages.

Before doing that, a few more items need to be added to the fractal code structure: the *cost of code in bits,* the *decoded error* for using this code, and a *flag* to tell whether a code been selected. So the new structure is the following:

```
1   typedef struct fractal_codes_10a
2   {
3       WORD    dx;          // D-origin x-component
4       WORD    dy;          // D-origin y-component
5       WORD    rx;          // R-origin x-component
6       WORD    ry;          // R-origin y-component
7       SHORT   q;           // intensity beta-value
8       WORD    s;           // spatial matrix index
9       LONG    err;         // error used in comparison
10      BIYE    bit;         // cost of in storage bit
11      BIYE    tag;         // flag of selecting status
12  }FRACODE;
```

The quadtree code selection will have the class of *quadtree selector:*

```
1   #define L             5   // number of layers
2   #define CHOSEN        1   // mark for a chosen code
3   #define NOT_CHOSEN    0   // mark for an unchosen code

4   class Quadtree{

        public :
5                   Quadtree();
6                   ~Quadtree();

7           INT     Quadtree_Selector( LONG lambda );

8           LONG    fcode_num[L];     // numbers of fractal codes
9           FRACODE *fcode[L];        // arrays of fractal codes

        private:

10          LONG    *ErrBuffer;       // a temporary error buffer of
                                      // the size fcode_num[0]
11  };
```

In this class, assume that there are `fcode_num[0]` fractal codes, `*fcode_[0]`, in the bottom, or first, layer (i.e., 1×1 blocks); `fcode_num[1]` fractal codes, `*fcode_[1]`, in the second layer (i.e., 2×2 blocks); `fcode_num[2]` fractal codes, `*fcode_[2]`, in the third layer (i.e., 4×4 blocks); `fcode_num[3]` fractal codes, `*fcode_[3]`, in the fourth layer (i.e., 8×8 blocks); and `fcode_num[4]` fractal codes, `*fcode_[4]`, in the top, or fifth, layer (i.e., 16×16 blocks). Thus, `fcode_num[i] = fcode_num[i+1]*4` , for `i` = 0, 1, 2, 3. Furthermore, *all codes are assumed to be in the quad order with respect to its above layer,* i.e., the fractal code `fcode[i+1][j]` and the union of four fractal codes, `fcode[i][4*j]`, `fcode[i][4*j+1]`, `fcode[i][4*j+2]`, and

0	1	2	3	4	5	6	7	8	9	10	...
25	26	27	28	29	30	31	32	33	34	35	...
50	51	52	53	54	55	56	57	58	59	60	...
75	76	77	78	79	80	81	82	83	84	85	...
100	101	102	103	104	105	106	107	108	109	110	...
125	126	127	128	129	130	131	132	133	134	135	...
...	...	...	...	...	...	...	...	...	...	...	...

top layer of a 400×400 image

0	1	4	5	8	9	12	13	16	17	20	...
2	3	6	7	10	11	14	15	18	19	22	...
100	101	104	105	108	109	112	113	116	117	120	...
102	103	106	107	110	111	114	115	118	119	122	...
200	201	204	205	208	209	212	213	216	217	220	...
202	203	206	207	210	211	214	215	218	219	222	...
...	...	...	...	...	...	...	...	...	...	...	...

4th-from-bottom, or 2nd-from-top, layer of a 400×400 image

0	1	4	5	16	17	20	21	32	33	36	...
2	3	6	7	18	19	22	23	34	35	38	...
8	9	12	13	24	25	28	29	40	41	44	...
10	11	14	15	26	27	30	31	42	43	46	...
400	401	404	405	416	417	420	421	432	433	436	...
402	403	406	407	418	419	422	423	434	435	438	...
...	...	...	...	...	...	...	...	...	...	...	...

3rd-from-bottom, or 3rd-from-top, layer of a 400×400 image

0	1	4	5	16	17	20	21	64	65	68	...
2	3	6	7	18	19	22	23	66	67	70	...
8	9	12	13	24	25	28	29	72	73	76	...
10	11	14	15	26	27	30	31	74	75	78	...
32	33	36	37	48	49	54	55	96	97	100	...
34	35	38	38	50	51	56	57	98	99	102	...
...	...	...	...	...	...	...	...	...	...	...	...

2nd-from-bottom, or 4th-from-top, layer of a 400×400 image

FIGURE 4.3.2 *Fractal codes in the* quad-scan order

`fcode[i][4*j+3]`, cover the identical image area on its support, for all $0 \leq$`j<` `fcode_num[i]`, and `i` $= 0, 1, 2, 3$.

Figure 4.3.2 shows a quad order example, in which we assume a 400×400 image allowing five layers with the 1×1 bottom layer. The top order is in the scan-line order; then each lower layer is ruled by the requirement of quad order, with a scan order among the four child blocks for each parent block. This specific quad order is called the *quad-scan order.*

By viewing two layers at a time from the bottom up as we explained in the previous subsection, you can optimize fractal codes for minimal total error. The

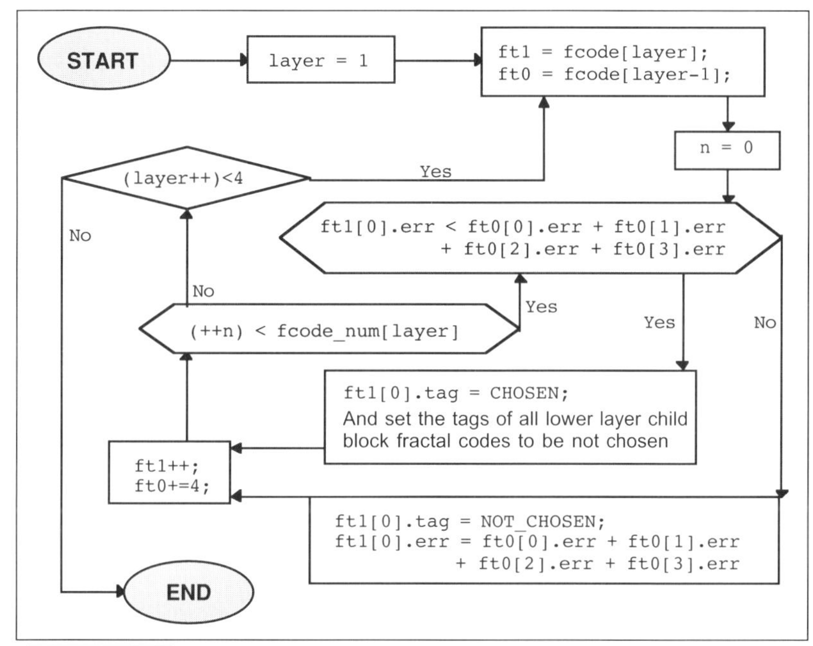

FIGURE 4.3.3 *Quadtree code selection algorithm*

algorithm is quite straightforward. The algorithm skeleton is drawn in Figure 4.3.3. The *C* program is implemented in the following code:

```
1    #define PARENT  2    // temporary mark for an unchosen code
                          // that some parent code is chosen

2    Quadtree::Quadtree_Selector(
3        LONG lambda           // distortion rate selected by user
4    ){
5        LONG    k,x,y;
6        LONG    layer;
7        FRACODE *ft,*fc;
8        LONG    *cerr,*perr;

         // Set bottom layer: error buffer to adjusted error
         //                   code flag to CHOSEN
9        cerr = ErrBuffer;
10       ft = fcode[0];
11       for(k=fcode_num[0]; k>0; k--){
12           *(cerr++) = ft->err + ft->.bit*lambda;
13           (ft++)->tag = CHOSEN;
14       }

         // Get higher layers:
15       for(layer=1;layer<L;layer++){
16           cerr =           // current error buffer pointer
17           perr = ErrBffr;  // previous error buffer pointer
18           ft = fcode[layer];
```

```
19          for(k=fcode_num[layer]; k>0; k--){
                // if new one is not better
20              if( (x = ft->err + ft->bit*lambda)
21                  >(y = perr[0]+perr[1]+perr[2]+perr[3]) ){
22                  *(cerr++) = y;
23                  (ft++)->tag = NOT_CHOSEN;
24              }
                // if new one is better
25              else{
26                  *(cerr++) = x;
27                  (ft++)->tag = CHOSEN;
28              }
29              perr += 4;
30          }
31      }

        // Reset lower layers:
32      for(layer=L-1; layer>0; layer--){
33          x = (layer==1) ? NOT_CHOSEN : PARENT;
34          ft = fcode[layer];
35          fc = fcode[layer-1];
36          for(k=fcode_num[layer]; k>0; k--){
37              switch(ft->tag){
38                  case PARENT:
39                      ft->tag = NOT_CHOSEN;
40                  case CHOSEN:
41                      fc[0].tag = fc[1].tag =
42                      fc[2].tag = fc[3].tag = s;
43                  default: break;
44              }
45          ft ++;
46          fc += 4;
47          }
48      }

49      return(OK);
50  }
```

The constructor and destructor of the class do nothing but allocate and deallocate a temporary memory buffer:

```
1   Quadtree::Quadtree( )
2   {
4       GetMemory(ErrBuffer,fcode_num[1]*sizeof(LONG));
5       return;
6   }

7   Quadtree::~Quadtree( )
8   {
9       FreeMemory(ErrBuffer);
10      return;
11  }
```

Actually, this buffer is unnecessary if we are allowed to overwrite the error item in the code structure of the fractal codes. We will leave this problem to the reader as an exercise.

TABLE 4.3.1 Lena *in quadtree partition*

λ	3×3	6×6	12×12	24×24	bpp	ratio	RMSE	PSNR
8	7520	2888	572	0	1.1239	7.12	3.0822	38.35
16	4576	2748	723	17	0.8254	9.69	3.6579	36.87
32	2680	2374	719	71	0.5982	13.37	4.4291	35.20
64	992	2000	754	112	0.3834	20.87	5.5030	33.32
128	196	1515	801	143	0.2718	29.43	6.5725	31.78
256	12	1001	765	187	0.2011	39.78	7.6119	30.50

TABLE 4.3.2 Golden Hill *in quadtree partition*

λ	3×3	6×6	12×12	24×24	bpp	ratio	RMSE	PSNR
8	14972	3061	63	0	1.8522	4.32	3.3790	37.56
16	8324	4451	131	0	1.3210	6.06	4.2616	35.54
32	3536	4868	326	0	0.8936	8.95	5.4048	33.48
64	1120	3696	770	0	0.5717	14.00	6.7610	31.53
128	156	2189	1131	19	0.3577	22.36	8.1583	29.90
256	0	968	1278	76	0.2377	33.66	9.5685	28.51

4.3.3 PERFORMANCE ON SAMPLE IMAGES

How does it work? Tables 4.3.1 and 4.3.2 show the statistics for *Lena* and *Golden Hill*. The bit-per-pixel rates and the compression ratios are simply calculated by assuming 26 bits per code: 16 bits for global reference address, 6 bits for block mean value, 3 bits for spatial symmetry, and 1 bit for quadtree. Futher reduction can be achieved by using entropy coding (see Chapter 12) and quantization (see Chapter 6) techniques. The actual image and its destination region partition diagram, which technically is called the *image template*, are illustrated in Figures 4.3.4 and 4.3.5.

4.3.4 TOP-DOWN SPLIT AND BOTTOM-UP MERGE

In reality, the compression speed is always an issue. As good trade-offs, the *top-down splitting selection algorithm* and the *bottom-up merging selection algorithm* are good algorithms to be used in practice.

The top-down splitting selection algorithm is shown in Figure 4.3.6. Initially, all codes are set to CHOSEN. We start from the top layer and the second-to-top layer, then move one layer lower in every step. In each step, two

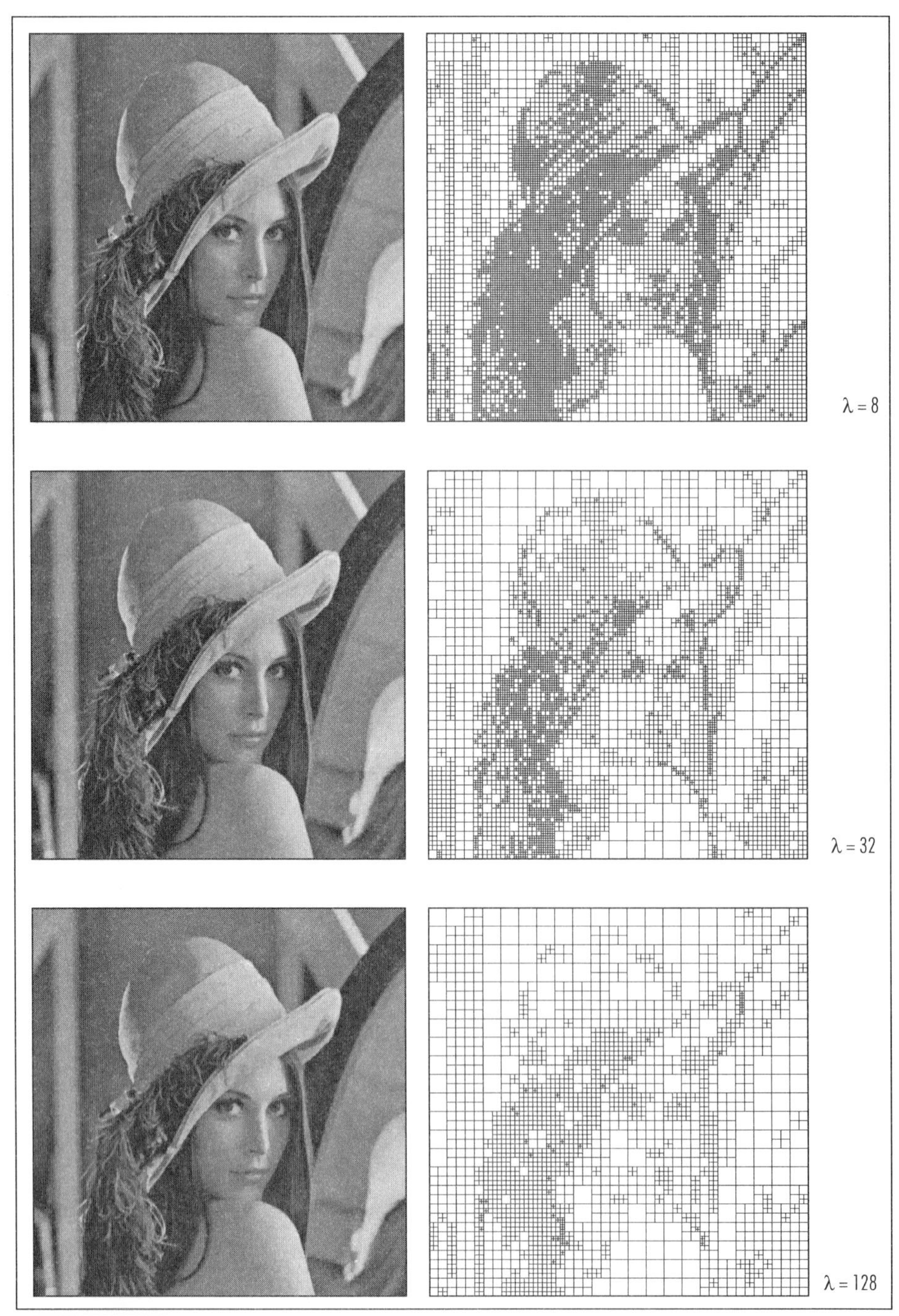

FIGURE 4.3.4 Lena *in quadtree partition*

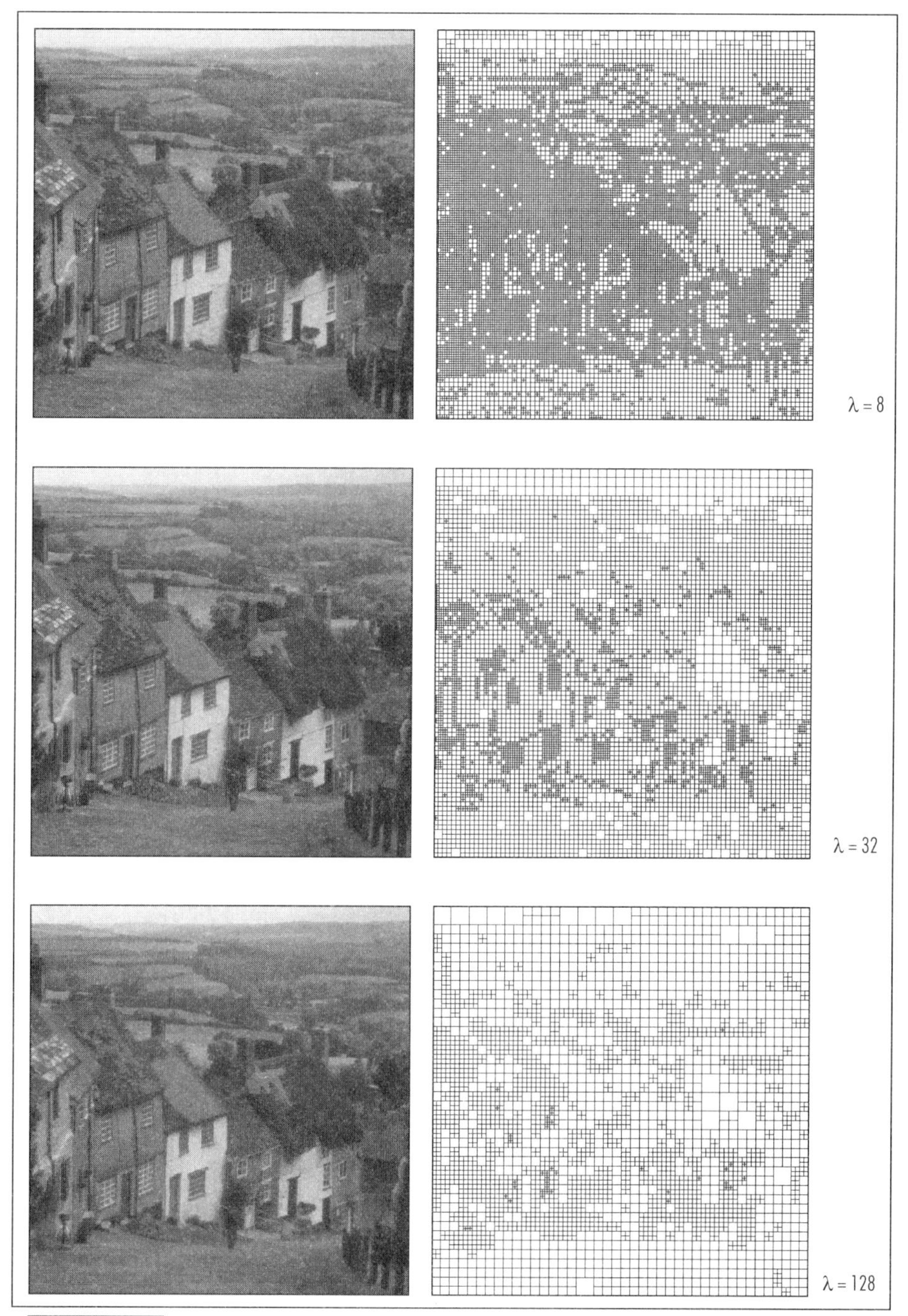

FIGURE 4.3.5 Golden Hill *in quadtree partition*

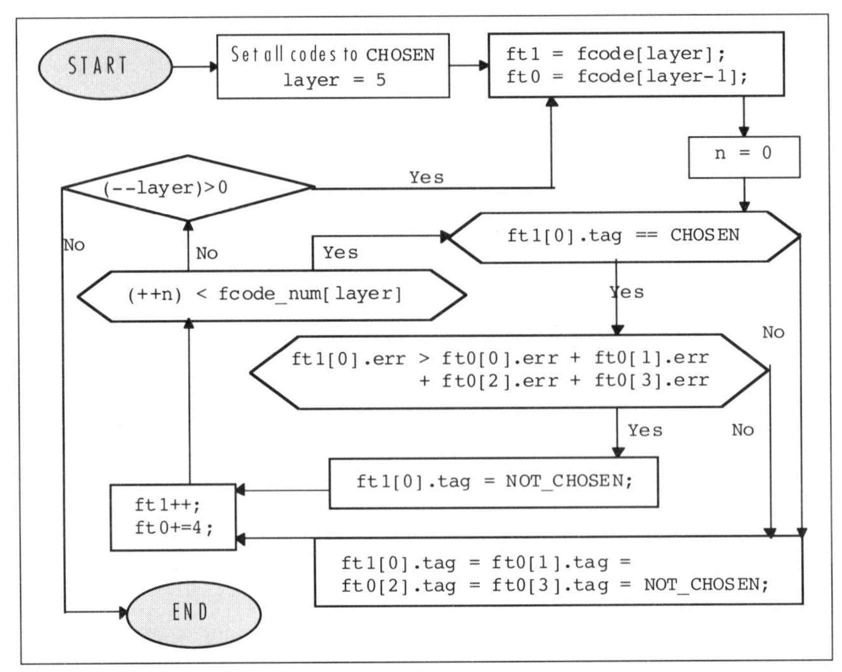

FIGURE 4.3.6 *Top-down code selection algorithm*

layers are compared at the same time. For each destination region code in the higher layer, if it has been set to NOT_CHOSEN, this implies that there must be some chosen code in an even higher layer that covers this area, so we will set all four lower-layer child codes to NOT_CHOSEN automatically. Otherwise, this code is compared with its four child codes, and either this code or the four children are set to NOT_CHOSEN, according to their errors. The advantage of this algorithm is that we can mix the searching in parallel to the code selection. In each step, the lower-layer codes are searched right before the comparison in that step. So for a higher-layer code marked with NOT_CHOSEN, the searching of its four lower-layer child codes can be skipped, for they will never be chosen. This will reduce the matching comparison time, especially if many codes in high layers are chosen.

Similar to the top-down algorithm, the bottom-up algorithm is shown in Figure 4.3.7. It starts from the bottom and the second-to-bottom layers. Again, each step involves two adjacent layers. For a given higher-layer code, if any of its lower-layer child codes has been set to NOT_CHOSEN, this implies that there must be an even lower chosen child code, and to avoid overwriting this choice, the higher-layer code will be set to NOT_CHOSEN. Otherwise, we will do the matching comparison. Just as before, the code searching and selecting are mixed by searching the higher-layer codes in each step prior to code comparison. For any code for which some child code has been set to NOT_CHOSEN, the searching of this code can be skipped according to the algorithm.

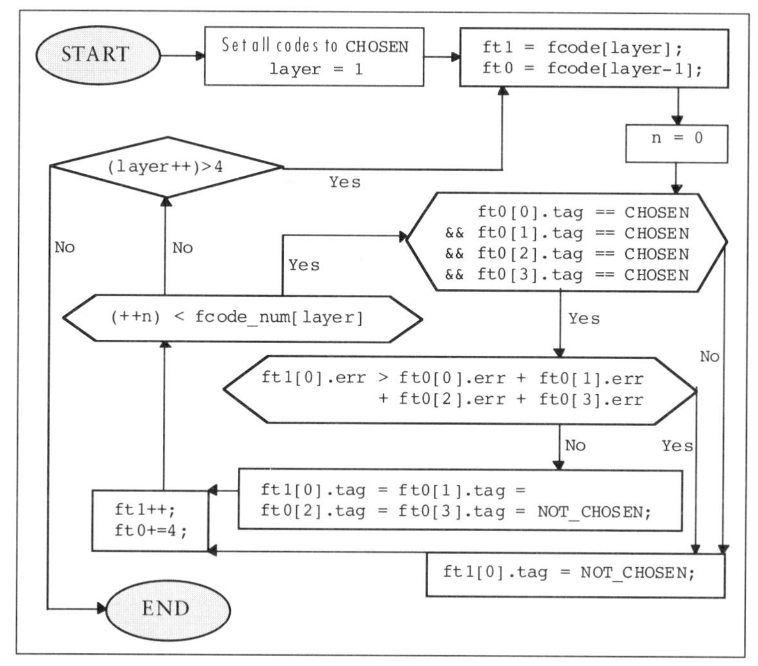

FIGURE 4.3.7 *Bottom-up code selection algorithm*

The top-down splitting selection algorithm performs better in high compression cases, because there are many large destination region codes that will be selected in the fractal representation data. When a code of large destination region is chosen, no further splitting will occur, and then many small destination region codes will be skipped in compression.

Conversely, the bottom-up merging selection algorithm performs better at the low compression and high quality end, for many small destination regions will never merge to a larger one, as most large destination regions will escape from being searched.

The performance of both algorithms is very close to the standard quadtree selector. Actually, it is convincing to say that in almost all real-world images, if a kth-layer code, in its destination region, is *better* than a set of $(k+2)$-layer codes, then there is a set of $(k+1)$-layer codes that is better than the set of $(k+2)$-layer codes. This property is called the *linearity of code transaction.* It is the key assumption for these two algorithms to achieve high performance.

There are some extreme cases. In those cases, the linearity of code transaction is broken, which makes the algorithms not work properly. For example, in the stripe case the splitting stops in the first step — we cannot get the small blocks even for a very small distortion rate. In a similar situation, for a middle range distortion rate, we never can merge the bottom layer, while in some higher layer, flat codes should be chosen. To avoid these cases, some verification routines need to be developed, which will be left to the reader as an open question. However,

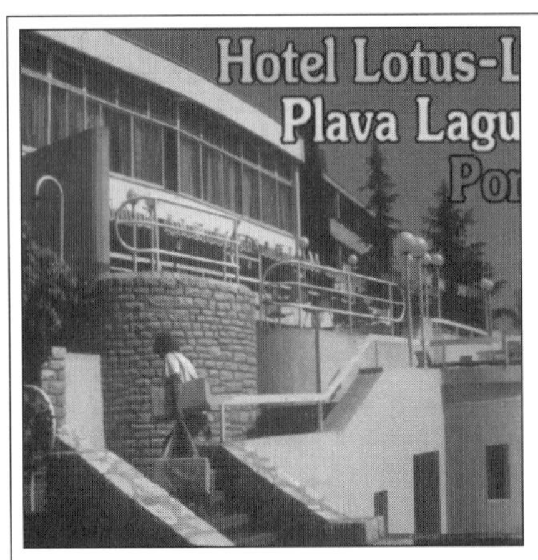

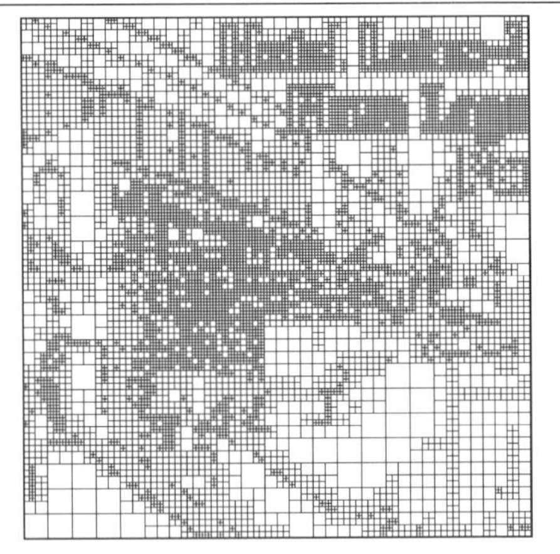

Top-down

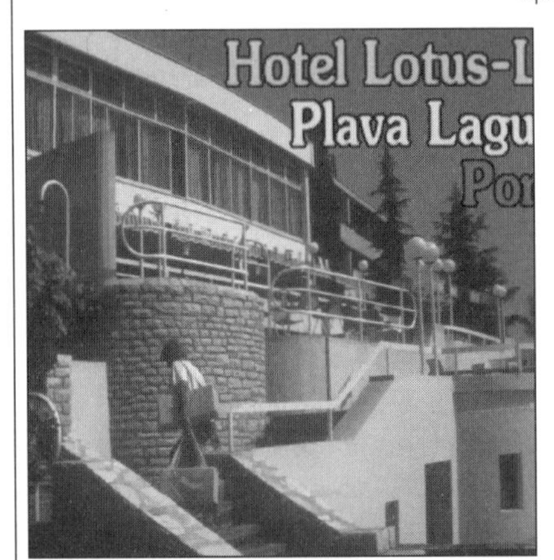

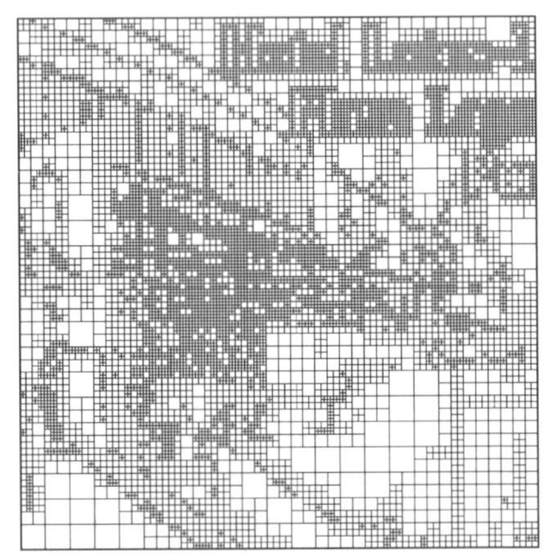

Standard

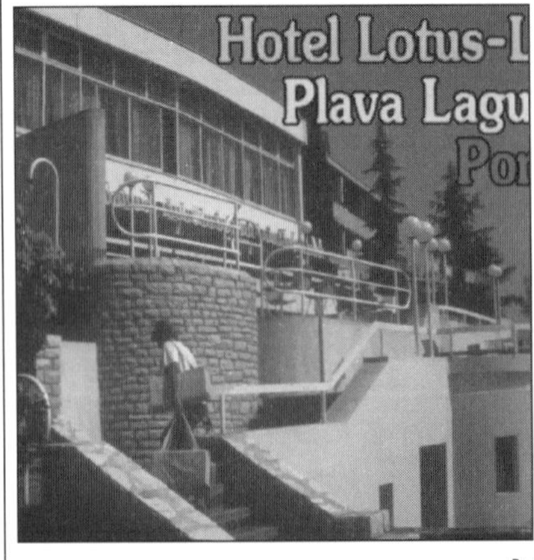

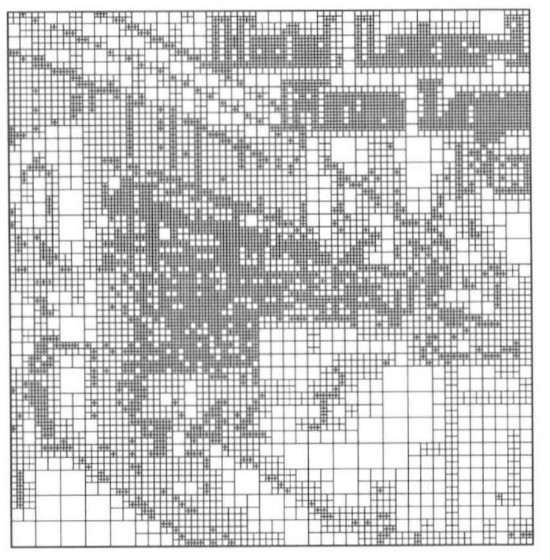

Bottom-up

FIGURE 4.3.8 Plava Laguna *in various variations of quadtree partition,* $\lambda = 32$

TABLE 4.3.3 Plava Laguna *in* 3×3–6×6–12×12–24×24 *quadtree partition*

λ	method	3×3	6×6	12×12	24×24	bpp	RMSE	PSNR	Time
8	*quadtree*	**13228** of 28224	**2293** of 7056	**364** of 1764	**0** of 441	*1.6259*	3.4932	37.27	100%
8	*top-down*	**13232** of 22368	**2284** of 7056	**366** of 1764	**0** of 441	*1.6256*	3.4960	37.26	94.8%
8	*bottom-up*	**13232** of 28224	**2296** of 7056	**363** of 937	**0** of 90	*1.6265*	3.4927	37.27	68.4%
32	*quadtree*	**7004** of 28224	**3053** of 7056	**415** of 1764	**37** of 441	*1.0757*	4.8431	34.43	100%
31.7	*top-down*	**7024** of 19136	**3028** of 6464	**420** of 1764	**37** of 441	*1.0757*	4.8461	34.42	89.9%
32.1	*bottom-up*	**7000** of 28224	**3054** of 7056	**419** of 1326	**36** of 140	*1.0757*	4.8452	34.42	76.7%
128	*quadtree*	**1528** of 28224	**3214** of 7056	**633** of 1764	**58** of 441	*0.5561*	7.9017	30.18	100%
127.4	*top-down*	**1536** of 14368	**3208** of 6112	**630** of 1764	**59** of 441	*0.5561*	7.9051	30.17	84.4%
128.6	*bottom-up*	**1528** of 28224	**3214** of 7056	**633** of 1668	**58** of 216	*0.5561*	7.9033	30.17	85.9%

without it, for most natural real-world images, as a randomly picked example as shown in Figure 4.3.8 and Table 4.3.3, the performances are simply as good as the complete quadtree selection. In fact, they are visually indistinguishable.

In Table 4.3.3, the first column is the distortion rate; the second column tells which code selection algorithm is used; the third, fourth, fifth, and sixth columns show how many codes are chosen versus how many codes have been searched for by the specified block sizes. The following columns give bit-per-pixel rate, L^2-error, and peak-signal-to-noise ratio. The last column is the search time with respect to full search.

4.3.5 PREDETERMINING PARTITION

From Table 4.3.3 we can see that more codes are searched for than are actually needed. If we can find quick invariants to classify a block region and to give a grade of likelihood of being selected, it may well save a lot of searching work. Particularly, it may be extremely helpful for designating special customized applications.

Block classification is not a new subject—it has been studied for texture matching and recognition, for VQ codebook classification, for image segmentation, and for other imaging research areas for many years. However, the goal here is slightly different, what we care about is how easy or complicated it is for a given block to find a good match to fit.

The research is still preliminary. The key is to find the proper invariant or combination of invariants, for example, the *average local variation*, defined by the average of square values given by the filter below:

1/8	1/8	1/8
1/8	−1	1/8
1/8	1/8	1/8

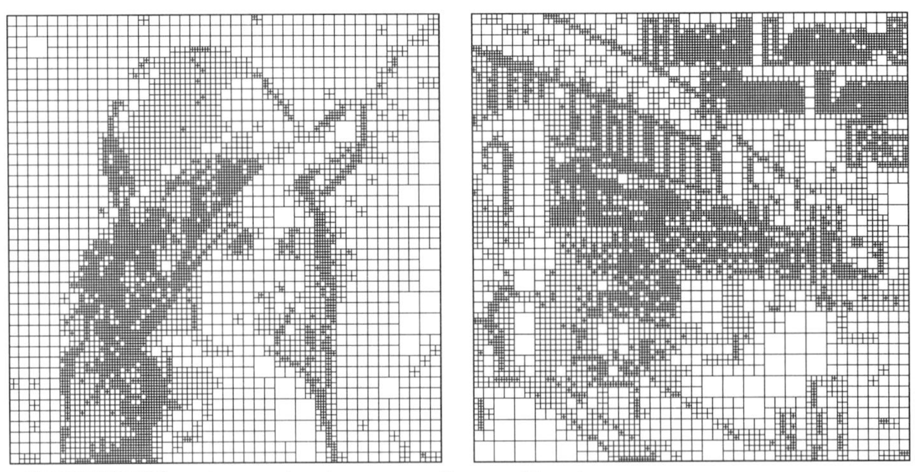

FIGURE 4.3.9 *Predetermined quadtree partition of* Lena *and* Plava Laguna

TABLE 4.3.4 *Predetermined quadtree partition*

	threshold	3×3	6×6	12×12	24×24	bpp	RMSE	PSNR
Lena	350.0	**3260**	**1609**	**890**	**65**	*0.5969*	4.8461	34.42
Plava Laguna	*502.5*	**7364**	**2587**	**533**	**31**	*1.0758*	5.5800	33.20

By setting adequate thresholds, the quadtree partition can be predetermined, as shown in Figure 4.3.9. Searching only the codes according to the given partition, the performance is surprisingly good: it decreases PSNR of less than 5% for cutting the speed to less than a quarter of the original compression time. The exact threshold setting and the detailed results of the example illustrated in Figure 4.3.9 is given in Table 4.3.4. It is interesting to compare it with the standard quadtree partition at $\lambda = 32$.

The local variation is the first invariant applied to predetermine the quadtree partition. More content and texture oriented invariants are expected to be applied in the near future. In fact, using visually based invariants not only reduces the compression time, but also, higher visual quality compression can be achieved.

4.4 MORE PARTITION METHODS

Other than the quadtree partition, various other image partitions have been implemented for various reasons. There are a few of them:

4.4.1 HORIZONTAL–VERTICAL PARTITION

An image partition method, using rectangles in place of squares, doing similar splitting as the quadtree splitting, is called the *horizontal–vertical partition*, or simply the *HV partition*. It starts from viewing the whole image as a rectangular piece. Next, we cut it (either vertically or horizontally) into two rectangles following a certain splitting criterion. Then, at every given moment, we go through all rectangles. For each of them, we decide either to leave as it is or to cut it (either vertically or horizontally) into two rectangles, according to the splitting criterion. We repeat this process until there is no more cutting.

The HV partition is a generalization of the quadtree partition, since every quadtree splitting can be decomposed into three HV splittings. Theoretically, it should give more; however, the increasing possibilities require more bits to distinguish them. Another problem of this partition lies with the *splitting criterion*. Surely, we can use the same method as we did in the quadtree case: try out all possible splittings of a given rectangle, choose the one that gives the most improvement in quality based on the *adjusted errors*, that is, the absolute error plus a bit rate penalty factor for a given distortion rate; and keep it unsplit if no more improvement can be found based on the adjusted errors. But this is unrealistic when we face the huge number of possible cuttings. Thus, what is an acceptable splitting criterion? Fisher and Menlove [F] gave a splitting criterion based purely on the rectangle's local variation.

Given a destination region $D=\{d_{ji}\}_{1\le i\le n,1\le j\le m}$, which is an $n\times m$ rectangle, the *biased horizontal differences* are defined as follows:

$$h_j=\frac{\min(j,m-j)}{m}\sum_{i=1}^{n}\left(d_{ij}-d_{i,j+1}\right), \tag{4.4.1}$$

for all $1\le j<m$; these are weighted differences of adjacent row averages. The *biased vertical differences* are defined similarly, using the formula

$$v_i=\frac{\min(i,n-i)}{n}\sum_{j=1}^{m}\left(d_{ij}-d_{i,j+1}\right), \tag{4.4.2}$$

for all $1\le i<n$; these are weighted differences of adjacent column averages. Each of these difference values corresponds to a rectangular horizontal or vertical splitting. For example, V_i corresponds to the vertical splitting that splits the rectangle into two; one has the first i columns and the other has the remaining $n-i$ columns. We split the rectangle where the absolute value of the difference is maximal among all possible choices.

The splitting may not be the best in the fractal sense; however, it tends to cut the image along some edge and discourages the generation of narrow rectangles.

Having been shown how to split a rectangle, when should we stop? Ideally, we should stop when there is no further improvement that can be gained in respect to code size distortion rate penalty. But this is not easy to compute. The implementation given by Fisher and Menlove adopts a predetermined error threshold. We keep a rectangle if its total collage error (not error average, but error sum), is less than the threshold. Their result is illustrated in Figure 4.4.1*a*.

A similar criterion can be used to apply triangular partition (Figure 4.4.1*b*).

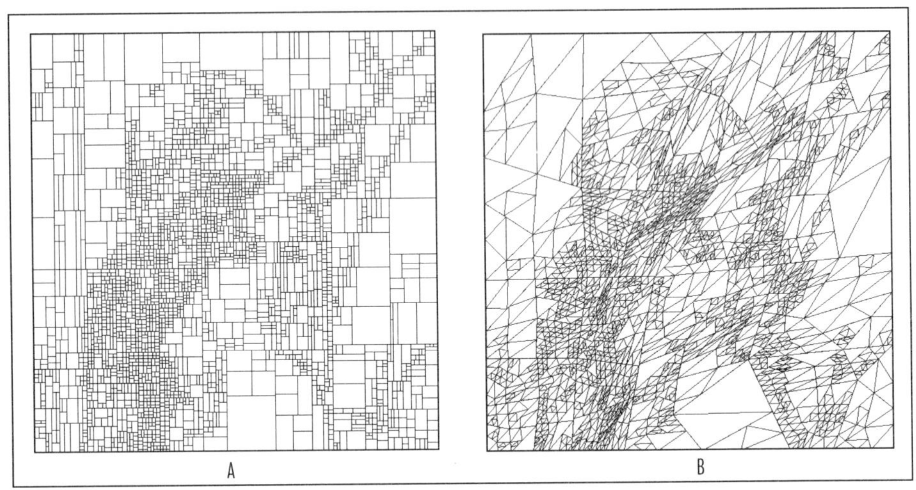

FIGURE 4.4.1 (A) HV partition; (B) Triangular partition. *(Both pictures courtesy of Y. Fisher [F])*

4.4.2 FUZZY HEXAGONAL PARTITION

The main visual defect of the quadtree method is the blockiness. Human eyes are extremely sensitive to those partition-joining lines, especially in high compression cases. Interesting work was done by Els Withers in 1989: the *fuzzy hexagonal partition*. In comparison with the weighted masking method introduced earlier, this method makes the joining line disappear, not through overlapping but through dithering. It preserves all the advantages of partitions.

Figure 4.4.2 gives the exact partition. The fuzzy *hexagon* is formed by 144 pixels (as is a 12×12 square block). It can be divided into three 48-pixel *fuzzy diamonds*. The code selection algorithm can be easily generated from the quadtree algorithm, so we leave it as an exercise.

However, the implementation is a little too complicated and does not cover a huge range of compression ratios. It is a good scheme for achieving good visual quality. We give no statistics, as the reason for this scheme was not for any mathematical measurement.

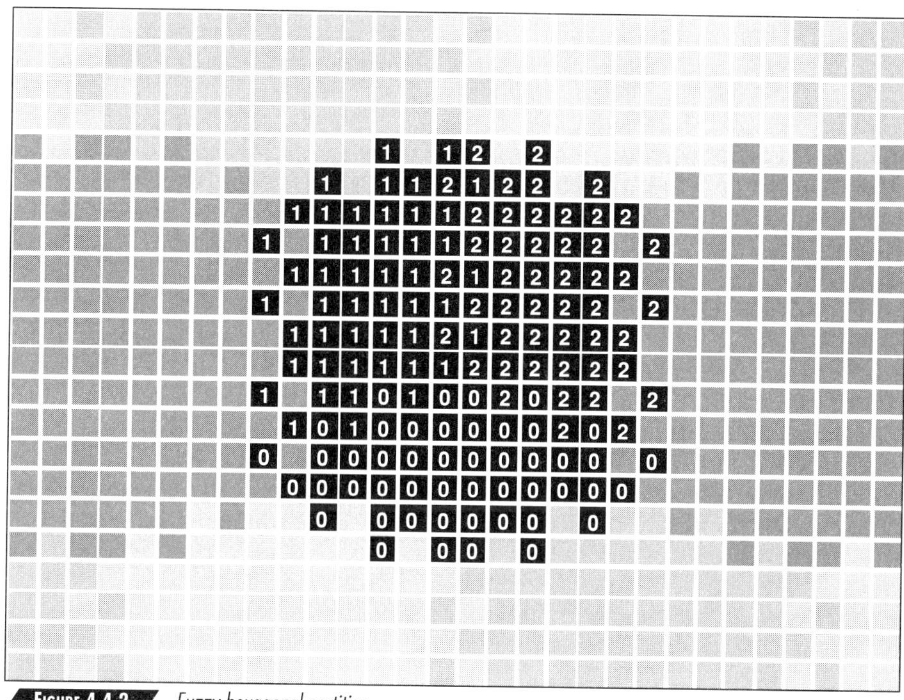

FIGURE 4.4.2 *Fuzzy hexagonal partition*

4.4.3 MIXED SQUARE PARTITION

The *mixed square partition* is an extension of a 3-layer quadtree partition. As you may notice, in almost all real-world image representations using the quadtree scheme, the difference between using three layers and using arbitrarily many layers is less than 5% of the data size. In fact, to cover the same area of image support, one needs to take exponentially more other smaller squares than larger ones.

Think of it in this way: You have an image covered by a 4-layer partition; each layer covers a quarter of the image, e.g., 16,000 codes for the bottom layer, 4000 codes for the second layer, 1000 for the third, and 250 for the top one, for a total of 21,250 codes. By splitting the top 250 codes into 1000 third-layer codes, we have a 3-layer partition of 22,000 codes. It increases the data by only 3.5%, but saves a whole layer of searching and comparison. Actually, by splitting those codes, we may find better matches to improve image quality slightly, so the real data size increases even less. When we analyze typical fractal image representations, we can see a principal layer, a higher layer for saving the data, and a lower layer for solving the hardship. The reason we sometimes keep more than three layers is for achieving various compression ratios and to allow for principal layer shifting.

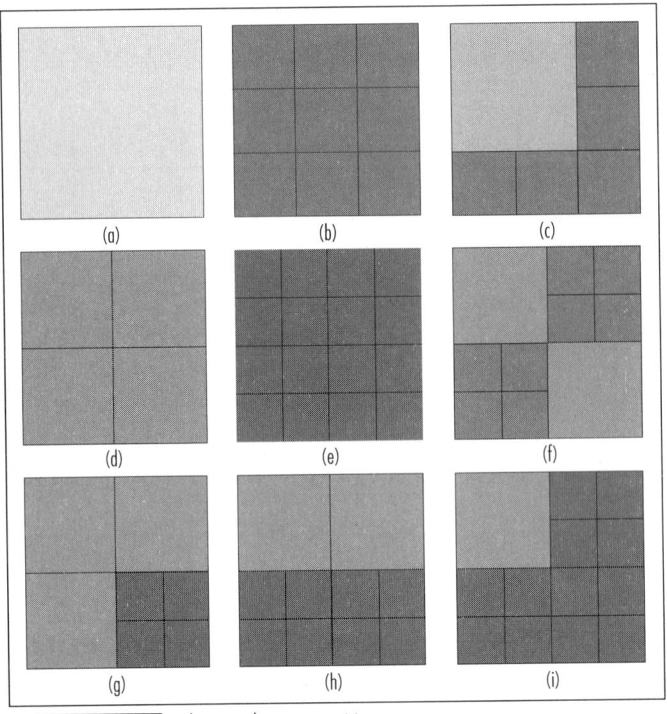

FIGURE 4.4.3 *The mixed square partition*

For a 3-layer partition, *top, bottom*, and *middle*—e.g., 12×12, 3×3, and 6×6—a top 12×12 region could have 17 different splittings, as shown in Figure 4.4.3: four rotations each of *g*, *h*, and *i*, two flips of *f*, and one each of *a*, *d*, and *e*.

The *mixed square partition* adds five more splttings to this list by allowing all possible 4×4 and 8×8 mixtures in a 12×12 area: four rotations of *c* and one of *b*. It smooths the splitting criterion and makes the code splitting not so brutal. Indeed, it adds two more numbers, 6 and 9, to the possible code numbers in a 12×12 area: 1, 4, 7, 10, 13, and 16.

The superiority of this partition in image quality is clear, because it allows a continuous change of code types without adding too much in splitting information. Still, it imposes the searching of two more square block sizes. One day we may revisit it, when there is a good partition predetermining routine available, or the searching speeds and times are no longer as critical an issue.

4.4.4 SPLIT AND MERGE

Adaptive partitioning methods have been studied by many. The idea is not only splitting regions but also allowing regions to merge.

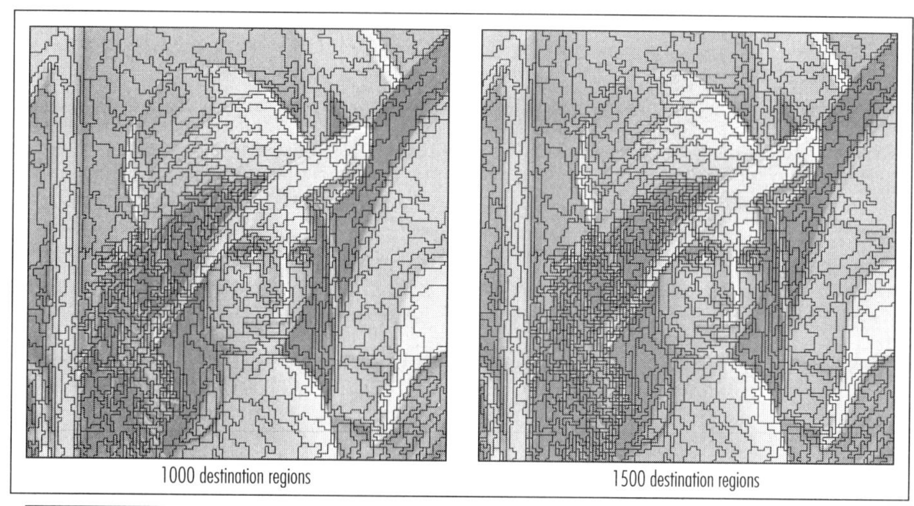

Figure 4.4.4 Lena *in* evolutionary partitioning *(Courtesy of D. Saupe and M. Ruhl, [SR])*

In the early years of Iterated Systems, Barnsley, Elton, and Withers ordered pixels along a Hilbert curve, an image partition being given by a partition of the curve into variable-length segments. The advantage of this method is that the variable size of destination regions allows maximization of performance, and irregular region boundaries eliminate the visual artifact caused by region boundary patterns, which is the primary artifact in most compression systems. The difficulty is the lack of effective searching techniques to fit this structure.

Other adaptive methods have been discussed in an overview given by Saupe, Hamzaoui, and Hartenstein [SHH], including Davoine *et al.*'s work using Delaunay triangulation [DBC], Thomas and Deravi's work in building large destination regions from atomic square blocks using heuristic search, and Saupe and Ruhl's adaptive partitioning using *evolutionary computation* [SR].

Saupe and Ruhl's evolutionary partition begins with a fixed size square block partition. Each destination block is sought, as in standard fractal coding, not for one, but for a list of acceptable fractal codes. Then these regions are merged randomly if they have codes from both lists that can be joined for both regions. Keep only the extendible ones for the list of the new region, and repeat the merge until no further new acceptable merge can be found. Finally, choose the best in each list to form a set of fractal codes. Figure 4.4.4 shows such partitions of *Lena* in 1000 and 1500 destination regions.

Like the Sun positioned in the Milky Way, like the Earth stumbling through its orbit around the Sun, I know, I can neither get closer to nor run away from you. Though I'm living within your rules, my instinct keeps telling me the ultimate truth that, I'm the center of the universe, you are important only because you are in my reach. . . . How often I think of you depends on how far you are away from me in a higher dimension.

— The Echo of Ego

Anonymous

5 SPATIAL TRANSFORMS

Better compressed image representation is demanded by many commercial applications. This requires fractal image representations to be efficient in their coefficient parameter storage. In other words, if there are two representations of the same image with indistinguishable visual qualities, the one that requires less storage capacity is preferred.

The next few chapters will give a detailed analysis of the storage of each fractal parameter of fractal elements. A fractal element consists of a destination region masking function and a transformation that can be decomposed into a spatial component and an intensity component. In any regular tiling image partition, a masking function is completely determined by its destination region set. As long as the destination regions are kept in a fixed order, with a fixed geometric shape, then the data used to characterize them will be extremely small. Therefore, the transformation coefficients take up most of the stored data. In this chapter our focus will be on the spatial component, which can be further decomposed into two steps: *translation* between the destination region origin and reference region origin, i.e., reference region origin address, or location, and *local deformation,* i.e., *s*-forms. This chapter will start with the reference region location.

5.1 COMPRESSION USING LOCAL SEARCHING

In the fractal codes given in our first sample encoder, *Codec 1.00,* the fixed image partition derives from the destination region origin locations. In contrast, the reference region origin addresses take more than half of the data file size. Is that necessary? Are they redundant? Could they be replaced by another efficient set of reference regions?

5.1.1 REFERENCE ORIGIN DISTRIBUTION

The first reaction is to plot their distribution and see whether any redundancy can be actually *seen.* Figure 5.1.1 shows the reference region origin distribution of *Lena* and *Golden Hill* represented by the uniform 4×4 square image partition. On these images, black dots represent where reference regions are allocated. They are quite uniformly distributed.

Are the reference region origin locations somewhat correlated to the destination region origin locations? Instead of drawing the reference region origin locations, the picture of the difference vectors between the reference region origin locations and their corresponding destination region origin locations is plotted as Figure 5.1.2. Looking at these relative reference region origin location maps, it is interesting to notice that

- Many reference regions are very close to their destination regions.
- More reference regions are either directly above or below their destination regions, or directly on the left or right of their destination regions.

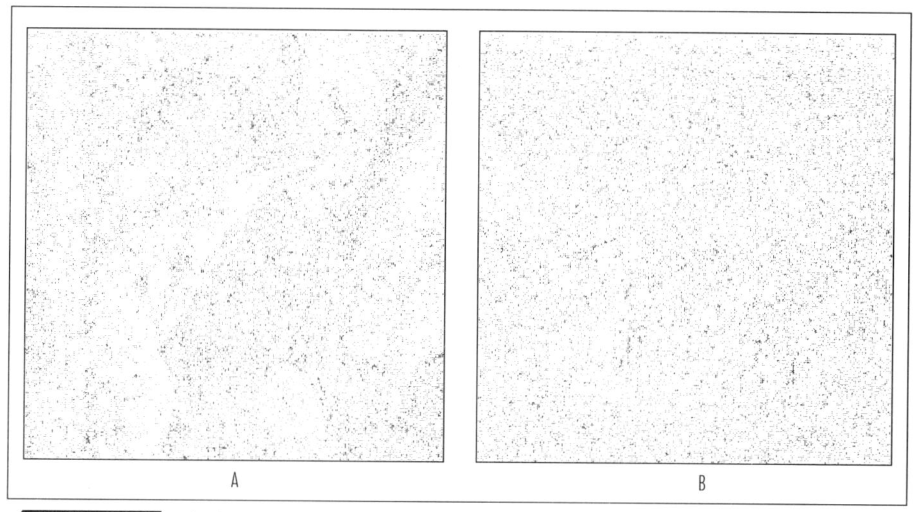

Figure 5.1.1 *Absolute reference region origin distribution: (A)* Lena; *(B)* Golden Hill

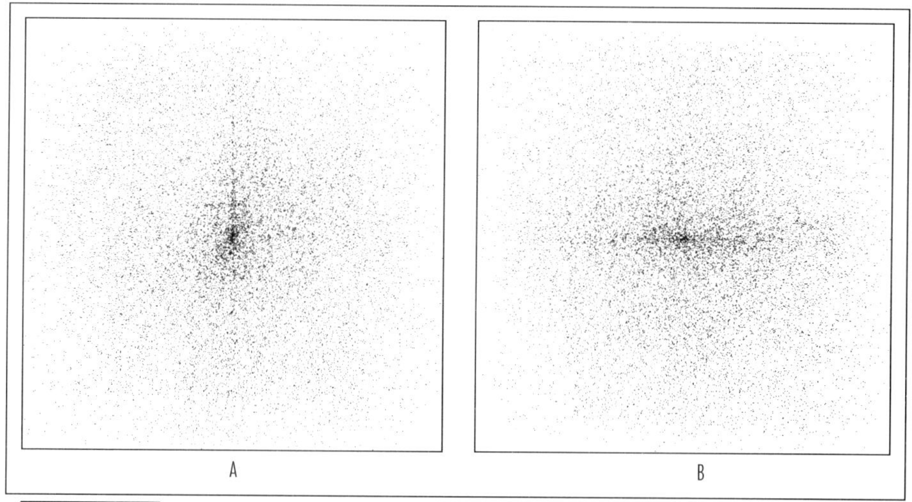

Figure 5.1.2 *Distribution of reference region origins relative to destination region origins: (A)* Lena; *(B)* Golden Hill

Based on these two facts, it is convenient to define the spatial distance as the sum of the absolute coordinates:

$$d((rx, ry), (dx, dy)) = |rx - dx| + |ry - dy|.$$

The local reference indices are drawn in Figure 5.1.3.

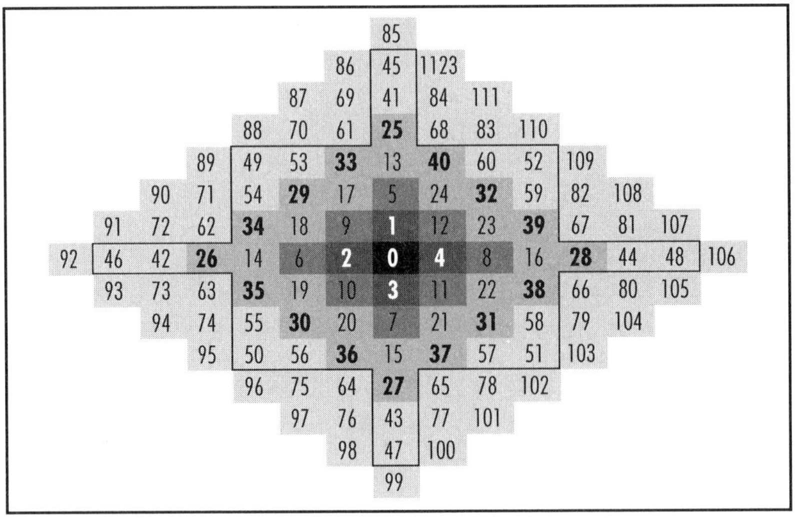

85
86 45 1123
87 69 41 84 111
88 70 61 **25** 68 83 110
89 49 53 **33** 13 **40** 60 52 109
90 71 54 **29** 17 5 24 **32** 59 82 108
91 72 62 **34** 18 9 1 12 23 **39** 67 81 107
92 46 42 **26** 14 6 2 0 4 8 16 **28** 44 48 106
93 73 63 **35** 19 10 3 11 22 **38** 66 80 105
94 74 55 **30** 20 7 21 **31** 58 79 104
95 50 56 **36** 15 **37** 57 51 103
96 75 64 **27** 65 78 102
97 76 43 77 101
98 47 100
99

FIGURE 5.1.3 *Local searching map*

TABLE 5.1.1 *Reference region origin distribution statistics*

relative radius	total reference	Lena			Golden Hill		
		reference	*percent*	*bits*	*reference*	*percent*	*bits*
r = 0	1	32	0.20 %	9.00	23	0.14 %	9.48
r = 1	4	32	0.19 %	11.00	22	0.13 %	11.54
r = 2, 3	20	37	0.23 %	13.11	36	0.22 %	13.15
r ∈ [4, 8)	88	127	0.78 %	13.47	69	0.42 %	14.35
r ∈ 8, 16)	368	249	1.52 %	14.56	197	1.20 %	14.90
r ∈ [16, 32)	1504	695	4.24 %	15.11	582	3.55 %	15.37
r ∈ [32, 64)	6080	1947	11.88 %	15.64	1867	11.40 %	15.70
r ∈ [64, 128)	24448	5867	35.81 %	16.05	6382	38.95 %	15.93
r ∈ [128, 512)	33023	7398	45.15 %	16.16	7206	43.98 %	16.20
total	65536	16384	100.00 %	15.94	16384	100.00 %	15.96

Using this distance, Table 5.1.1 contains statistics that show the redundancy of the reference region origin address information in our fractal representation. For simplicity, all of our sample fractal representations use the uniform 4×4 square image partitions. The reader should be able to generate them in similar general cases.

The column *total reference* lists the number of reference region origins allocated in the ring area given by the radius. For each image the *reference* column is the number of fractal codes whose reference region origins are in that ring area, column *percent* is the percentage of fractal codes in the ring in comparison with the total number of fractal codes, and column *bits* is the average bit number for a datum in that ring in an entropy calculation, i.e.,

$$bits = \log_2 total\_range - \log_2 percent.$$

The reality concluded from Table 5.1.1 is less optimistic than the impression got from Figure 5.1.2. Globally the reference region distribution is not as redundant as hoped. There is only a 0.3% gain using entropy coding packing scheme.

However, by using this knowledge, it is natural to build the next experiment: a compression algorithm that encourages fractal codes having locally addressed reference region origins by giving some distortion error penalty to the globally addressed codes.

5.1.2 LOCAL REFERENCE REGIONS

Considering Table 5.1.1, assuming 6 bits for the intensity offset and 3 bits for the spatial form, a fractal code will take from 18 to 26 bits. Now these bit numbers can be used as the code sizes in the code selection criterion presented in Section 4.1.2. In the code selection criterion, the *distortion rate* λ is the parameter that determines the trade-off between the increasing bit size and the improvement in quality. That is, for a chosen fixed λ, the bit size penalty of each code, defined to be the product of the code size and λ, is added to the code collage error prior to the code selection.

For example, in an L^1-metric, if for some destination region a code of radius 1 with an error 200 and a code of radius 80 of error 150 are presented, which one should be chosen? It depends on which of the numbers $200 + 18\lambda$ and $150 + 25\lambda$ is smaller. Thus, the radius 1 code will be chosen if $\lambda \geq 8$, and the radius 80 code will be selected when $\lambda < 8$.

In the newly compressed fractal representation, the statistics could be different from the previous one. Actually, for $\lambda > 0$, the types of codes with fewer bits are preferred. Their numbers increase when λ increases, and the new statistics will be in their favor. Now, let us compress again, using these new statistics, i.e., using the same distortion rate λ. Repeating this step, an annealing procedure is generated. It ultimately deduces the best selection of codes for that distortion rate.

In general, in an L^P-metric, $p \geq 1$, the code selection formula is not so straightforward, since the global error is not the sum of local errors. However, the p radical root part can be left out. The equivalent metric, defined as the p-power of the absolute differences, has local additivity. Its code selection criteria are used in image compression L^P-metric cases. As a matter of fact, in all unspecified discussions, an L^2-metric is assumed.

From the above discussion, we see that the bit size penalty of the best *relative reference region origin* depends upon the image and its distortion rate. But in general, there is no luxury of annealing for every image in every λ. After numerous experiments, a fixed bit size table is decided upon to avoid the time-consuming annealing. The performance seems to be very similar. Table 5.1.2 lists our choices.

TABLE 5.1.2 Range bits for local codes

radius	$r = 0$	$r = 1$	$r = 2$	$r = 3, 4$	$r \in [5, 7]$	$r \geq 8$
code size	13	15	17	19	20	26

Using this table as the cost penalty in our fractal code selection, the results were favorable, as shown in Tables 5.1.3 and 5.1.4. In local searching, the only spatial form used is the identity.

The first column is the distortion rate λ used in local searching. The following five columns are the numbers of reference regions found in the specified radii. The sixth column tells the estimated number of *bits per pixel* based on the fractal code entropy calculation. Then the next three columns are the L^1 and L^2 *distortion errors*, and the *peak signal-to-noise ratio*.

TABLE 5.1.3 *Local searching of* Lena

λ	r = 0	r = 1	r = 2	3, 4	5,6,7	bpp	L(1)	L(2)	PSNR
0	**36**	**32**	**19**	**52**	**105**	1.6236	2.1160	3.3663	37.59
1	**137**	**174**	**129**	**179**	**249**	1.6100	2.1217	3.3692	37.58
4	**1,779**	**1824**	**808**	**632**	**720**	1.4248	2.2731	3.4599	37.35
9	**4,366**	**3101**	**998**	**591**	**718**	1.2077	2.5281	3.6642	36.85
16	**6,438**	**3210**	**906**	**486**	**667**	1.0776	2.7405	3.8750	36.37
25	**7,889**	**3116**	**765**	**432**	**576**	0.9957	2.9170	4.0759	35.93
36	**8,938**	**3026**	**670**	**401**	**506**	0.9345	3.0751	4.2749	35.51
49	**9,732**	**2971**	**587**	**329**	**462**	0.8876	3.2272	4.4822	35.10
64	**10,386**	**2861**	**505**	**276**	**431**	0.8524	3.3609	4.6717	34.74
81	**10,965**	**2705**	**478**	**245**	**405**	0.8215	3.4941	4.8725	34.37
100	**11,409**	**2600**	**438**	**255**	**366**	0.7971	3.6048	5.0547	34.06
144	**12,112**	**2443**	**348**	**250**	**327**	0.7576	3.8282	5.4320	33.43
200	**12,701**	**2298**	**283**	**208**	**300**	0.7251	4.0193	5.7877	32.88

TABLE 5.1.4 *Local searching of* Golden Hill

λ	r = 0	r = 1	r = 2	3, 4	5,6,7	bpp	L(1)	L(2)	PSNR
0	**25**	**23**	**15**	**46**	**55**	1.6255	2.8470	4.1647	35.74
1	**55**	**76**	**70**	**101**	**126**	1.6211	2.8490	4.1652	35.74
4	**455**	**514**	**233**	**212**	**274**	1.5826	2.8818	4.1837	35.70
9	**1319**	**2239**	**449**	**414**	**578**	1.4997	2.9983	4.2675	35.53
16	**2468**	**1894**	**732**	**642**	**852**	1.3830	3.2234	4.4666	35.13
25	**3739**	**2575**	**897**	**772**	**947**	1.2643	3.4991	4.7470	34.60
36	**4963**	**2982**	**969**	**789**	**973**	1.1648	3.7751	5.0530	34.06
49	**6014**	**3412**	**946**	**741**	**878**	1.0797	4.0635	5.3980	33.48
64	**7157**	**3494**	**906**	**670**	**773**	1.0059	4.3585	5.7675	32.91
81	**8172**	**3529**	**805**	**591**	**639**	0.9446	4.6260	6.1136	32.40
100	**9009**	**3497**	**743**	**491**	**584**	0.8942	4.8669	6.4393	31.95

In both sample images the image visual quality decreases dramatically when the distortion rate is around 25, i.e., the penalty between the concentric code (the code with radius 0) and a global code is 300 ($=25\cdot(25-13)$). The error per pixel will be the square root of 18.75 ($=300/16$), i.e., about 4. When the distortion rate is less than or equal to 4, i.e., the global penalty is no more than 48, there are no visual losses in either case. When the distortion rate is higher than 49, i.e., the penalty is greater than 588, then both images fade away (cf. Figures 5.1.4 and 5.1.5).

Conclusively, an *average penalty error per pixel* is set to 5, i.e., the distortion rate is no more than 60, and the global penalty is no more than 720 in this case. In practice, distortion rates between 4 and 60 are interesting.

FIGURE 5.1.4 *Local searching—partial* Lena *[(136, 64)–(392, 320)]: (A)* $\lambda = 8$; *(B)* $\lambda = 32$; *(C)* $\lambda = 64$; *(D)* $\lambda = 128$.

FIGURE 5.1.5 *Local searching—partial* Golden Hill *[(128 ,218)– (384, 474)]: (A)* $\lambda = 8$; *(B)* $\lambda = 32$; *(C)* $\lambda = 64$; *(D)* $\lambda = 128$

Study Tables 5.1.3 and 5.1.4 for the distortion rates 9, 16, and 25. The distribution ratio of the five local layers is not regular, but an approximate ratio,

$$4 : 3 : 1 : 1 : 1,$$

is good enough to use in the penalty setting.

The ratio between the local and global is not constant; it varies from 1:5 to 5:1.

Since there are 1, 4, 8, 28, and 72 locations in each local layer respectively, the bit numbers of each location, calculated using the formula

$$\text{bit number} = \log_2 \frac{\text{total count}}{\text{individual count}}, \tag{5.1.1}$$

are 1.320, 3.737, 6.322, 8.129, and 9.492 bits. In practice, the penalties 2, 4, 6, 8, and 9 are used respectively. In Tables 5.1.3 and 5.1.4 other parameters are assumed to have 11 bits.

5.1.3 COMPRESSION ALGORITHM

The compression algorithm is similar to the one implemented in Chapter 3, except that the searching core needs to be modified by adding the local searching. The skeleton of the scheme is drawn in Figure 5.1.6, and the exact implementation will be an exercise. The *adjusted error* is defined as the actual error plus the penalty, which is the code bit size multiplied by the given distortion rate.

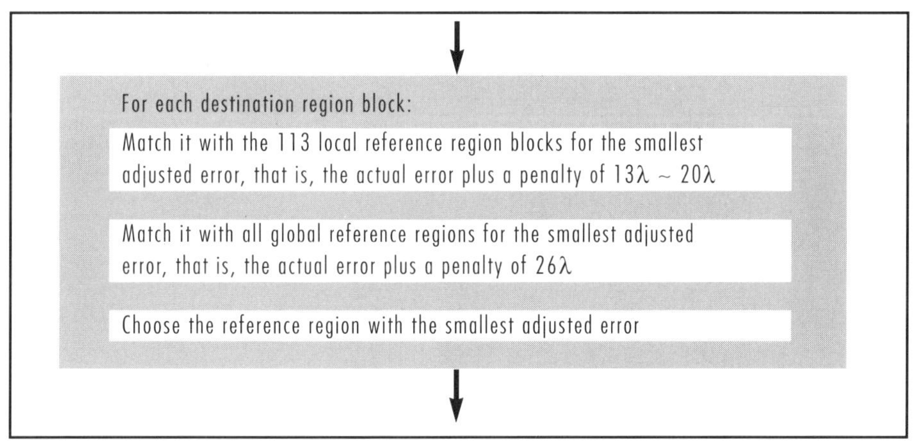

FIGURE 5.1.6 *Local searching compression algorithm*

The local searching of a destination block is much faster than its corresponding global searching since there are many fewer block matches involved. For example, for a 512×512 grayscale image, compressed using a uniform 4×4 destination region partition, in a global searching with 8 standard spatial forms, there are $(128\times128)\times(253\times253)\times8=8{,}389{,}787{,}648$ block comparisons. And in the local searching algorithm the number of comparisons is much smaller, reduced to $(128\times128)\times113\times8=14{,}811{,}136$ block comparisons. In reality, these numbers are equivalent to the compression speed. This represents 8 hours vs 1 minute on a 100MHz Pentium IBM-PC.

In practice, local searching has been used to improve both performance and compression speeds. If the local codes are searched first and the codes that have errors smaller than some predetermined threshold are selected without further global searching, then the global searching and comparison time can be greatly reduced. A local fractal code is said to be *good enough* with respect to some *threshold* $\varepsilon\geq0$ if the adjusted error of the code minus the global code penalty is less than or equal to the threshold ε. The compression algorithm using this strategy has been called the *early-kick-out scheme*. In Table 5.1.5 the performance

TABLE 5.1.5 *Local searching thresholds in a 4× 4 system*

	Lena λ = 20			Golden Hill λ = 20		
Threshold ε	*globally searched*	*globally chosen*	*L(2) distortion*	*globally searched*	*globally chosen*	*L(2) distortion*
0	**16,384**	**4465**	4.4549	**16,384**	**9683**	6.6476
400	**8,926**	**4465**	4.4549	**14,477**	**9683**	6.6476
576	**5,030**	**4446**	4.4563	**10,553**	**9639**	6.6490
784	**3,525**	**3487**	4.6431	**7,248**	**7215**	6.7602
1024	**2,813**	**2805**	5.8410	**6,293**	**6288**	6.8750

statistics are collected for this algorithm. A reasonable threshold, e.g., 576 (equivalent to an average error of $\sqrt{576/4\times4}=6$) in the illustrated case, improves compression speed without degenerating any image quality.

5.1.4 LOCAL CODES VS GLOBAL CODES

As both local and global reference blocks are used, it is interesting to see their final collage error and decoded error distributions as shown in Figure 5.1.7. Notice that the local searching tends to take care of most low-error blocks. In Figure 5.1.8, the similar graphs of local and global searching code errors are given with respect to the block variation instead of block error.

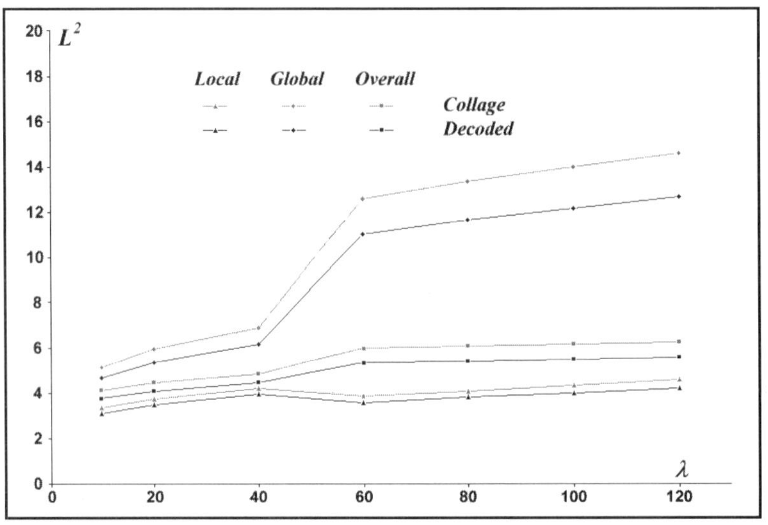

FIGURE 5.1.7 *Error distributions of local and global reference codes of* Lena *in a 4× 4 system*

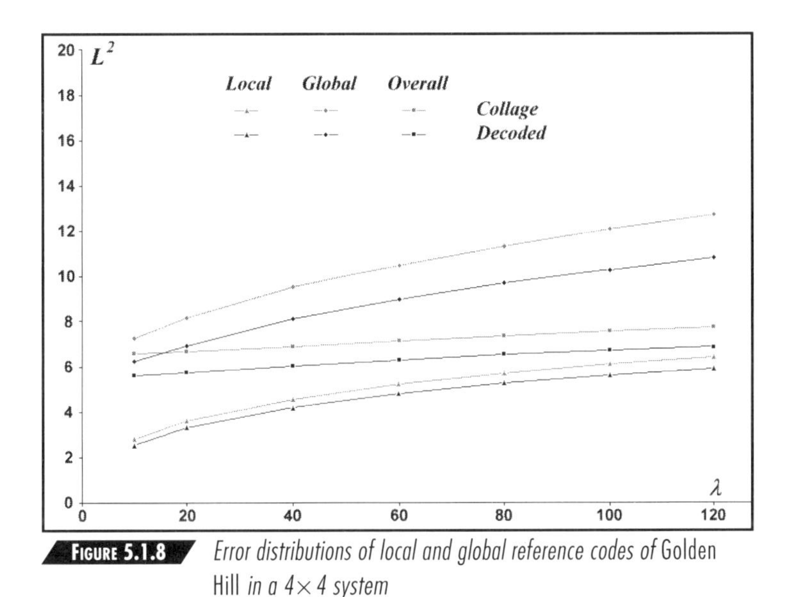

FIGURE 5.1.8 *Error distributions of local and global reference codes of* Golden Hill *in a* 4×4 *system*

5.2 REFERENCE SPATIAL FORMS

The spatial form index is the other parameter in spatial transformation of a fractal code. It has different statistics in both local and global codes.

5.2.1 SPATIAL FORMS IN LOCAL SEARCHING

Among the eight indexed standard spatial forms the trivial one has been used far more than the others, particularly in the local case. Looking through all fractal representations across all distortion rates for both images *Lena* and *Golden Hill* (Tables 5.2.1 and 5.2.2), the spatial form distributions are very similar.

Besides the fact that the trivial spatial form is used far more than the others, the last four *s*-forms tend also to be used less, about a 2:3 ratio in comparison with the first four in the global case.

So an entropy coding scheme will pack the trivial *s*-form with fewer bits than the other *s*-forms. By calculating these bit numbers according to the *s*-form distribution of a fractal image representation, using these bit numbers to determine the code-selecting criterion for a given distortion rate, and recompressing the image, an annealing process that is similar to the local reference origin annealing is established.

In fact, the total entropy decreases even by assuming that the trivial *s*-form uses one bit fewer than the others, as shown in Tables 5.2.3 and 5.2.4. The annealing process concludes that *the only spatial form that should be used in local searching is the trivial one.*

The performance comparison of using eight *s*-forms vs the trivial one is given in Tables 5.2.5 and 5.2.6. The images obtained by using only the trivial *s*-form not only take less local searching time by a factor of 8, but also are visually indistinguishable and numerically with smaller errors.

TABLE 5.2.1 *Standard spatial form distribution of* Lena *for* $\lambda = 16$

Isometry	total	0	1	2	3	4	5	6	7
Local	**11707**	5822	818	1366	536	1008	547	579	1031
Global	**4677**	821	664	706	707	484	384	431	480

TABLE 5.2.2 *Standard spatial form distribution of* Golden Hill *for* $\lambda = 16$

Isometry	total	0	1	2	3	4	5	6	7
Local	**6588**	2710	942	668	463	490	357	387	571
Global	**9796**	1480	1340	1307	1347	1049	1077	1104	1092

TABLE 5.2.3 *Standard spatial form distribution of* Lena *for* $\lambda = 16$ *with encouragement*

Isometry	total	0	1	2	3	4	5	6	7
Local	**12100**	9093	693	889	552	237	180	178	278
Global	**4284**	743	607	642	652	437	359	398	446

TABLE 5.2.4 *Standard spatial form distribution of* Golden Hill *for* $\lambda = 16$ *with encouragement*

Isometry	total	0	1	2	3	4	5	6	7
Local	**7325**	4389	795	679	521	249	208	195	289
Global	**9059**	1356	1230	1218	1254	973	996	1036	996

TABLE 5.2.5 *Local searching performance of* Lena

λ	*s*-form	$r = 0$	$r = 1$	$r = 2$	3, 4	5,6,7	bpp	$L(1)$	$L(2)$	PSNR
16	**8**	6438	3210	906	486	667	1.0776	*2.7405*	*3.8750*	*35.08*
15	1	5535	3174	773	412	431	1.0725	*2.7400*	*3.8603*	*35.11*

TABLE 5.2.6 *Local searching performance of* Golden Hill

λ	*s*-form	$r = 0$	$r = 1$	$r = 2$	3, 4	5,6,7	bpp	$L(1)$	$L(2)$	PSNR
16	**8**	2468	1894	732	642	852	1.3830	*3.2234*	*4.4666*	*33.85*
17	1	2397	1540	421	353	456	1.3898	*3.2193*	*4.4638*	*33.85*

Exercise: Improve *Codec 1.00* by adding the local searching of the trivial identity s-form to create a new compression system: *Codec 2.00*.

In summary:

- The local searching algorithm provides an effective technique in reducing fractal image representation data size.
- The only spatial form that should be used in local searching is the trivial one.
- The last four range spatial isometries in the standard eight spatial forms have lower frequencies in comparison to the first four, so in some compressors they can be omitted for practical reasons.

5.2.2 INCORPORATING QUADTREE PARTITIONS

By combining local searching and quadtree partitions a simple, and decent fractal image compression system is built. In fact, for any given distortion rate λ, a penalty is added to the collage error of each code. The quadtree partition is performed directly by comparing these adjusted errors.

Figure 5.2.1 and Figure 5.2.2 are the performance graphs on the standard testing images—the first graph shows the mixture of 3×3, 6×6, 12×12, and 24×24 blocks; the second graph shows the mixture of 4×4, 8×8, and 16×16 blocks. Figure 5.2.3, Figure 5.2.4, and Color Plates 1, 8, 9, and 16 show the exact visual results.

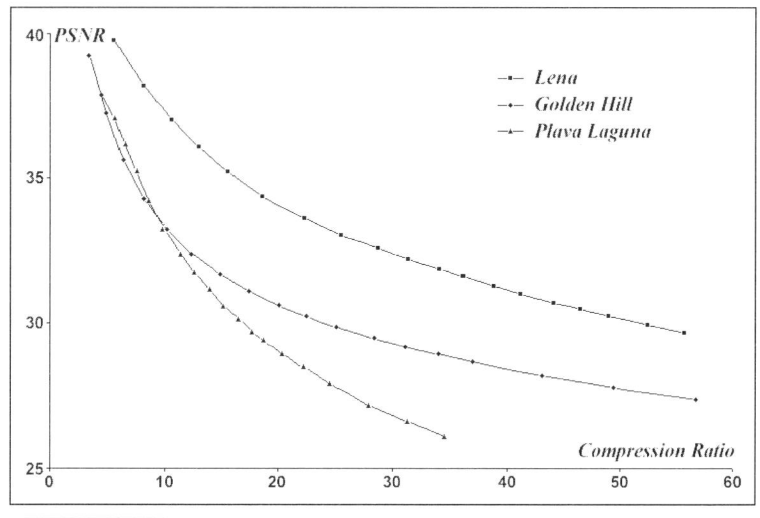

FIGURE 5.2.1 *Compression performance of a quadtree system using 3×3, 6×6, 12×12, and 24×24 blocks*

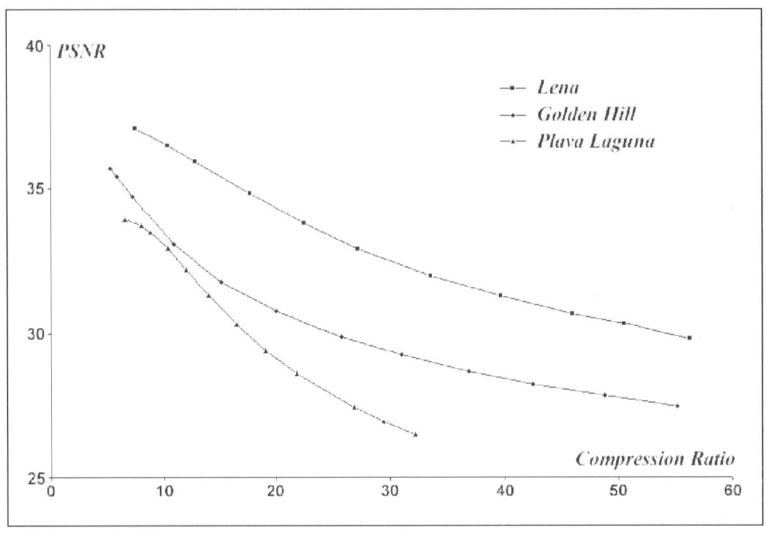

FIGURE 5.2.2 *Compression performance of a quadtree system using 4× 4, 8× 8, and 16× 16 blocks*

File size:	**8320** bytes
Bit rate:	**0.2539** bpp
Compression:	**31.5 : 1**
rmse:	**6.2440**
PSNR:	**32.21**

FIGURE 5.2.3 Lena *in baseline fractal compression system: using 3× 3, 6× 6, 12× 12, and 24× 24 blocks*

File size:	**13056** bytes
Bit rate:	**0.3985** bpp
Compression:	**20.4 : 1**
rmse:	**7.4865**
PSNR:	**30.65**

Figure 5.2.4 Golden Hill *in baseline fractal compression system: using 3×3, 6×6, 12×12, and 24×24 blocks*

5.2.3 Spatial scaling factors

After reading the collage theorem and then looking at the implementation, one may wonder *why 2 to 1 spatial contraction should be chosen*. Actually, no such request has been stated in the condition of the theorem. The first natural choice could be 1-to-1—what is wrong with that?

In a fractal model, a fractal element provides a way to identify two similar objects, destination object and reference object, and refers to the destination object from the reference object. Thus, it is allright to use 1-to-1—but we need to avoid one problem: not to let the reference form a dead cycle, particularly not refer a region to itself. In the searching algorithm introduced in *Codec 1.00* and *Codec 2.00*, if we switch from 2-to-1 to 1-to-1, a block could end up by referring to itself, which is against the intention of the fractal model of identifying different similar objects.

For example, in Figure 5.2.5*a* a simple 8×8 image (a crossing line), which could be a piece of a larger image, is given. Using 1-to-1, a simple fractal element could be the one whose reference region is the same as its destination region, which is the whole block, with γ-value 3/4 and β-value 35. The decoded block is shown in Figure 5.2.5*b*. The collage error, in an L^2-metric, is an average

200	200	200	60	20	60	200	200
200	200	100	40	40	100	200	200
200	200	60	20	60	200	200	200
200	100	40	40	100	200	200	200
200	60	20	60	200	200	200	200
100	40	40	100	200	200	200	200
60	20	60	200	200	200	200	200
40	40	100	200	200	200	200	200

A

140	140	140	140	140	140	140	140
140	140	140	140	140	140	140	140
140	140	140	140	140	140	140	140
140	140	140	140	140	140	140	140
140	140	140	140	140	140	140	140
140	140	140	140	140	140	140	140
140	140	140	140	140	140	140	140
140	140	140	140	140	140	140	140

B

FIGURE 5.2.5 *Spatial 1-to-1 study*

of 22.23, and the decoded error is 64.48 ($< 22.23 \cdot 4$). The decoded image is purely flat and contains no information of the crossing line; therefore it is useless. Again, by letting the γ-value close to 1, the collage error can be reduced to less than any given positive number, but the decoded image is the same flat block.

By using 2-to-1 reference this problem is automatically avoided. However, to some images 1-to-1 type fractal codes could be very useful if *self-referencing deadly cycles* are avoided, since often there are different similar objects of the same scale found in the same image. One way to avoid these cycles is to make sure that the fractal codes are put in a certain order, e.g., the scan line order. Thus, the reference region of any 1-to-1 fractal code is always covered by the union of destination regions of all preceding fractal codes.

Exercise: Add 1-to-1 spatial transforms to the current compression system.

What about the other spatial contraction factor? It should definitely not be larger than 1, since the less detailed object needs to be represented by a more detailed one and not the other way around. For a noninteger ratio it involves the image resampling convention. A decoder needs to perform in sync with an encoder. The same sampling rule for converting the reference region into the resolution of the destination region needs to be established for both the encoder and the decoder. An inappropriate conversion could generate some unexpected artifact in the process. Figures 5.2.6 and 5.2.7 show some choice of 3-to-2 and 3-to-1 sampling filters respectively.

Exercise: Add 3-to-2 and 3-to-1 spatial forms to the current compression system.

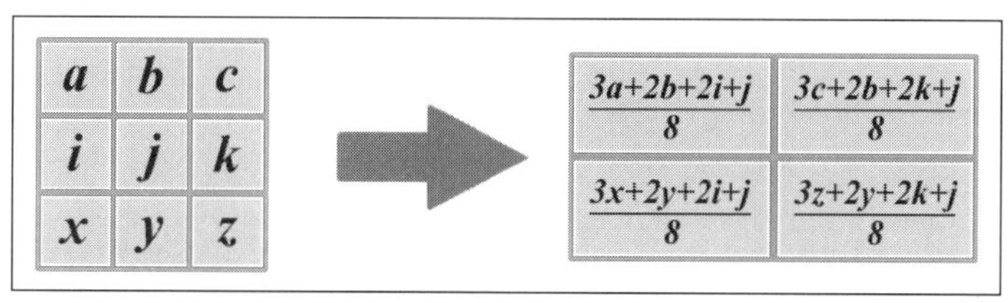

FIGURE 5.2.6 3-to-2 resampling formula

FIGURE 5.2.7 3-to-1 resampling formula

Will allowing large spatial contraction factors help image compression? The previous exercise will partially answer this question.

5.2.4 MORE SPATIAL FORMS

In most constructive examples, general spatial affine transformations seem to be necessary. The image similarities in the real-world image are often skewed and deformed. It is difficult to find the optimal balance point between more spatial forms and fewer complex computations. In this subsection some additional spatial forms are given to improve the image quality and the compression ratio further.

Take the uniform 4×4 block partition as the sample case. The first additional set of spatial forms is a set of skewing transforms of the following form:

$$\begin{pmatrix} 2 & 2\xi \\ 0 & 2 \end{pmatrix}, \text{ for } \xi \in \left\{-1,\ -\frac{7}{8},\ -\frac{3}{4}, \dots,\ -\frac{1}{8},\ 0,\ \frac{1}{8}, \dots,\ \frac{7}{8},\ 1\right\}. \quad (5.2.2)$$

Thus, the reference block is a parallelogram as shown in Figure 5.2.8.

Again, in this case one needs a subsampling convention. The most natural convention is the *area weighted average interpolation*. A simpler approximation is the *neighborhood pixel bilinear interpolation*. This method will be explained with illustrative example. The general sampling formula is left as an exercise.

Given the skewing spatial form of $\xi=3/8$ for a 4×4 block, the reference region of the s-form of the pixel of the square $[2,3]\times[2,3]$ is highlighted in Figure 5.2.8, which is the parallelogram: $(5.5, 4)-(7.5, 4)-(6.25, 6)-(8.25, 6)$.

FIGURE 5.2.8 Sample skewing spatial form: ξ=3/8

Thus, the center of the reference "pixel" is (6.875, 5). As a result, using the notation illustrated in Figure 5.2.8, the *area weighted average interpolation* gives the value

$$\frac{5}{64}a+\frac{1}{4}b+\frac{11}{64}c+\frac{1}{192}d+\frac{23}{96}e+\frac{47}{192}f+\frac{1}{96}g$$

$$=0.0781a+0.2500b+0.1719c+0.0052d+0.2396e+0.2448f+0.0104g, \quad (5.2.3)$$

and the *neighborhood pixel bilinear interpolation* gives the value

$$\frac{5}{16}b+\frac{3}{16}c+\frac{5}{16}e+\frac{3}{16}f=0.3125\,b+0.1875\,c+0.3125\,e+0.1875\,f. \quad (5.2.4)$$

Those skewing spatial forms give nice curves that the square spatial forms cannot generate, as illustrated in Figure 5.2.9. These sets of spatial forms improve image visual quality by giving geometric patterns more precision.

Similar to the skewing s-forms that are skewed in the horizontal direction, a set of vertical skewing forms can be defined as well:

$$\begin{pmatrix} 2 & 0 \\ -2\eta & 2 \end{pmatrix}, \text{ for } \eta\in\left\{-1,\ -\frac{7}{8},\ -\frac{3}{4},\cdots,\ -\frac{1}{8},\ 0,\ \frac{1}{8},\cdots,\ \frac{7}{8},\ 1\right\}. \quad (5.2.5)$$

A combined set of 64 additional spatial forms is suggested in the following:

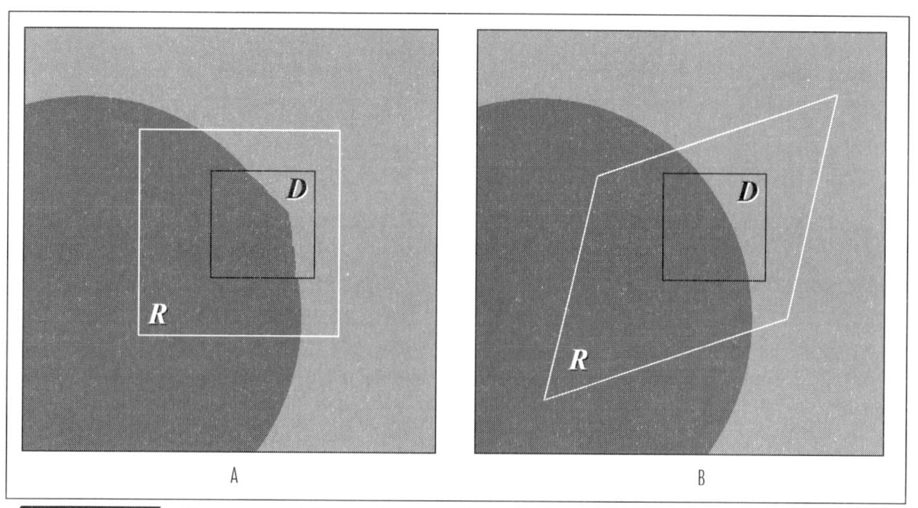

FIGURE 5.2.9 Skewing spatial form study: (A) Without skewing s-forms; (B) With skewing s-forms.

$$\begin{pmatrix} 2 & 2\xi \\ -2\eta & 2 \end{pmatrix}, \text{ for } \xi, \eta \in \left\{ \frac{1}{8}, \frac{1}{4}, \frac{3}{8}, \frac{1}{2}, \frac{5}{8}, \frac{3}{4}, \frac{7}{8}, 1 \right\}. \tag{5.2.6}$$

Other than the slow compression speeds, though they can be improved by using various techniques presented in this book, the performance improves remarkably.

5.3 SEED REFERENCE IMAGE

After dividing fractal codes into local codes and global codes, the selection of global codes appears quite randomly—it often has nothing to do with the fractal property. The randomness is partially due to the imperfect image comparison metric model and the lack of effective reference block candidates. Considering the large bit size of global codes, these codes are very likely to be alternated with other descriptions using fewer bits, unless the complexity of the regions covered by corresponding codes are higher.

The remainder of this chapter analyzes the global reference block distribution with the hope that some gain in bit rate can be achieved by indexing the reference location addresses. As a result, adding fixed customized or universal seed reference images becomes an effective alternative to the global codes—especially for a variety of applications. The techniques of using universal seed image is equivalent to some variations of the *vector quantization* schemes.

5.3.1 INDEXED REFERENCE REGIONS

Given an image fractal representation, it is interesting to investigate which fractal codes are global and which are local. Figure 5.3.1 illustrates the destination regions in the pictures *Lena* and *Golden Hill* that are covered by the global fractal codes (where a pixel is marked to black if it is covered by a global code and to white otherwise). Figure 5.3.2 illustrates the global reference regions of the same fractal representations.

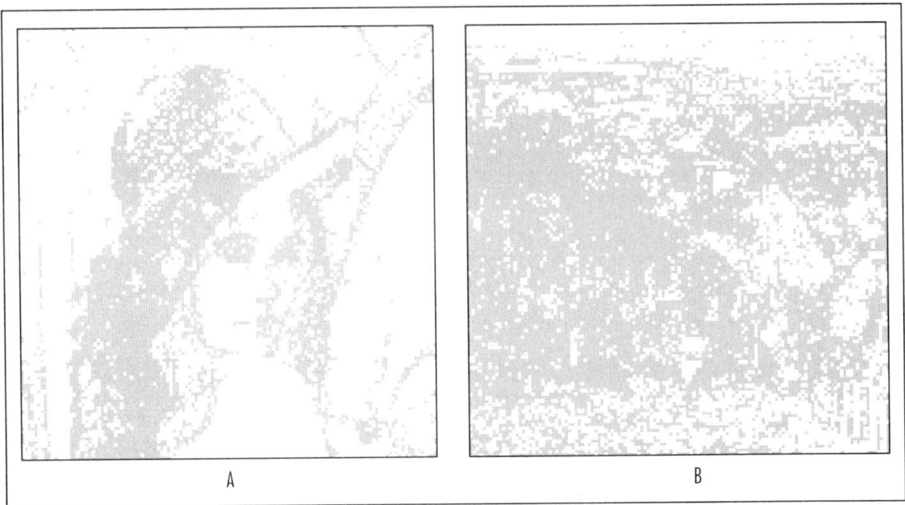

FIGURE 5.3.1 *Destination regions covered by global fractal codes: (A)* Lena; *(B)* Golden Hill

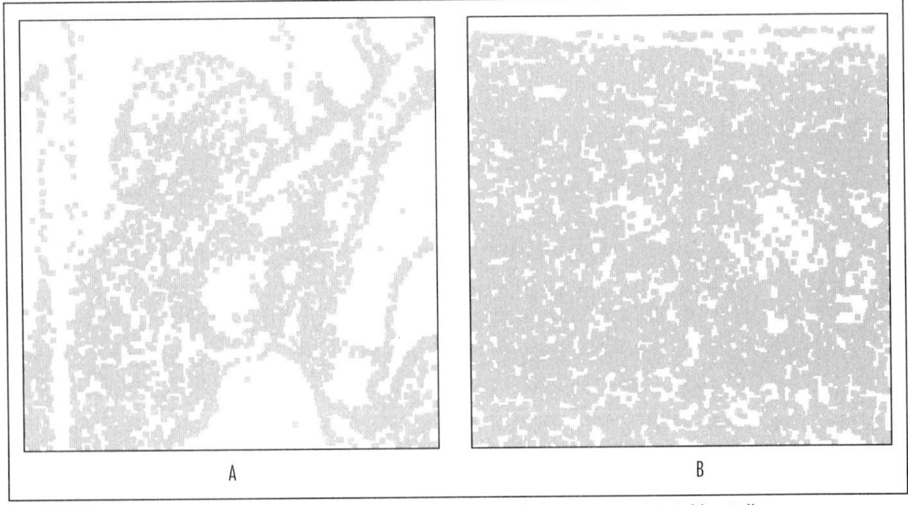

FIGURE 5.3.2 *Reference regions used by global fractal codes: (A)* Lena; *(B)* Golden Hill

From Figure 5.3.2*a* one can tell that there is room to improve the global reference address bit rate, and from Figure 5.3.2*b* the situation seems to be the opposite. In comparing Figure 5.3.2 to Figure 5.3.1, an interesting phenomenon is observed: they look like each other. This is not an accident, particularly after various other images are verified. Based on this, a new global reference address coding technique is designed by indexing the global reference blocks. It improves compression performance, especially to those images with extremely unbalanced complexities. This technique is called the *global indexed reference technique*.

Given a fractal representation that has a mixture of local and global codes, a binary global code region map can be obtained by marking the global destination region as *black* and local destination region as *white*. Expanding the black dots on the global code map to their neighborhood dots within radius r, for some positive number r, the new map is called the *global r-map*. Precisely, the global r-map is a binary image that marks all pixels that have a distance (to some global destination region) less than r to black, and other pixels to white.

Given a positive number m, called the *matching number, the (r,m)-reference list* consists of the complete list of possible reference regions containing at least m black points from the global r-map. An index to that list can be assumed by both coding and decoding systems. In extreme cases, the pair (1, 1) gives the list of reference regions that intersect some global destination region. This list gives the complete reference set when r is large enough.

By choosing the proper number pair (r, m), the global reference regions can be indexed with a few bits fewer. The fractal codes that violate the list can always be replaced by some new codes. Applying to the image *Lena*, a 2-bit gain can be expected from each global code. However, it is useless in the case of the *Golden Hill* image, as shown in Figure 5.3.2*b*.

What is the right pair of parameters (r, m) to be used? Experiments show that it is image dependent and distortion rate dependent. The correct way is to find those pairs that suit the particular image by using an *annealing process* as described below.

After a fractal presentation is given, a large enough pair (r, m) can be set so that all the global codes are valid. Now, let the radius number r be reduced by 1 or the matching number m be increased by 1. For any invalid global fractal codes, i.e., whose reference regions are not in the new (r, m) reference list, the best valid choices will replace them. If the ratio between the increasing error and code data size reduction is smaller than the designated distortion rate, then repeat this procedure. Otherwise, use the previous number pair as the final choice.

Exercise: Give a complete C implementation of the global indexed reference algorithm.

5.3.2 FRACTAL TRANSFORMATION TEMPLATE

To improve image quality, a separate global reference image is introduced as an *image template*. In specific real-world applications there are always some special properties that are shared within an application and not universally applied to general images. For example, a collection of oil paintings, an atlas of city maps, or a database of fingerprints. Furthermore, for images obtained by using the same process, such as those scanned from Kodak Royal Gold 100 prints or digitized from Fuji Velvia 50 slides or obtained directly from a Kodak/Nikon DCS 460 digital camera, some common properties can be found. These common properties are where the further redundancies reside.

FIGURE 5.3.3 *Fractal representation using FTT*

The approach for further improvement is to add a piece of a representative image that has the shared properties within that specific application as an extra piece of reference image.

Such a reference image piece is called the *fractal transformation template* (FTT). This image piece is installed together with the application system before any images are compressed. The FTTs can be manually or automatically composed from a gallery of images or selected from a collection of FTT candidates.

This method is application specific. Figure 5.3.3 shows an example. Even with a quite arbitrarily small FTT, the picture obtained using FTT obviously has higher image quality.

Because the FTT reference image is given as a part of the compression and decompression system, the decompression is not iterative. In this case, the collage errors are the same numbers as those of the decoded errors.

Actually, in some applications, higher resolution FTTs can be sent ahead of time, such as letter fonts; the decoded image using FTT could even achieve "better than original" results, such as a fax copy recovered by FTT fonts.

5.3.3 UNIVERSAL SEED IMAGE

As previously shown, the shared reference image can dramatically improve decoded image quality for a specified group of images. What happens if a universally shared reference image is provided? It may not help to improve every image, but there are some advantages in using it:

- *Fast compression.* A predetermined searching tree structure can be built in advance that accelerates the compression speed (which will be presented in Chapter 7).
- *No error spread.* Since the seed image is given, there is no need for iteration. The matching errors, i.e., collage errors, are the actual decoded errors.
- *Fast decompression.* Of course, no need for iteration.

Comparing the idea of universal seed image with the *vector quantization technique* (VQ), it is interesting to see that though they are approached differently, yet practically they are similar. The next subsection presents the basics of vector quantization, especially where it concerns obtaining a universal seed image.

5.3.4 VECTOR QUANTIZATION

Given a fixed square block dimension $n \times n$, each image piece can be viewed as a vector of n^2 entries, one entry for each pixel. The L^2-distance between two image pieces is indeed identically equal to the L^2-distance between the two vectors associated to them. Thus, all reference blocks from the universal seed image can be viewed as a codebook, for each reference block is a code vector indexed by its location. Reciprocally, if a codebook is given, a seed image can be composed by putting one codeblock after another, and then assuming that the valid reference addresses are in the grid of multiples of the block width n. Therefore, a seed image is equivalent to a codebook of image blocks.

A well-known algorithm for designing a VQ codebook is given by Linde, Buzo, and Gray, called the *generalized Lloyd–Max algorithm*, or briefly, the *LBG algorithm*. The LBG algorithm requires an initial codebook and the set of all blocks that is represented, which are called *training blocks*. The algorithm proceeds as follows:

1. *Clustering*. Map every training block into its nearest code block, and calculate the overall distortion error. Quit if the overall error is lower than a predetermined threshold, or if the *error improvement*—which is defined as the difference between the previous overall error and the current overall error—is smaller than a predetermined threshold.
2. *Updating*. For each code block form the set of all training blocks that map to this code block—which is called *the cluster of this code block*. Find the *best representation block of the cluster*, which is defined to be the block that gives the minimal total error by summing the error between this representation block and each training block in the cluster. (In the case of L^2-distance, the representation block is the mean of all training blocks in the cluster.) And replace the code block by the representation block.
3. Apply Step 2 to every code block, then go back to Step 1.

Note that every time a code block is replaced by a new representation block, the overall distortion error decreases. And each changing of code block also may cause some code block cluster changes, since some training blocks from this cluster could find another code block closer than the new representation block, and some training blocks from other clusters could be closer to this new representation block than to their current code block.

In any case, the LBG algorithm converges because its overall distortion decreases monotonically. However, it converges only to a local optimal codebook. Generally, numerous local optima exist, and many of them yield poor performance. Therefore, choosing an initial codebook is important. Good performance is usually obtained by starting from a good initial codebook.

What is the best way to generate a good initial codebook? Equitz's *pairwise-nearest-neighborhood (PNN) algorithm* [E2] is an appropriate choice. The PNN algorithm sets the initial codebook to the whole training set—each training block is a code block. Then repeat the following procedures:

1. *Merging*. Map every training block into its nearest code block. Merge the two closest code blocks into one. The combined cluster is defined to be the union of the clusters of both code blocks.

2. *Updating*. Choose the best representation block of the combined cluster, which is defined to be the block that gives the minimal total error, as previously described. Quit if you reach the desired codebook size. Otherwise, go back to the first step.

Again, when L^2-distance is used, in the updating step, the new code blocks are exactly the mean of all training blocks in that cluster.

In Chapter 7 some extremely fast but less optimized cluster splitting algorithms will be given. These algorithms are good for generating initial codebooks.

In creating a universal seed image, the set of training blocks will be the picture block pieces from all images—the size of this set is gigantic. In this case, both above algorithms become impractical, and so some evolution is required. In the following subsection an evolved adaptive algorithm will be presented for generating a universal seed image.

5.3.5 SEED IMAGE CONSTRUCTION ALGORITHM

Given a huge string of training blocks (in our case, the $n \times n$ D-blocks that have no adequate local fractal codes) from all images, let K be the number of code blocks desired. This block number K should be in the range from 1,000 to 1,000,000.

At each step, we have the following:

k the number of code blocks;

C_i for $i = 0, 1, \cdots, k-1$, the ith code block;

σ_i for $i = 0, 1, \cdots, k-1$, the number of processed training blocks represented by the code block C_i;

m $= \min_{0 \le i < j < k} \left\| C_i - C_j \right\|$, the minimal distance between ith and jth code blocks;

T a training block;

$d(T)$ $= \min_{0 \le i < k} \left\| T - C_i \right\|$, the distance of T to the closest code block;

$i(T)$ the index for one of the closest code blocks, i.e., $\left\| T - C_{i(T)} \right\| = d(T)$.

A full loop through all training blocks is given by processing each training block below. For a training block T,

1. If $k < K$, add T as a code block to the codebook, $C_k \leftarrow T$ and $\sigma_k \leftarrow 1$, increase k by 1, update the number m.
2. Else, if $d(T) < 2m$, add T to the $i(T)$th cluster by updating

$$C_{i(T)} \leftarrow \frac{\sigma_{i(T)} \cdot C_{i(T)} + T}{\sigma_{i(T)} + 1} \text{ and } \sigma_{i(T)} \leftarrow \sigma_{i(T)} + 1, \tag{5.3.1}$$

update the number m.

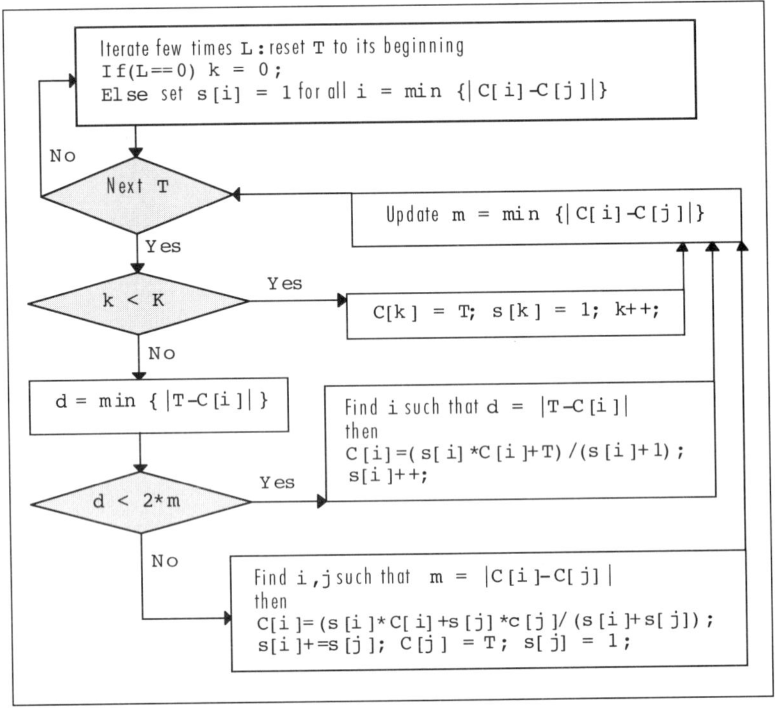

FIGURE 5.3.4 *Seed image construction algorithm*

3. And else, find i and j such that $m = \left\| C_i - C_j \right\|$, merge C_j to C_i by

$$C_i \leftarrow \frac{\sigma_i \cdot C_i + \sigma_j \cdot C_j}{\sigma_i + \sigma_j} \text{ and } \sigma_i \leftarrow \sigma_i + \sigma_j, \tag{5.3.2}$$

add T as a code block to the codebook, $C_j \leftarrow T$ and $\sigma_j \leftarrow 1$, update the number m, and done.

For a high compression performance, two full loops are recommended. In the beginning of the second loop, the numbers σ_i, $i = 0, 1, \cdots, k-1$, are set to 1.

After few more full loops, the numbers σ_i, $i = 0, 1, \cdots, k-1$, give exactly the probabilities of the code blocks C_i, $i = 0, 1, \cdots, k-1$, respectively, which can be used directly for indexing and entropy coding.

Figure 5.3.4 shows the skeleton of this algorithm. One more question:

Problem: Given a predetermined seed image, is there a fast way to match an arbitrary block with the best code?

A partial answer will be given in Chapter 7.

At last the horizon seems open once more, granting even that it is not bright; our ships can at last put out to sea in face of every danger; every hazard is again permitted to the discerner; the sea, our sea, again lies open before us; perhaps never before did such an "open sea" exist.

—Joyful Wisdom

Friedrich Nietzsche

Brightness and Contrast

6

When an object is moved from one place to another, to a viewer, in the viewing plane, not only can the shape of the object be deformed and scaled, but its color may also be changed because of the surrounding lighting. This color change (in a fractal image model) is reflected in the other half of the transformation of a fractal element: the *intensity component*. In the PIFS fractal modeling presented in Section 3.3, the intensity component has three parameters: the *intensity brightness adjustment* β*-value*, the *intensity contrast adjustment* γ*-value*, and the *intensity lighting adjustment t-form*. In this chapter, the β*-value* and γ*-value* will be discussed in depth. A new parameter, the *absolute luminance*—that is, the *intensity mean*, which therefore will be called the *mean value*—will be introduced as an alternative choice to the β-value, because it behaves efficiently for data storage.

More intensity adjustment terms can be added to the intensity component of a fractal code. The lighting tilt *t*-form is such a term that has a clear physical interpretation. Other additional terms, e.g., polynomial terms and discrete orthogonal vectors, which have been studied in research, will be briefly covered. One special choice of additional intensity terms is the set of the standard DCT base vector, which will be presented as a part of the JPEG still image standard.

6.1 BRIGHTNESS PARAMETERS

The first term of the color intensity adjustment is the *brightness adjustment*, called β*-value*, which is the correction, or shift, or translation, between the total color intensity mapped from the reference region and the total actual color intensity of the destination region. Technically, it is also called the *order zero term*, since it is exactly the constant term if you expand the color intensity component into a polynomial of spatial and intensity variables.

6.1.1 INTENSITY CORRECTION

With compression as the focus, this chapter explains how to store the intensity parameters efficiently. In almost all current effective working implementations, the *brightness adjustment* β-values (also called the *intensity correction*), take the most bits of intensity parameters in fractal image data. Figures 6.1.1 and 6.1.2 show the β-value distribution graphs of *Lena* and *Golden Hill* in fractal representations generated by *Codec 1.00* and *Codec 2.00* with distortion rate 16.

The set of β-values of a given image can be viewed as a 2-dimensional array. It will be packed using a standard simple 2-dimensional data decorrelating algorithm: the *2-dimensional DPCM* (*Differential Pulse Coding Modulation*).

The 2-dimensional DPCM algorithm is illustrated in Figure 6.1.3. A 2-dimensional image data array is ordered in the scan line order. Each datum is

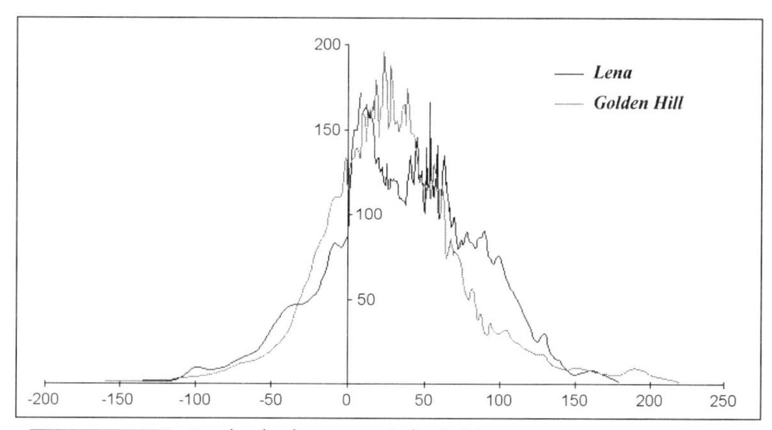

FIGURE 6.1.1 β-value distribution using Codec 1.00

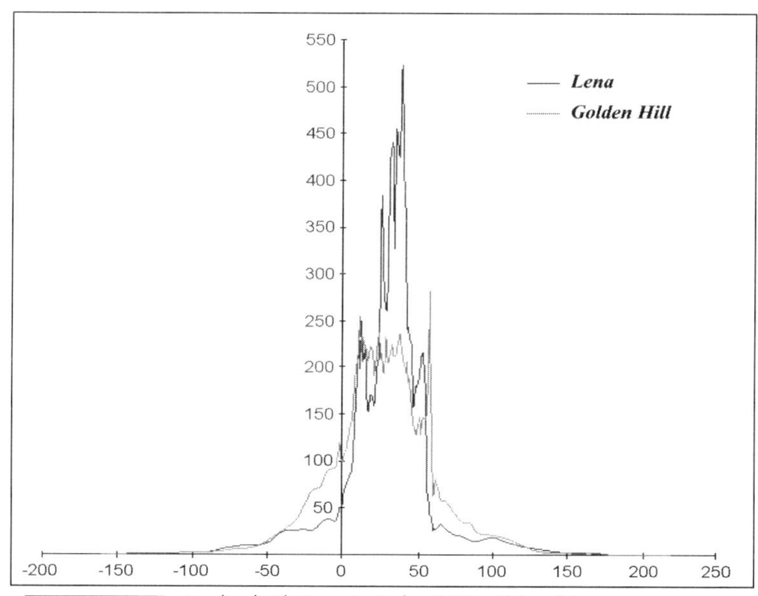

FIGURE 6.1.2 β-value distribution using Codec 2.00 with $\lambda = 16$

a	b	c	d	...
i	j	k	l	...
p	q	r	s	...
⋮	⋮	⋮	⋮	⋱

→

a	b-a	c-b	d-c	...
i-a	j-b	k-c	l-d	...
p-i	q-j	r-k	s-l	...
⋮	⋮	⋮	⋮	⋱

FIGURE 6.1.3 A 2-dimensional DPCM

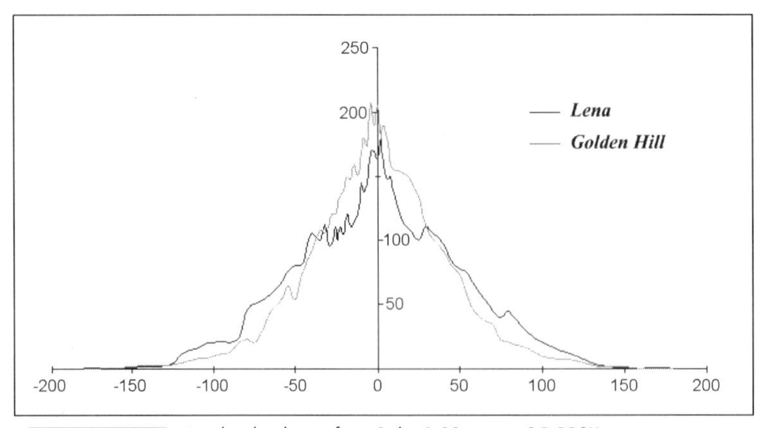

FIGURE 6.1.4 β-value distribution from Codec 1.00 using a 2-D DPCM

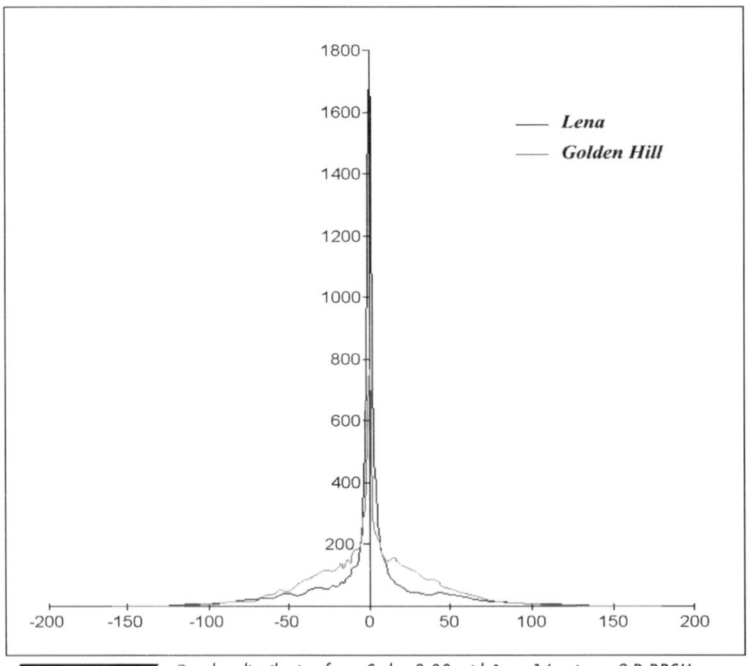

FIGURE 6.1.5 β-value distribution from Codec 2.00 with $\lambda = 16$ using a 2-D DPCM

subtracted from the corresponding datum above in the previous row, except for those in the first row. Each datum in the first row is subtracted from its preceding datum, except for the very first datum. The first datum is stored.

Figures 6.1.4 and 6.1.5 shows the graphs when the same data strings from Figures 6.1.1 and 6.1.2 are packed using the 2-dimensional DPCM.

From these figures two facts are concluded:

- *local searching reduces the entropy of β-values*, since *Codec 2.00* graphs have significantly lower entropies than the corresponding *Codec 1.00* ones; and
- the β-*values are more efficient when packed in the decorrelated difference form.*

Table 6.1.1 shows not only the exact entropy calculations, but also the results of packing the DPCM differences of β-value by using the most commonly adopted *probability distribution functions: Gaussian distribution, Laplacian distribution*, and *gamma distribution.* The exact definitions of *entropy* and these distribution functions are covered in Chapter 12.

TABLE 6.1.1 *Average β-value bit rate using an entropy codec*

β	Lena		Golden Hill	
Codec	*1.00*	*2.00 & λ=16*	*1.00*	*2.00 & λ= 16*
Entropy	7.64770	6.59744	7.55659	7.04236
Difference Entropy	7.56643	5.76247	7.42813	6.93850
Gaussian Function	7.60356	6.83305	7.46961	7.08069
Laplacian Function	7.65788	6.46473	7.46925	7.00280
Gamma Function	7.98088	6.04140	7.77343	7.13262

The brightness adjustment β-value is used for obtaining the correct color brightness by performing an additive correction. What if we store directly the *total absolute luminance* of a destination region instead of its brightness correction? The next subsection will investigate this problem.

6.1.2 INTENSITY MEAN VALUE

The average absolute luminance of a destination region is called the *mean value* of this region. In the digital case, the *mean value* of a destination region is the sum of all intensity values of the pixels in this region divided by the pixel number in this region.

In the instance of a fractal image model, observe that the set of mean values of this model depends only on the masking functions. In the models that use image partitioning, the mean value of a destination region is completely independent of the fractal transformation chosen for this region. As a result, the mean values can be quantized and packed independently from the compression process, in contrast to the fact that the β-values are actually determined in the compression process. In fact, two fractal representations of an image, compressed using *Codec 1.00* and *Codec 2.00* respectively, will have different sets of β-values and the same set of mean values.

Again, applying the 2-dimensional DPCM method to the 2-dimensional mean array, the distribution of mean value differences is charted in Figure 6.1.7. Table 6.1.2 gives the same statistics as in Table 6.1.1 with positive results. The mean value distributions not only have low entropies but also behave in very standard distributions. The difference between such an actual entropy and the bit rate obtained from a standard distribution function—Gaussian, Laplacian and gamma function—is less than 0.1 bit. Overall, the mean values take 15–20% less storage space than their equivalent β-values.

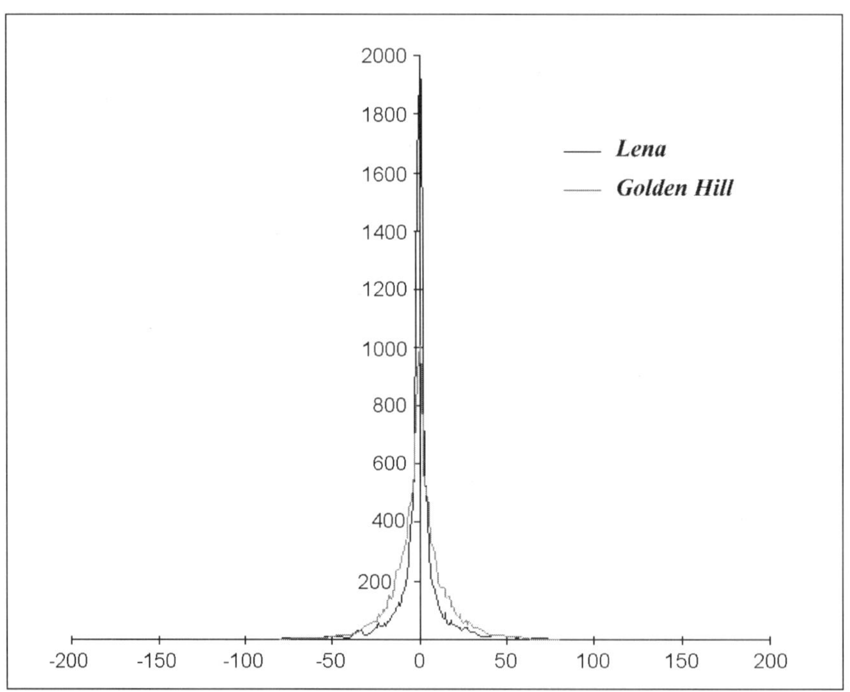

FIGURE 6.1.6 *Mean value difference distribution using* Codec 1.00

TABLE 6.1.2 *Average mean value bit rate using an entropy codec*

m	Lena	Golden Hill
Entropy	7.38883	7.36342
Difference Entropy	5.07433	5.77052
Gaussian Function	5.57568	5.86041
Laplacian Function	5.41538	5.78346
Gamma Function	5.16442	5.82233

6.1.3 GENERALIZED COLLAGE THEOREM

How does the switch from β-value to mean value affect our theoretical base? *The collage theorem* will be updated in this subsection.

A *partitioned iterated function system with means* $\tilde{W}$ is a PIFS whose β-values are replaced by mean values. Using similar notation as in Section 3.3.2, let $\tilde{W} = \left\{\Phi_i = (\mu_i, \tilde{T}_i)\right\}_{i=1,2,\cdots,m}$, where $\tilde{T}_i$ has the affine parameters

$$\tilde{T}_i = \left((x_D^i, y_D^i), (x_R^i, y_R^i), \sigma_i = \begin{pmatrix} s_{00}^i & s_{01}^i \\ s_{10}^i & s_{11}^i \end{pmatrix}, \tau_i = 0, \gamma_i, m_i \right). \tag{6.1.1}$$

The associated Markov process, in this case, is decomposed into two steps:

Step 1: $f \mapsto U(f)$:

$$U(f)(x,y) = \begin{cases} \gamma_i \cdot f\left(\sigma_i^{-1}\begin{pmatrix} x - x_D^i \\ y - y_D^i \end{pmatrix} + \begin{pmatrix} x_R^i \\ y_R^i \end{pmatrix}\right), & \text{if } (x,y) \in \mu_i^{-1}(1) \subset \Re^2, \text{ for } 1 \le i \le m; \\ 0, & \text{otherwise.} \end{cases} \tag{6.1.2}$$

Step 2: $U(f) \mapsto \tilde{W}(f)$:

$$\tilde{W}(f)(x,y) = \begin{cases} U(f)(x,y) + m_i - \gamma_i \cdot \dfrac{\iint_{D_i} U(f)(\xi,\eta)\, d\xi\, d\eta}{\iint_{D_i} d\xi\, d\eta}, & \text{if } (x,y) \in \mu_i^{-1}(1) \subset \Re^2, \\ & \text{for some } 1 \le i \le m; \\ 0, & \text{otherwise;} \end{cases} \tag{6.1.3}$$

for any $(x,y) \in \Re^2$ and $f \in \mathcal{P}(\Re^2)$.

Actually, the β-value is calculated in each iteration as

$$\beta_i = m_i - \gamma_i \cdot \frac{\iint_{D_i} U(f)(\xi,\eta)\, d\xi\, d\eta}{\iint_{D_i} d\xi\, d\eta}, \tag{6.1.4}$$

for each $1 \le i \le m$.

Consider a special case by beginning with a partitioned IFS *W*: Let $\{\beta_i\}_{i=1,2,\cdots,m}$ denote the collection of β-values of *W*, and *A* the attractor image of *W*. Then construct a PIFS with means $\tilde{W}$ from the PIFS *W* by replacing the β-values $\{\beta_i\}_{i=1,2,\cdots,m}$ with the attractor means

$$m_i = \frac{\iint_{(x,y)\in D_i} A(x,y)\,dx\,dy}{\iint_{(x,y)\in D_i} dx\,dy}, \tag{6.1.5}$$

for all $i = 1, 2, \cdots, m$. The sequence

$$P,\ \tilde{W}(P),\ \tilde{W}^2(P),\ \cdots,\ \tilde{W}^n(P),\ \cdots, \tag{6.1.6}$$

for any image P, clearly converges to the same attractor A of the PIFS W. Indeed, this can be deduced from the fact that

$$d(\tilde{W}(P), A) \le d(W(P), A), \tag{6.1.7}$$

for any image P, which is obvious from the definitions.

This discussion actually proves the existence of the PIFS with means stated in the next theorem:

GENERALIZED ENCODING THEOREM:
Given any PIFS W with a contraction factor p, $0 < p < 1$, there exists one and only one PIFS with means $\tilde{W}$ such that both models have the same collection of masking functions and converge to the same attractor image.

Furthermore, for any image P,

1. if $d(P, \tilde{W}(P)) \le \varepsilon$ for some $\varepsilon > 0$, then $d(P, A) \le \dfrac{\varepsilon}{1-p}$;
2. $d(\tilde{W}^n(P), A) \le p^n$, for all integers n.

The uniqueness is a trivial verification. The proof of the two properties is the same as the proof of the collage theorem by using the fact that

$$d(\tilde{W}(P), A) \le d(W(P), A). \tag{6.1.8}$$

Exercise: Give a proof of the above theorem in detail.

The theorem implies that PIFS with means is a more general model than PIFS, and to use means could possibly give a better (and certainly not worse) fractal image model than to use β-values.

What happens if use the mean of the original image O instead of the attractor's? The first observation is that the sequence

$$P,\ \tilde{W}(P),\ \tilde{W}^2(P),\ \cdots,\ \tilde{W}^n(P),\ \cdots, \tag{6.1.9}$$

for an image P, may *never* converge. However, the next theorem will assure that the PIFS with means $\tilde{W}$ still can be used to represent an image.

GENERALIZED COLLAGE THEOREM:
Given a partitioned iterated function system with means

$$\tilde{W} = \{ (D, R, \sigma, \tau, \gamma, m) \} \tag{6.1.10}$$

having the contractivity factor p less than 1, and given an image O, a so-called original image, if

1. the mean values of $\tilde{W}$ are actually the mean values from the image O, i.e.,

$$m_i = \frac{\iint_{(x,y)\in D_i} O(x, y)\, dx\, dy}{\iint_{(x,y)\in D_i} dx\, dy} \tag{6.1.11}$$

   for all $i = 1, 2, \cdots, m$, and

2. $d(\tilde{W}(O), O) < \varepsilon$, for some ε, then, for any image P,

$$d(\tilde{W}^n(P), O) < \frac{\varepsilon}{1-p}, \tag{6.1.12}$$

   for any large enough integer n.

Proof: Again, a PIFS W will be constructed as a bridge to the proof. This PIFS W is obtained from $\tilde{W}$ by exchanging the means to the collection of β-values $\{\beta_i\}_{i=1,2,\cdots,m}$ so that $W(O) = \tilde{W}(O)$, i.e., they are given by the formula

$$\beta_i = \frac{\iint_{(x,y)\in D_i} O(x, y)\, dx\, dy}{\iint_{(x,y)\in D_i} dx\, dy} - \gamma_i \cdot \frac{\iint_{(x,y)\in D_i} U(O)(x, y)\, dx\, dy}{\iint_{(x,y)\in D_i} dx\, dy} \tag{6.1.13}$$

for all $i = 1, 2, \cdots, m$. Repeatedly use the formula

$$d(\tilde{W}(P), O) \le d(W(P), O), \tag{6.1.14}$$

for any image P, which is the direct result of the first condition. Then,

$$\begin{aligned} d(\tilde{W}^n(P), O) &\le d(W(\tilde{W}^{n-1}(P)), O) \\ &\le d(W(\tilde{W}^{n-1}(P)), W(O)) + d(W(O), O) \\ &\le p \cdot d(\tilde{W}^{n-1}(P), O) + \varepsilon \\ &\le p \cdot (p \cdot d(\tilde{W}^{n-2}(P), O) + \varepsilon) + \varepsilon \\ &\le p^n \cdot d(\tilde{W}^0(P), O) + p^{n-1} \cdot \varepsilon + \cdots + p \cdot \varepsilon + \varepsilon \\ &\le p^n \cdot d(P, O) + \frac{d(\tilde{W}(O), O)}{1-p}. \end{aligned}$$

The second condition concludes the theorem ♦

In practice, all our experiments always show that

$$d(\tilde{W}^n(P), O) \le d(W^n(P), O), \tag{6.1.15}$$

for large n. Is this always true? It is a question for the reader.

From now on, the mean values will be used in most fractal image PIFS models, and henceforth PIFS will mean a PIFS with either β-values or means.

6.1.4 COMPRESSION USING MEANS

From the above discussion, it is conclusive that one should use mean values for better compression performance. However, it will cause some necessary changes in the decompression procedure. In every step of the iteration, instead of adding the correction, the correction needs to be calculated first by differentiating between the current intensity value and the designated intensity value. This increases the computational complexity in each iteration, but in return, the number of iterations is reduced a great deal. Chapter 8 will discuss the decompression techniques in detail. And Table 6.1.3 shows that the number of iterations is dramatically reduced by using mean values.

Comparing Table 6.1.3 and the corresponding decompression table in Chapter 3 side by side, the quality of an m-decoder after 4 iterations is both visually and numerically as good as a q-decoder after 10 iterations, and the quality of an m-decoder after 8 iterations is as good as a q-decoder after 20 iterations.

If the original mean data are used instead of the decoded attractor, convergence is no longer the case. However, the result is always better—somewhere between the attractor and the original image—as shown in Table 6.1.4.

Having performed the same experiments with many other images, notice that the decoded images do not improve after five iterations, and all of them have a slightly improved quality performance (in comparison with using β-values). Consequently, the default iteration number will drop from 16 to 5.

TABLE 6.1.3 *Iteration and convergence using mean values of the attractors*

L^2	Lena			Golden Hill		
i	*original*	*attractor*	*previous*	*original*	*attractor*	*previous*
0	48.016402	47.803839		49.891805	49.595829	
1	9.313221	8.481671	47.236881	10.382104	9.210134	48.919721
2	4.433032	2.822330	8.275642	5.389032	3.298798	8.790541
3	3.421712	.632966	2.705847	4.215694	.716302	3.168183
4	3.375076	.287388	.607858	4.167173	.303169	.692886
5	3.371716	.196568	.279768	4.164997	.205654	.300229
6	3.370957	.147354	.189715	4.164862	.155883	.202476
7	3.371010	.112741	.146341	4.164570	.119827	.151943
8	3.370927	.090436	.114053	4.164164	.095484	.115118
9	3.371030	.071309	.091422	4.164281	.075240	.090310
10	3.370755	.053667	.070799	4.164249	.059786	.073418
11	3.370579	.045721	.059594	4.164221	.045134	.058561
12	3.370724	.037263	.050631	4.164218	.032973	.044495
13	3.370605	.027896	.043057	4.164164	.024316	.033716
14	3.370511	.022183	.033885	4.164135	.018113	.026851
15	3.370626	.012040	.025240	4.164101	.008286	.019137
16	3.370577	.005167	.013102	4.164073	.000000	.008286
17	3.370583	.000000	.005167			

TABLE 6.1.4 *Iteration and convergence using mean values of the originals*

L^2	Lena		Golden Hill	
i	*original*	*previous*	*original*	*previous*
0	48.016402		49.891805	
1	11.496596	46.617727	12.037807	48.404100
2	6.815158	10.079661	7.849318	9.611075
3	3.749522	5.157243	4.728427	5.564765
4	3.396842	1.508483	4.209134	1.919773
5	3.364696	0.682158	4.157121	0.786823
6	3.361183	0.533536	4.152344	0.557941
7	3.362913	0.486842	4.151988	0.497280
8	3.362926	0.461244	4.151451	0.465315
9	3.361212	0.441506	4.150711	0.446070
10	3.361645	0.430198	4.151849	0.428095
11	3.361931	0.419995	4.153813	0.414541
12	3.361653	0.414003	4.150401	0.411605
...				
99	3.360668	0.396114	4.152402	0.409602
100	3.361808	0.400400	4.152022	0.409803

6.1.5 MEAN QUANTIZATION

Mean values depend only upon image partitions. In a lossy compression the *mean difference quantization* is introduced, which is a very useful technique in reducing the mean data size while preserving good visual quality. Research has shown that *the eye is more sensitive to the luminance change in the smooth homogeneous region than in the texture high-variance region.*

Thus, the mean differences will be quantized in such a way that the low differences are more accurate and high differences less precise.

Tables 6.1.5 and 6.1.6 are two highly recommended quantization tables used in some research projects. They are called the *fine quantization table* and the *coarse quantization table*, respectively.

TABLE 6.1.5 *The mean value fine quantization table*

representative	values	representative	values
0	0		
1	1	−1	−1
2	2	−2	−2
3	3, 4	−3	−3, −4
6	5, 6, 7	−6	−5, −6, −7
9	8, 9, 10	−9	−8, −9, −10
12	11, 12, 13	−12	−11, −12, −13
15	14, 15, 16, 17	−15	−14, −15, −16, −17
5k, *k*=4 ... , 50	5*k*−2, ... , 5*k*+2	−5k, *k*=4, ... , 50	−5*k*+2, ... , −5*k*−2
255	253, 254, 255	−255	−253,−254, −255

TABLE 6.1.6 *The mean value coarse quantization table*

representative	values	representative	values
0	0		
1	1, 2	−1	−1, −2
4	3, 4, 5	−4	−3, −4, −5
7	6, 7, 8,	−7	−6, −7, −8,
10	9, 10, 11, 12	−10	−9, −10, −11, −12
5k, *k*=3,4,5,6	5*k*−2, ... , 5*k*+2	−5k, *k*=3,4,5,6	−5*k*+2, ... , −5*k*−2
37+7k, *k*=0,1,2,3,4	34+7*k*, 35+7*k*, ... , 40+7*k*	−37−7k, *k*=0,1,2,3,4	−34−7*k*, −35−7*k*, ... , −40−7*k*
74+9k, *k*=0,1,2,3,4	70+9*k*, 71+9*k*, ... , 78+9*k*	−74−9k, *k*=0,1,2,3,4	−70−9*k*, −71−9*k*, ... , −78−9*k*
121+11k, *k*=0,1,2,3,4	116+11*k*, 117+11*k*, ... , 126+11*k*	−121−11k, *k*=0,1,2,3,4	−116−11*k*, −117−11*k*, ... , −126−11*k*
180+15k, *k*=0,1,2,3	173+15*k*, 174+15*k*, ... , 187+15*k*	−180−15k, *k*=0,1,2,3,4	−173−15*k*, −174−15*k*, ... , −187−15*k*
240	233, 234, ... , 255	−240	−233,−234, ... ,−255

The fine quantization table quantizes 511 mean differences into 111 representatives. The coarse quantization table reduces them to 57 items.

The entropy table 6.1.2. can be improved further to Table 6.1.7.

The numerical improvement of the images *Lena* and *Golden Hill* is shown in Tables 6.1.8 and 6.1.9. The visual improvement is more significant. So far, all images compressed with the same distortion rate, regardless of the quantization, are visually identical.

TABLE 6.1.7 *Average mean value bit rate using an entropy codec*

m	Lena		Golden Hill	
Quantization	*fine*	*coarse*	*fine*	*coarse*
Difference Entropy	4.25410	3.74377	4.51182	4.10565
Gaussian Function	5.07650	4.54432	4.72316	4.39536
Laplacian Function	4.57618	4.05627	4.53917	4.17781
gamma Function	4.33056	3.81804	4.60905	4.19215

TABLE 6.1.8 *Compression comparison of* Lena *in a* 4×4 *block fractal system*

no quantization			fine quantization			coarse quantization		
λ	*PSNR*	*bpp*	λ	*PSNR*	*bpp*	λ	*PSNR*	*bpp*
8	37.03	1.1983	5	37.21	1.1801	4	37.08	1.2137
15	36.52	1.0027	9	36.97	0.9916	8	36.86	0.9960
33	35.63	0.8063	18	36.51	0.8080	16	36.45	0.8044
112	33.85	0.6008	51	35.44	0.6029	42	35.55	0.6025

TABLE 6.1.9 *Compression comparison of* Golden Hill *in a* 4×4 *block fractal system*

no quantization			fine quantization			coarse quantization		
λ	*PSNR*	*bpp*	λ	*PSNR*	*bpp*	λ	*PSNR*	*bpp*
17	35.06	1.3596	8	35.32	1.3450	6	35.21	1.3514
27	34.51	1.2043	17	35.02	1.2031	15	34.97	1.2088
48	33.53	1.0022	33	34.37	1.0017	30	34.37	1.0053
86	32.23	0.8004	61	33.34	0.8024	57	33.41	0.8011

From the statistics in Tables 6.1.8 and 6.1.9, and actual visual testing, it can be concluded that some quantization table should always be used in a lossy image compression. For a high compression ratio, the coarse quantization table is recommended.

We did more tests. The results are always consistent. For the same data size the quantized ones are obviously better than the nonquantized ones. Yet, for the same quality distortion rate, the differences among those different quantization tables are visually unnoticeable.

In summary, in a fractal lossy compression system, mean values should be used in place of brightness adjustment β-values, and mean values should always pack in quantized differences.

6.2 CONTRAST FACTOR

The next important color intensity parameter is the *contrast adjustment*, also called the *intensity scaling factor*, or simply *γ-value*. It is the *order one term* in the intensity variable in the intensity transform of a fractal element.

6.2.1 CONTRAST SCALING FACTOR

So far, only constant intensity scaling γ-values have been used, e.g., $\gamma=3/4$. In this subsection, the variable γ-value case will be examined, and the comparison between using variable γ-value and using constant γ-value will be discussed.

Let $W=\{(D, R, \sigma, \tau, \gamma, \beta/m)\}$ be a PIFS fractal representation of an image P. The L_2 *collage error* of a fractal code is computed as

$$\textit{collage error} = \left(\frac{\int_D \left(P(x,y) - W(P)(x,y)\right)^2 \, dx\, dy}{\int_D dx\, dy} \right)^{1/2}$$

$$= \sqrt{\frac{\sum_{j=1}^{|D|} \left(d_j - \gamma \cdot r_j - \beta\right)^2}{|D|}} \tag{6.2.1}$$

in the digital case, where d_j is an image sample pixel value in some destination region D, r_j is the pixel value in the corresponding location of the reference region R, and $N=|D|$ is the total number of sample pixels in the destination region D.

The mean values of the destination region and the reference region are denoted by

$$m = \frac{1}{N}\sum_{j=1}^{N} d_j \text{ and } m_R = \frac{1}{N}\sum_{j=1}^{N} r_j . \tag{6.2.2}$$

By taking a few partial derivatives to minimize error, the intensity translation will be

$$\beta = m - \gamma \cdot m_R, \tag{6.2.3}$$

and the intensity scaling factor will be

$$\gamma = \frac{\sum_{j=1}^{N} (d_j - m)(r_j - m_R)}{\sum_{j=1}^{N} (r_j - m_R)^2} = \frac{< \mathbf{d} - m, \mathbf{r} - m_R >}{< \mathbf{r} - m_R, \mathbf{r} - m_R >}, \tag{6.2.4}$$

and consequently the collage error will be

$$\varepsilon = \sum_{j=1}^{N} \left((d_j - m) - \gamma (r_j - m_R) \right)^2 = < \mathbf{d} - m, \mathbf{d} - m > - \frac{< \mathbf{d} - m, \mathbf{r} - m_R >^2}{< \mathbf{r} - m_R, \mathbf{r} - m_R >}. \tag{6.2.5}$$

From the collage theorem, the decoded error is bound by the collage error, with a factor of $1/(1-\gamma)$. As a consequence, proportional penalty values should be assigned to large γ-values that are close to 1. What are the right penalties? This is a complicated question. Before we even consider this problem, the first question is how many accurate bits of the γ-value should be given. The simplest way to test this is to decompress an image using close but different γ-values. Table 6.2.1 shows the decoded errors of an image that is compressed using the constant γ-value 3/4 and decompressed using various γ-values in a PIFS with means.

TABLE 6.2.1 *Decompressing a $\gamma = 3/4$ image using various γ-values*

	Lena		Golden Hill	
γ	*rmse*	*PSNR*	*rmse*	*PSNR*
1/4	8.7213	29.32	9.4793	28.60
1/2	5.7074	33.00	6.5354	31.83
5/8	4.2454	35.57	5.0604	34.05
11/16	3.6726	36.83	4.4779	35.11
23/32	3.4738	37.31	4.2726	35.52
47/64	3.4071	37.48	4.2017	35.66
3/4	**3.3611**	**37.60**	**4.1519**	**35.77**
49/64	3.3546	37.62	4.1317	35.81
25/32	3.3694	37.58	4.1393	35.79
13/16	3.4867	37.28	4.2479	35.57
7/8	4.0761	35.93	4.8383	34.44
1	6.3184	32.12	7.3014	30.86
5/4	9.3508	28.71	10.9888	27.31

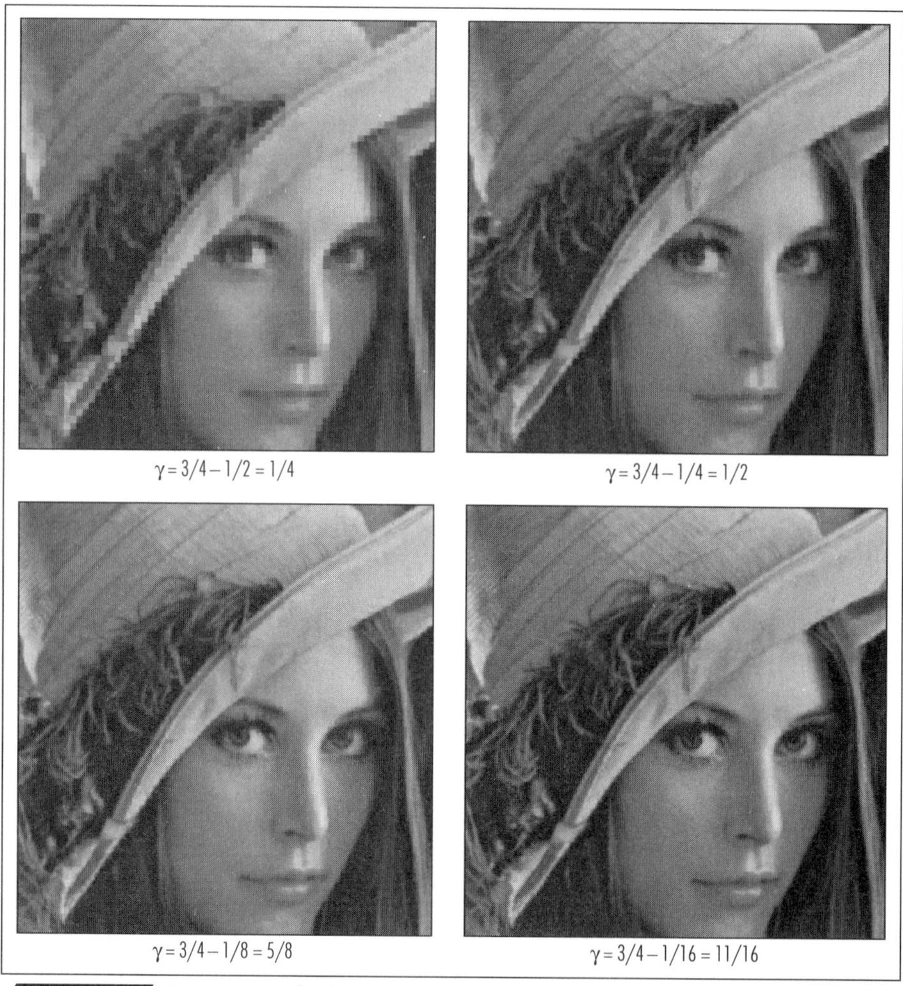

FIGURE 6.2.1 *Lena compressed with* $\gamma = 3/4$ *and decompressed with various values of* γ

The data in Table 6.2.1, make it clear that there is no reason to choose γ-value in steps finer than 1/32. Surprisingly, the best γ-value to decompress an image is not the same γ-value used for compression, but a value that is a slightly higher value. (In the sample case, the image *Lena* is compressed using $\gamma = 3/4$, and the best decompression γ-value is $\gamma = 49/64$). Is this phenomenon caused by an integer rounding error of the implementation? No. The fact is more profound: *The PIFS image model has a tendency to flatten an image. By slightly increasing the contrast factor, some of these losses are actually compensated for, and a better overall picture results.*

Figure 6.2.1 *(continued)* Lena *compressed with* $\gamma = 3/4$ *and decompressed with various values of* γ

Figure 6.2.1 visually displays the typical behavior of the γ-value: *a small* γ *smooths the image, and a large* γ *sharpens the texture*. In conclusion, this parameter reflects the contrast changes.

Notice that in the last two γ values the restriction on $|\gamma| < 1$ is broken. This puts the convergence at risk. However, the destination region intensity mean value assures that the decoded image could not be too distant from the original one. That actually is the most important advantage in using the mean as the main intensity parameter. In fact, in the case of using intensity mean values, the penalty for a contrast scaling factor close to 1 can be ignored.

6.2.2 DISCRETE CONTRAST SCALING FACTORS

Compress both sample images: *Lena* and *Golden Hill*, using the best contrast factor γ-value for each fractal code without receiving any penalty. The histograms of the γ-value distribution for both images are shown in Figure 6.2.2 and Figure 6.2.3. The used γ-values spread all over the places.

Comparing an image compressed and decompressed using a variable γ-value and the same image compressed and decompressed using a constant γ-value, there is no visual difference, as displayed in Figure 6.2.4. In fact, the additional γ-values produce more accurate local contrast, but no additional textures are created. As human eyes are particularly sensitive to geometric patterns and textures, it is worthwhile spending additional bits in that direction. Consequently, *there is no reason to use more than a few constant γ-values.*

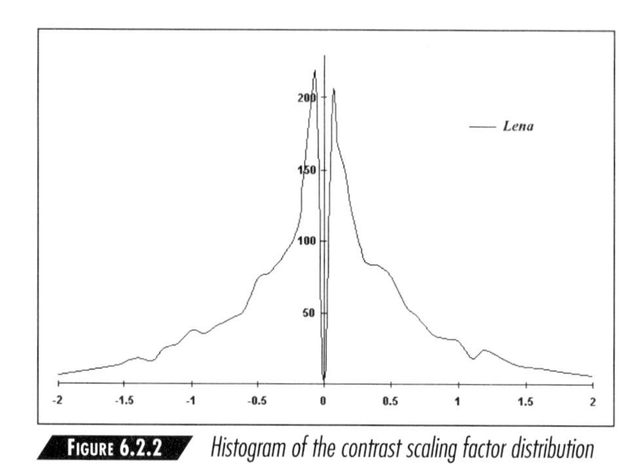

FIGURE 6.2.2 *Histogram of the contrast scaling factor distribution for* Lena

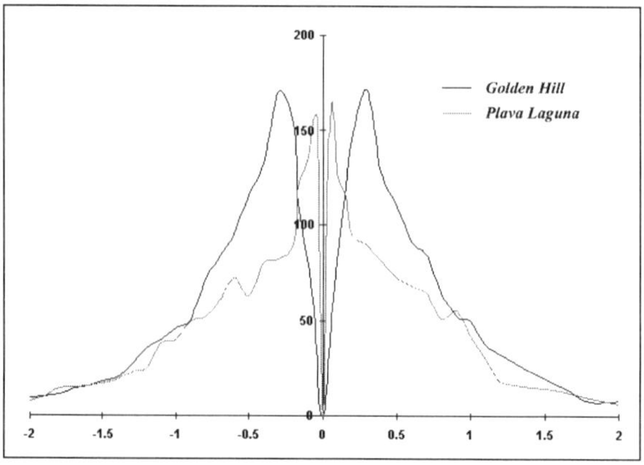

FIGURE 6.2.3 *Histograms of the contrast scaling factor distribution for* Golden Hill *and* Plava Laguna

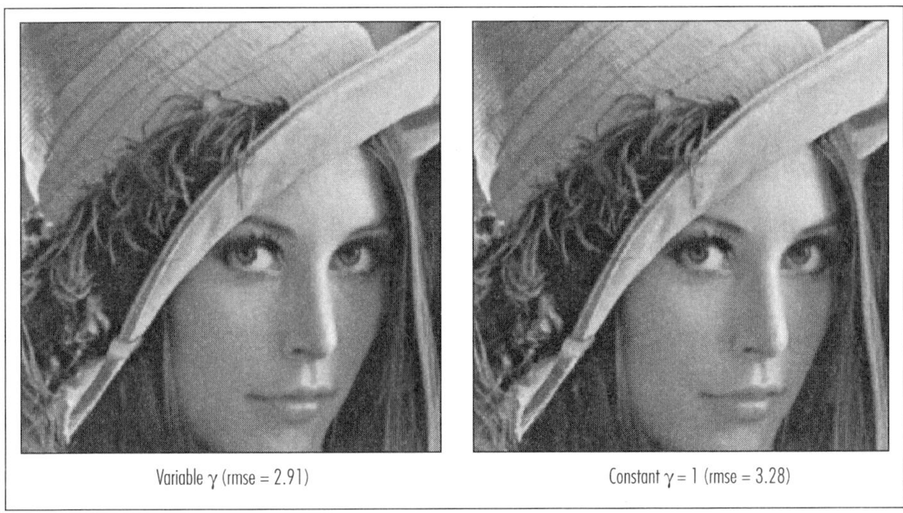

FIGURE 6.2.4 *Variable γ-value vs. constant γ-value*

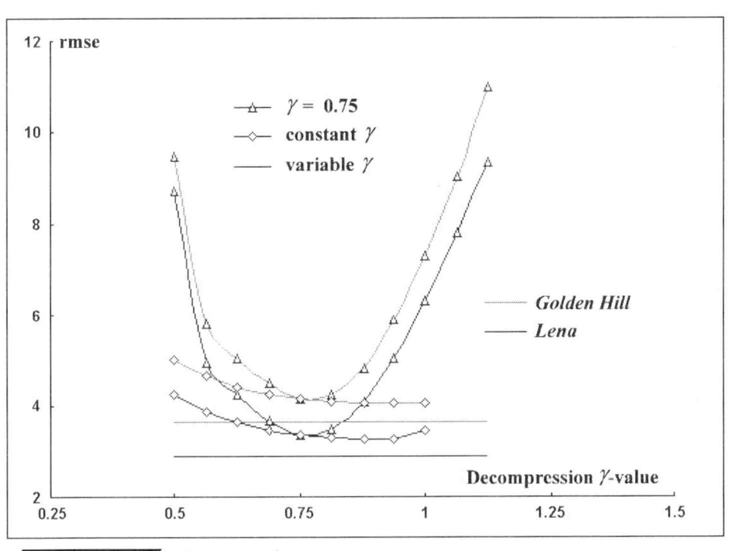

FIGURE 6.2.5 *The choice of γ*

In summary, all the above statistics are plotted in Figure 6.2.5:

- The *straight horizontal lines*, the *best* fractal representations. They are compressed using variable γ-'s and decompressed in the same way.
- The *cup curves minimizing at* $\gamma = 3/4$, the fractal representations compressed using the fixed $\gamma = 3/4$ and decompressed in different fixed constant γ's.
- The *asterisk sequences*, the fractal representations compressed using the fixed constant γ's and decompressed in the same fixed γ's respectively.

TABLE 6.2.2 *Difference between decoded error and collage error*

rmse	Lena			Golden Hill		
γ	*Collage*	*Decoded*	*Differ*	*Collage*	*Decoded*	*Differ*
1/2	3.7807	4.2684	0.4877	4.4683	5.0345	0.5662
9/16	3.5329	3.8880	0.3551	4.2471	4.6626	0.4155
5/8	3.3587	3.6348	0.2761	4.0892	4.4093	0.3201
11/16	3.2369	3.4624	0.2255	3.9739	4.2526	0.2787
3/4	3.1542	3.3611	0.2069	3.8897	4.1519	0.2622
13/16	3.0988	3.3000	0.2012	3.8328	4.0982	0.2654
7/8	3.0649	3.2698	0.2049	3.7954	4.0715	0.2761
15/16	3.0554	**3.2685**	0.2131	3.7764	4.0704	0.2940
1	3.0580	3.2790	0.2210	3.7627	**4.0665**	0.3038
All	2.6335	2.9082	0.2747	3.3103	3.6471	0.3368

For all four pairs of parallel curves, the upper one in each pair is always for *Golden Hill* and the lower one is for *Lena*.

If only one γ-value will be used, which constant γ-value should be chosen? There is an additional factor that needs to be considered: the difference between the *collage error* (which is the comparison error in compression) and the *decoded error* (which is the actual error after decompression). This has been illustrated in Table 6.2.2. The smaller the difference, the better the fractal representation converges. Actually, this was the reason why $\gamma=3/4$ was chosen as the default choice in our various implementations.

6.2.3 CONTRAST SCALING FACTORS IN A MIXED SEARCHING SCHEME

In the previous chapter, the local reference region addresses are coded differently from the global ones. Therefore, there is no penalty cost in having different constant contrast γ-values for the locally addressed and globally addressed fractal codes. The goal is to find those constants. Again, the statistics method is used by testing various images.

Table 6.2.3 shows the comparison results for *Lena* and *Golden Hill* using different combinations of local and global γ-values. The exact procedure is as follows: first compress the image with both local and global γ-values set at 3/4 and error distortion rate 16, obtain the ratio between the number of local fractal codes and the number of global fractal codes (e.g., about 3:2 for *Lena* and 3:7 for *Golden Hill*), and then compress the image in each γ-values combination in the same ratio by adjusting decimally the error distortion rate.

It is expected that the result will be image dependent. The conclusion is better than expected: Most images have yielded the same result as shown in Table 6.2.3: *It is best to use* $\gamma=1/2$ *for the local fractal codes and* $\gamma=7/8$ *for the global fractal codes.* Moreover, no significant improvement has been found by using any other choice of the γ-value pairs with any images.

TABLE 6.2.3 Combination of the local γ-value and the global γ-value

rmse	Lena				Golden Hill			
L \ G	3/4	7/8	15/16	1	3/4	7/8	15/16	1
7/16	3.7505	3.6750	3.6785	3.6975	4.3631	4.2963	4.2989	4.3059
1/2	3.7442	**3.6615**	3.6696	3.6828	4.3580	**4.2928**	4.2948	4.2991
9/16	3.7392	3.6629	3.6660	3.6828	4.3670	4.3030	4.3073	4.3097
5/8	3.7520	3.6729	3.6771	3.6926	4.3802	4.3189	4.3222	4.3258
3/4	3.7965	3.7192	3.7229	3.7383	4.3801	4.3162	4.3219	4.3302
7/8	3.8268	3.7982	3.7975	3.8185	4.4198	4.3791	4.3873	4.4000

Intuitively, for a local fractal code it is very likely that the reference portion of the image has the same slope as the destination region. Shrink the reference region down—this explains why $\gamma = 1/2$ should be the best choice for local fractal codes.

6.3 GENERAL INTENSITY TRANSFORMATIONS

In the previous chapter, the VQ algorithm was presented as an extreme case of fractal compression by allowing a special spatial component. Here, JPEG DCT will be studied as an extreme case of a special intensity component of a general fractal image model.

So far, in the implemented fractal model, all intensity transforms in all destination regions take the simplest form:

$$z_D \leftarrow \gamma \cdot z_R + \beta, \tag{6.3.1}$$

where z_D is the intensity value at a sample point of the destination region, and z_R is the intensity value at the corresponding point in the reference region (shown in Figure 6.3.1). The parameters of this transform have been interpreted as the brightness and contrast adjustments.

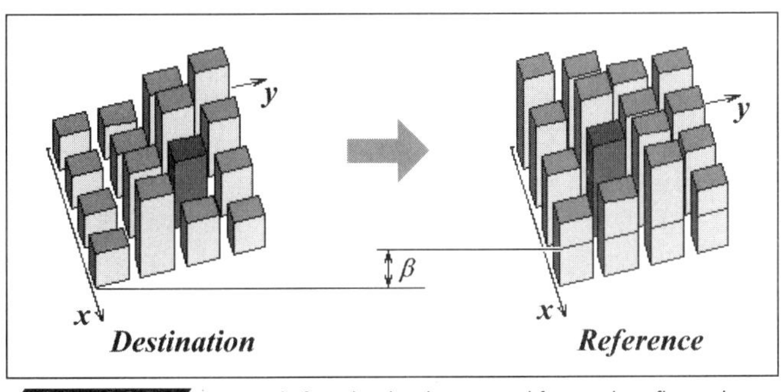

FIGURE 6.3.1 A sample fractal code whose spatial form is the y-flip, and whose intensity map is $z_D \leftarrow \gamma \cdot z_R + \beta$

In this section, formula 6.3.1 of the intensity component of a fractal code will be extended by adding more terms to this expression.

6.3.1 INTENSITY IN SPATIAL POLYNOMIALS

First, additional terms were implemented (called *tilts*) in an Iterated Systems research project in early 1991. They are the two linear terms of intensity transform in spatial variables:

$$z_D \leftarrow \gamma \cdot z_R + \beta + \xi \cdot x + \eta \cdot y. \tag{6.3.2}$$

The pair (ξ, η) is called a pair of *tilt parameters*, or a *tilt form*. Near a light source there is a constant luminance decay that depends upon the location of the light source. The *tilt form* models this situation perfectly. Generally, the tilt forms characterize precisely the background light changing, and they are not for enhancing an object's detail. By tilting a block, the block edge artifact is greatly reduced, and L^2-distortion reduces accordingly. Nevertheless, no other visual improvement is achieved.

Second-order and third-order terms have been studied by Dudbridge and Monro [D3] as *Bath transforms*. In that case the intensity component has the formula

$$z_D \leftarrow \gamma \cdot z_R + P(x, y), \tag{6.3.3}$$

where $P(x, y)$ is a polynomial in spatial variables x and y:

$$P(x, y) = \beta + \xi \cdot x + \eta \cdot y + ax^2 + by^2 + cxy + ex^3 + fy^3 + \cdots. \tag{6.3.4}$$

The physical meaning of high-order terms have not been justified.

6.3.2 ORTHOGONAL BLOCK VECTORS

In the discrete case, each term in the polynomial is equivalent to a constant monomial block. For example, for an 8×8 block, the constant term, the x term, the xy term, and the y^2 term are equivalent to the formulas given below:

$$1 = \begin{pmatrix} 1&1&1&1&1&1&1&1 \\ 1&1&1&1&1&1&1&1 \\ 1&1&1&1&1&1&1&1 \\ 1&1&1&1&1&1&1&1 \\ 1&1&1&1&1&1&1&1 \\ 1&1&1&1&1&1&1&1 \\ 1&1&1&1&1&1&1&1 \\ 1&1&1&1&1&1&1&1 \end{pmatrix}, \quad xy = \begin{pmatrix} 49&35&21&7&-7&-21&-35&-49 \\ 35&25&15&5&-5&-15&-35&-35 \\ 21&15&9&3&-3&-9&-15&-21 \\ 7&5&3&1&-1&-3&-5&-7 \\ -7&-5&-3&-1&1&3&5&7 \\ -21&-15&-9&-3&3&9&15&21 \\ -35&-25&-15&-5&5&15&25&35 \\ -49&-35&-21&-7&7&21&35&49 \end{pmatrix}, \tag{6.3.5}$$

$$x = \begin{pmatrix} -49 & -35 & -21 & -7 & 7 & 21 & 35 & 49 \\ -49 & -35 & -21 & -7 & 7 & 21 & 35 & 49 \\ -49 & -35 & -21 & -7 & 7 & 21 & 35 & 49 \\ -49 & -35 & -21 & -7 & 7 & 21 & 35 & 49 \\ -49 & -35 & -21 & -7 & 7 & 21 & 35 & 49 \\ -49 & -35 & -21 & -7 & 7 & 21 & 35 & 49 \\ -49 & -35 & -21 & -7 & 7 & 21 & 35 & 49 \\ -49 & -35 & -21 & -7 & 7 & 21 & 35 & 49 \end{pmatrix}, \; y^2 = \begin{pmatrix} -49 & -49 & -49 & -49 & -49 & -49 & -49 & -49 \\ -25 & -25 & -25 & -25 & -25 & -25 & -25 & -25 \\ -9 & -9 & -9 & -9 & -9 & -9 & -9 & -9 \\ -1 & -1 & -1 & -1 & -1 & -1 & -1 & -1 \\ 1 & 1 & 1 & 1 & 1 & 1 & 1 & 1 \\ 9 & 9 & 9 & 9 & 9 & 9 & 9 & 9 \\ 25 & 25 & 25 & 25 & 25 & 25 & 25 & 25 \\ 49 & 49 & 49 & 49 & 49 & 49 & 49 & 49 \end{pmatrix}. \quad (6.3.6)$$

Geometrically, these monomial image blocks are shown in Figure 6.3.2, where the pixel values are remapped as –100 for black, 0 for gray, and 100 for white.

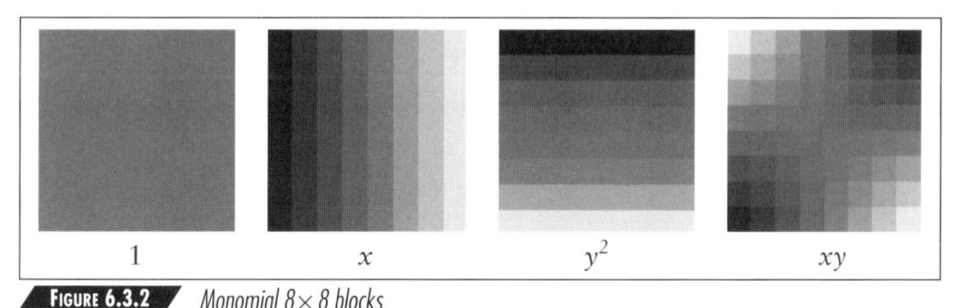

FIGURE 6.3.2 *Monomial 8×8 blocks*

Therefore, in general, the intensity transform can be written in the form

$$z_D \leftarrow \gamma \cdot z_R + \sum_i \alpha_i B_i \,, \qquad (6.3.7)$$

where the B_i, called *base blocks*, are selected constant blocks. This expression was first studied by Øien and Lepsøy [F]. For efficiency of representation, the set of blocks

$$\{z_R, B_0, B_1, \cdots, B_{k-1}\}\,, \qquad (6.3.8)$$

as vectors, is orthogonal, or at least linearly independent.

What is the best way to choose these base blocks? It is simple to choose a fixed set of base blocks, and in this case the block z_R should be ignored, for it changes in different destination blocks. So the next step is to select an orthonormal basis $\{B_0, B_1, \cdots, B_{k-1}\}$ for the 8×8-dimensional vector space. Consequently, $k = 64$ for a complete basis.

Certainly, the trivial basis is $\mathbf{E} = \{E_{ij}\}_{0 \le i, j < 8}$, where $E_{ij}(s, t) = \delta_{(s-i)}\delta_{(t-j)}$, for all $0 \le i, j, s, t < 8$, in which case it is equivalent to store the pixel values. Apparently, this is not what is desired. What is needed is a basis that packs energy well. In other words, most image blocks can be well approximated by a combination of very few image blocks. For instance, applying this requirement to the flat block, the result is that the equation

$$\sum_{s,t} B_i(s,t) = 0 \tag{6.3.9}$$

should hold for most base blocks. Any block that satisfies equation (6.3.9) will be called an *AC* (*alternating-current*) block, and if it does not satisfy the equation, it will be called a *DC* (*direct-current*) block.

6.3.3 THE HAAR–WALSH–HADAMARD TRANSFORM

It is not difficult to see mathematically that the most natural basis, for simplicity and symmetry, is the Haar–Walsh–Hadamard basis. For an 8×8 block, it can be well described by the matrix

$$H_8 = \frac{1}{\sqrt{8}} \begin{pmatrix} 1 & 1 & 1 & 1 & 1 & 1 & 1 & 1 \\ 1 & 1 & 1 & 1 & -1 & -1 & -1 & -1 \\ 1 & 1 & -1 & -1 & -1 & -1 & 1 & 1 \\ 1 & 1 & -1 & -1 & 1 & 1 & -1 & -1 \\ 1 & -1 & -1 & 1 & 1 & -1 & -1 & 1 \\ 1 & -1 & -1 & 1 & -1 & 1 & 1 & -1 \\ 1 & -1 & 1 & -1 & -1 & 1 & -1 & 1 \\ 1 & -1 & 1 & -1 & 1 & -1 & 1 & -1 \end{pmatrix}. \tag{6.3.10}$$

This is a real orthonomal matrix, that is, $H_8 = H_8^T = H_8^{-1}$. *The Haar–Walsh–Hadamard basis* is defined to be

$$\mathbf{H} = H_8 \cdot \mathbf{E} \cdot H_8 = \{H_{ij} = H_8 E_{ij} H_8\}_{0 \le i,j < 8}, \tag{6.3.11}$$

as pictured in Figure 6.3.3, where 1/8 is set to white and –1/8 to black. In this basis all blocks are AC blocks, except for the block in the upper left corner, H_{00}, the only DC block.

The nice thing about this basis is that the decomposition can be calculated easily. In fact, for any given block

$$B = (b_{ij}) = \sum_{i,j} a_{ij} H_{ij}, \tag{6.3.12}$$

it is straightforward to prove that

$$(a_{ij}) = H_8 B H_8, \text{ and } B = H_8 (a_{ij}) H_8. \tag{6.3.13}$$

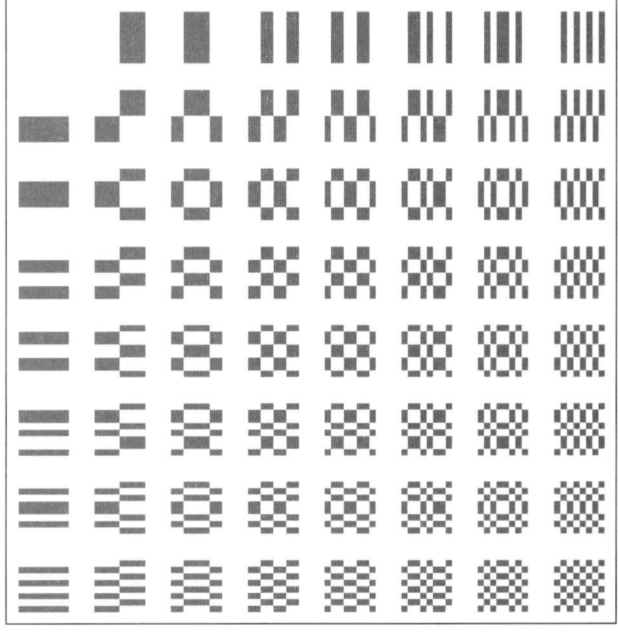

Figure 6.3.3 Haar–Walsh–Hadamard basis

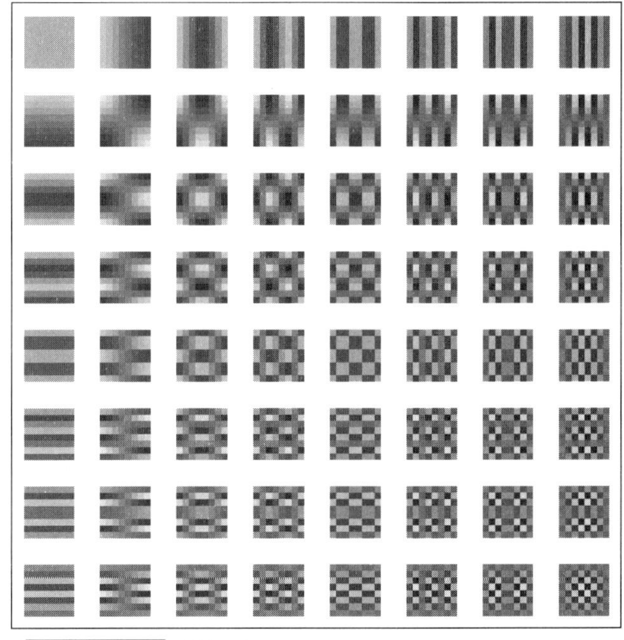

Figure 6.3.4 DCT basis

This triple matrix multiplication is the both a forward and a backward Haar–Walsh–Hadamard transform, which really makes the calculation of the decomposition coefficients a_{ij} handy.

6.3.4 THE DCT TRANSFORM

Applying the Haar–Walsh–Hadamard basis to real images, close to block edges visible artifacts are apparent—in particular for highly compressed pictures.

All Haar–Walsh–Hadamard base blocks have high-contrast geometric shapes. A coded block is a combination of a few of them, which will have some clear geometric shapes near the block edges that may not match with the edge shapes generated by its adjacent blocks.

A better alternative is found, the *discrete cosine transform* (DCT) basis $\mathbf{D} = \{D_{ij}\}_{0 \le i,j < 8}$, for its moderate geometric shape near the block edges, as shown in Figure 6.3.4. Similarly, in this basis all blocks are AC blocks but for the block D_{00}, which is the block mean.

The decomposition coefficients a_{ij} still can be calculated using simple transformations. Again, for $B = (b_{ij}) = \sum_{i,j} a_{ij} D_{ij}$, the forward transform (FDCT) is defined as

$$a_{ij} = c_i c_j \sum_{s,t=0}^{7} b_{st} \cos\left[\frac{(2s+1)i\pi}{16}\right] \cos\left[\frac{(2t+1)j\pi}{16}\right], \tag{6.3.14}$$

and the inverse transform (IDCT) is defined as

$$b_{ij} = \sum_{s,t=0}^{7} c_s c_t a_{st} \cos\left[\frac{(2i+1)s\pi}{16}\right] \cos\left[\frac{(2j+1)t\pi}{16}\right], \tag{6.3.15}$$

where

$$c_0 = \frac{\sqrt{2}}{4} \text{ and } c_i = \frac{1}{2} \text{ for } i = 1, 2, \cdots, 7. \tag{6.3.16}$$

The DCT has become by far the most widely used transformation for image compression.

6.3.5 THE JPEG BASELINE STANDARD

When the contrast adjustment value γ is set to zero, the spatial transform is irrelevant; the result of using the DCT basis became the JPEG (Joint Photographic Experts Group) image compression standard in 1990.

For different requirements of various applications, the JPEG DCT-based still image coding standard operates in four different modes: *sequential*, *progressive*, *lossless*, and *hierarchical*. The baseline mode is sequential, which is formed by 8×8 blocks in an array in scan line order.

An image is always partitioned into 8×8 blocks. Given an 8×8 image block $B = (b_{ij})_{0 \le i, j < 8}$, the exact algorithm for generating data codes is as follows. The sample block given in Rabbani and Jones's book [RJ],

$$B = \begin{pmatrix} 139 & 144 & 149 & 153 & 155 & 155 & 155 & 155 \\ 144 & 151 & 153 & 156 & 159 & 156 & 156 & 156 \\ 150 & 155 & 160 & 163 & 158 & 156 & 156 & 156 \\ 159 & 161 & 162 & 160 & 160 & 159 & 159 & 159 \\ 159 & 160 & 161 & 162 & 162 & 155 & 155 & 155 \\ 161 & 161 & 161 & 161 & 160 & 157 & 157 & 157 \\ 162 & 162 & 161 & 163 & 162 & 157 & 157 & 157 \\ 162 & 162 & 161 & 161 & 163 & 158 & 158 & 158 \end{pmatrix}, \tag{6.3.17}$$

will be used for illustration here as well.

1. *FDCT*. Get DCT coefficients using FDCT (6.3.14). Thus, on the sample:

$$A = \left(a_{ij}\right)_{0 \le i, j < 8} = \begin{pmatrix} 1260 & -1 & -12 & -5 & 2 & -2 & -3 & 1 \\ -23 & -17 & -6 & -3 & -3 & 0 & 0 & -1 \\ -11 & -9 & -2 & 2 & 0 & -1 & -1 & 0 \\ -7 & -2 & 0 & 1 & 1 & 0 & 0 & 0 \\ -1 & -1 & 1 & 2 & 0 & -1 & 1 & 1 \\ 2 & 0 & 2 & 0 & -1 & 1 & 1 & -1 \\ -1 & 0 & 0 & -1 & 0 & 2 & 1 & -1 \\ -3 & 2 & -4 & -2 & 2 & 1 & -1 & 0 \end{pmatrix}. \tag{6.3.18}$$

2. *Quantization*. All coefficients are normalized (quantized) by a user-defined normalization array $Q = (q_{ij})_{0 \le i, j < 8}$. That is,

$$\tilde{A} = \left(\tilde{a}_{ij} = \left[\frac{a_{ij}}{q_{ij}} \right] \right)_{0 \le i, j < 8}. \tag{6.3.19}$$

The typical normalization arrays used by JPEG for luminance screen and chrominance screen are

$$Q_Y = \begin{pmatrix} 16 & 11 & 10 & 16 & 24 & 40 & 51 & 61 \\ 12 & 12 & 14 & 19 & 26 & 58 & 60 & 55 \\ 14 & 13 & 16 & 24 & 40 & 57 & 69 & 56 \\ 14 & 17 & 22 & 29 & 51 & 87 & 80 & 62 \\ 18 & 22 & 37 & 56 & 68 & 109 & 103 & 77 \\ 24 & 35 & 55 & 64 & 81 & 104 & 113 & 92 \\ 49 & 64 & 78 & 87 & 103 & 121 & 120 & 101 \\ 72 & 92 & 95 & 98 & 112 & 100 & 103 & 99 \end{pmatrix} \tag{6.3.20}$$

and

$$Q_C = \begin{pmatrix} 17 & 18 & 24 & 47 & 99 & 99 & 99 & 99 \\ 18 & 21 & 26 & 66 & 99 & 99 & 99 & 99 \\ 24 & 26 & 56 & 99 & 99 & 99 & 99 & 99 \\ 47 & 66 & 99 & 99 & 99 & 99 & 99 & 99 \\ 99 & 99 & 99 & 99 & 99 & 99 & 99 & 99 \\ 99 & 99 & 99 & 99 & 99 & 99 & 99 & 99 \\ 99 & 99 & 99 & 99 & 99 & 99 & 99 & 99 \\ 99 & 99 & 99 & 99 & 99 & 99 & 99 & 99 \end{pmatrix} \tag{6.3.21}$$

respectively. Thus applying Q_Y to the sample:

$$\tilde{A} = \begin{pmatrix} 79 & 0 & -1 & 0 & 0 & 0 & 0 & 0 \\ -2 & -1 & 0 & 0 & 0 & 0 & 0 & 0 \\ -1 & -1 & 0 & 0 & 0 & 0 & 0 & 0 \\ 0 & 0 & 0 & 0 & 0 & 0 & 0 & 0 \\ 0 & 0 & 0 & 0 & 0 & 0 & 0 & 0 \\ 0 & 0 & 0 & 0 & 0 & 0 & 0 & 0 \\ 0 & 0 & 0 & 0 & 0 & 0 & 0 & 0 \\ 0 & 0 & 0 & 0 & 0 & 0 & 0 & 0 \end{pmatrix}. \tag{6.3.22}$$

3. *DC Coding.* The only top-left DC coefficient a_{00}, as the quantized mean value of the current block, is differentiated from the previous block mean, then coded using a specific Huffman table.

   The numbers 0, ±1, ±2, ... are divided into categories. As shown in Table 6.3.1, the kth category contains 2^k numbers: $\pm 2^{k-1}$, $\pm(2^{k-1}+1)$, ..., $\pm(2^{k-1}-1)$, which have k bit Huffman subcodes: `s000...0`, `s100...0`, `s010...0`, `s1100...0`, ..., `s111...1`, where s is a sign bit, 0 for negative and 1 for positive.

   The category indices are coded by the specified Huffman table.

For the baseline system, up to two separated Huffman tables can be specified in the file header information, as one table for the luminance image and the another table for chrominance image in general. A typical JPEG system uses Huffman tables shown in Table 6.3.2. This table and Table 6.3.3 are taken from Pennebaker and Mitchell's JPEG book [PM].

In the sample, if the previous block has DC value 74, the value that needs to be coded is $5 = 79 - 74$, in category 3, which has code `100`. The number 5 has the subcode `100`. Hence, the final DC code will have 6 bits: `100100`.

TABLE 6.3.1 *Coefficients categories*

category	coefficients range	subcode word	code length
0	0		0
1	−1, 1	1,0	1
2	−3, −2, 2, 3	11, 01, 00, 10	2
3	±4, ±5, ±6, ±7	s00, s10, s01, s11	3
4	±8, ±9, ... , ±15	s000, s100, ..., s111	4
5	±16, ±17, ... , ±31	s000, s100, ..., s111	5
6	±32, ±33, ... , ±63	s0000, s1000, ..., s1111	6
7	±64, ±65, ... , ±127	s00000, s10000, ..., s11111	7
8	±128, ... , ±255	s000000, ..., s1111111	8
9	±256, ... , ±511	s0000000, ..., s111111111	9
10	±512, ... , ±1023	s00000000, ..., s1111111111	10

TABLE 6.3.2 *Huffman table for DC difference*

	luminance		chrominance	
category	*code word*	*code length*	*code word*	*code length*
0	00	2	00	2
1	010	3	01	2
2	011	3	10	2
3	100	3	110	3
4	101	3	1110	4
5	110	3	11110	5
6	1110	4	111110	6
7	11110	5	1111110	7
8	111110	6	11111110	8
9	1111110	7	111111110	9
10	11111110	8	1111111110	10

TABLE 6.3.3 *Partial Huffman table luminance AC coefficients*

	luminance		chrominance	
run-zero / category	*code word*	*code length*	*code word*	*code length*
EOB	1010	4	00	2
16 ZRL	11111111001	11	1111111010	10
0/1	00	2	01	2
0/2	01	2	100	3
0/3	100	3	1010	4
0/4	1011	4	11000	5
0/5	11010	5	11001	5
0/6	1111000	7	111000	6
0/7	11111000	8	1111000	7
0/8	1111110110	10	111110100	9
0/9	...	16	1111110110	10
0/10	...	16	111111110100	12
1/1	1100	4	1011	4
1/2	11011	5	111001	6
1/3	1111001	7	11110110	8
1/4	111110110	9	11110110	9
1/5	11111110110	11	11111110110	11
1/6	...	16	111111110101	12
...	...	...	...	...
2/1	11100	5	11010	5
2/2	11111001	8	11110111	8
2/3	1111110111	10	1111110111	10
2/4	111111110100	12	111111110110	12
2/5	...	16	111111111000010	15
...	...	...	...	...
3/1	111010	6	11011	5
3/2	111110111	9	11111000	8
3/3	111111110101	12	1111111000	10
3/4	...	16	111111110111	12
...	...	...	...	...
4/1	111011	6	111010	6
4/2	1111111000	10	111110110	9
...	...	...	...	...
5/1	1111010	7	111011	6

** All skipped code words have a length of 16 bits.*

TABLE 6.3.3 (continued) Partial Huffman table luminance AC coefficients

	luminance		chrominance	
run-zero / category	code word	code length	code word	code length
5/2	11111110111	11	1111111001	10
...	...	...	...	...
6/1	1111011	7	1111001	7
6/2	111111110110	12	11111110111	11
...	...	...	...	...
7/1	11111010	8	1111010	7
7/2	111111110111	12	11111111000	11
...	...	...	...	...
8/1	111111000	9	11111001	8
8/2	111111111100000	15	...	16
...	...	...	...	...
9/1	111111001	9	111110111	9
10/1	111111010	9	111111000	9
11/1	1111111001	10	111111001	9
12/1	1111111010	10	111111010	9
13/1	11111111000	11	11111111001	11
14/1	...	16	11111111100000	14
15/1	...	16	111111111000011	15
...	...	...	...	...

* All skipped code words have a length of 16 bits.

4. *AC Coding*. The other DCT AC coefficients are coded in the zigzag 1-dimensional order shown in Figure 6.3.5. Zeros are patched to the next nonzero entry with a run-length number, and the nonzero entry is packed as a combination of its category and its remainder, similar to DC coefficients. When all nonzero coefficients are stored, an EOB (end of block) symbol follows.

   Applying this to the sample, the 1-dimensional AC string is

$$0, -2, -1, -1, -1, 0, 0, -1, 0, 0, 0, 0, 0, 0, 0, 0, \ldots 0. \tag{6.3.23}$$

   Grouping zeros to its next nonzero entry, the coding string becomes

$$1/-2,\ 0/-1,\ 0/-1,\ 0/-1,\ 2/-1,\ \text{EOB}. \tag{6.3.24}$$

   A typical AC Huffman table is shown in Table 6.3.3. It gives the following binary code string of 26 bits:

$$1101100\ 000\ 000\ 000\ 111000\ 1010. \tag{6.3.24}$$

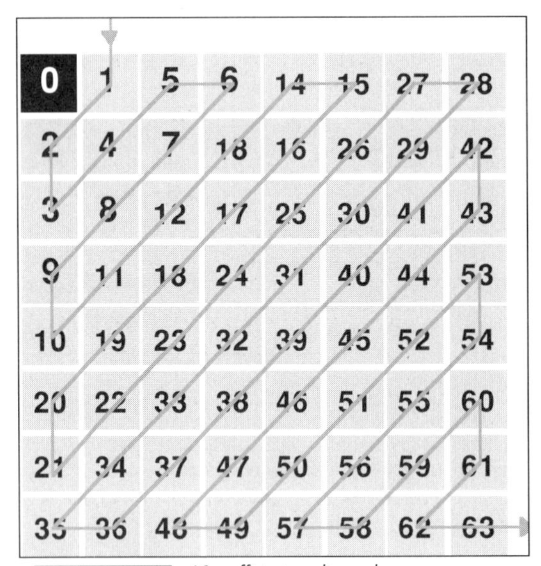

FIGURE 6.3.5 *AC coefficients coding order*

Compression ratio: **32:1**
Bit rate: **0.25 bpp**
Image file size: **8192 bytes**

FIGURE 6.3.6 Lena *compressed in JPEG DCT*

So after all, the sample 8×8 block is coded in 32 bits, a 16:1 compression. Applying IDCT, the recovered and decoded block looks like the following:

$$B' = \begin{pmatrix} 144 & 146 & 149 & 152 & 154 & 156 & 156 & 156 \\ 148 & 150 & 152 & 154 & 156 & 156 & 156 & 156 \\ 155 & 156 & 157 & 158 & 158 & 157 & 156 & 155 \\ 160 & 161 & 161 & 162 & 161 & 159 & 157 & 155 \\ 163 & 163 & 164 & 163 & 162 & 160 & 158 & 156 \\ 163 & 163 & 164 & 164 & 162 & 160 & 158 & 157 \\ 160 & 161 & 162 & 162 & 162 & 161 & 159 & 158 \\ 158 & 159 & 161 & 161 & 162 & 161 & 159 & 158 \end{pmatrix} \qquad (6.3.25)$$

for which the RMSE from the original block is 2.26.

Applying the technology to real world images, the blockiness and mosquito-like aliasing are the typical artifacts in high compression ratios.

Figure 6.3.6 shows a 0.25 bpp (32:1) JPEG DCT compressed *Lena*. The L^2-distortion is 7.34 (or PSNR 30.81). Some of the latest research has shown promise for improvement by adding postprocessing.

In the empty immensity of earth, sky, and water, there she was, incomprehensible, firing into a continent.

—Heart of Darkness

Joseph Conrad

Clustering Searching 7

The first fractal image compression algorithm was a slow, exhaustive search algorithm. It takes about 8 hours to compress a 512×512 grayscale image using the algorithm provided in Chapter 3 on a Pentium 100MHz PC. Based on the same algorithm, Iterated Systems's first hardware board speeded up this process and did the compression in 5 minutes. In this chapter, a practical software solution will be provided. It reduces the compression time to seconds with an unnoticeable reduction in quality. The technique that will be introduced here, first used in fractal imaging by Iterated Systems research scientists, is called the *Heckbert clustering algorithm*.

Among all of the digital image compression techniques, fractal image compression stands out with its own inherent advantages—*resolution independence* and *fast decompression*—due to the following two fundamental facts: the fractal transformations represent images from their own structurally similar pieces, and the asymmetric decompression algorithm is an extremely simple copying procedure.

However, like many other searching and sorting algorithms, the fractal compression scheme is much slower than most computational compression techniques. The unbearably slow rate of early fractal compression algorithms has been the main reason that many applications use alternative methods, even though the compression time seems less critical than the decompression time because an image compressed once can be used forever. In an industry that produces thousands of images, compression speed and monetary cost are two sides of the same equation. A reasonably fast compression scheme is not just a bonus but a necessity in the commercial market.

This book presents an encouraging solution to this problem: a revolutionary algorithm to reduce the compression time from 8 hours to 25 seconds. With the compression speed improved by a thousand times (faster than Iterated Systems's first-generation hardware board), the quality reduction is insignificant: there is no visual difference, and less than about a 5% peak signal-to-noise error increase.

The method of speeding up a searching or sorting algorithm is as follows: to classify reference objects into groups and to compare each target object with only one of the groups that has the highest possibility of finding a good match. Thus, to have an optimum speed, the number of objects in a group need to be small, and consequently, the number of groups becomes inevitably large. Some early approaches that classify the image regions into the categories of smooth, edge, texture, etc. were not successful because the number of groups was too small, and a high percentage of regions did not fall into any of the categories.

The software fractal image compression presented here is not based on a predetermined classification but adopts an image-dependent classification using Heckbert clustering. The first implementation was initiated by John Elton. Various implementations were studied at Iterated Systems. The art, in this case, is how to balance among the classification, the designation, and the searching to achieve both speed and performance.

7.1 VECTORS AND CLUSTERS

Let $W=\{ (D, R, \sigma, \tau, \gamma, \beta/m) \}$ be a PIFS fractal model of some given image P described in Chapter 3, Section 3.3, where D is a destination region, R a reference region, σ an s-form, τ a t-form, γ a γ-value, and β a β-value that can be replaced by a mean value m as was discussed in Chapter 6. To simplify the discussion, the tilting form is assumed to be the trivial "nontilted" one, $\tau=0$.

7.1.1 IMAGE REGION VECTORS

In the digital case, a destination region is really a set of pixels. Putting them into a specific order, one can assume that $D=\{d_i\}_{1\le i\le N}$. Their corresponding points in the reference region under the spatial component of the transformation of the fractal element, which may fall outside of the image sample pixel grid, will be denoted by $R=\{r_i\}_{1\le i\le N}$. Therefore, the *collage error* in L^2-metric of a fractal element can be computed as

$$\textit{collage error} = \sqrt{\frac{\sum_{i=1}^{N}\left(P(d_i)-\gamma\cdot P(r_i)-\beta\right)^2}{N}}, \tag{7.1.1}$$

where, in the digital case, $P(d_i)$ is the pixel value at the point $d_i \in D$, $P(r_i)$ is the pixel value of the reference region image piece after transform and resampling to the same destination region sample grid, and N is the total number of sample pixels in the destination region.

In the uniformly distributed image partition case, the number N is a constant over the image. Thus, a destination region image piece $P(D)=\{P(d_i)\}$ can be viewed as a vector in the N-dimensional vector space $\Re^N$. Similarly, the corresponding reference region image piece $P(R)=\{P(r_i)\}$ can be viewed as a vector in $\Re^N$ as well.

The mean values of the destination region D and the reference region R are denoted by

$$m=\frac{1}{N}\sum_{i=1}^{N}P(d_i) \text{ and } m_R=\frac{1}{N}\sum_{i=1}^{N}P(r_i). \tag{7.1.2}$$

The collage error becomes the exact distance between the two vectors $P(D)-m$ and $\gamma(P(R)-m_R)$ in the N-dimensional vector space $\Re^N$. The vectors $P(D)-m$

and $\gamma(P(R) - m_R)$ are called the *zero-sum vectors* of the destination region D and the reference region R, respectively.

Considering the set of all normalized reference region zero-sum vectors, the searching problem can be readdressed in the following way:

Problem 1: Given a finite set Ω of zero-sum vectors in $\Re^N$, for any zero-sum vector $\mathbf{d}$ in $\Re^N$ find a fast algorithm to sort a vector $\mathbf{r}$ from Ω such that $d(\mathbf{d}, \mathbf{r}) = d(\mathbf{d}, \Omega)$.

When speed is a higher priority, Problem 1 can be expanded:

Problem 2: Given a finite set Ω of vectors in $\Re^N$, for any vector $\mathbf{d}$ in $\Re^N$ find a fast algorithm to sort a vector $\mathbf{r}$ from Ω such that $d(\mathbf{d}, \mathbf{r}) = d(\mathbf{d}, \Omega)$ is "very small." That is, be prepared to choose not necessarily the "best" but a "good enough" matching.

7.1.2 CLUSTER STRUCTURE

Definition 1: Given a finite set of vectors Ω in the set $\Re^N$, a *cluster structure*, or a *clustering*, of the set Ω is a finite collection $\{\Omega_k\}_{0 \le k < M}$ of M nonempty subsets of Ω and a map

$$c: \Re^N \longrightarrow \{0, 1, \cdots, M-1\} \tag{7.1.3}$$

such that $\bigcup_{0 \le k < M} \Omega_k = \Omega$ and $c^{-1}(k) \ne \varnothing$, for all $0 \le k < M$. The map c is called the *classification map*. Each subset Ω_k is called a *cluster*, for $k = 0, 1, \cdots, M-1$.

Definition 2: The *diameter of a cluster* Ω_k, $k = 0, 1, \cdots, M-1$, is defined by

$$\delta(\Omega_k) = \sup_{\mathbf{d}, \mathbf{r} \in \Omega_k} d(\mathbf{d}, \mathbf{r}). \tag{7.1.4}$$

Definition 3: The *speed of the clustering* is defined by

$$\sigma(\Omega) = \frac{\max\limits_{0 \le k < M} \{|\Omega_k|\} - 1}{|\Omega|}. \tag{7.1.5}$$

The clustering is said to be *instantaneous* if $\sigma(\Omega) = 0$.

Definition 4: The *error of the clustering* is defined by

$$\varepsilon(\Omega) = \sup_{\mathbf{d} \in \Re^N} \left(d(\mathbf{d}, \Omega_{c(D)}) - d(\mathbf{d}, \Omega) \right). \quad (7.1.6)$$

The clustering is said to be *perfect* if $\varepsilon(\Omega) = 0$.

Clearly, the trivial clustering formed by the single set Ω itself is a perfect clustering with speed 1.

The clustering speed and clustering error are two important concepts introduced for measuring how good a given clustering structure is. The goal is to minimize both of these values.

7.2 CLUSTERING ALGORITHMS

In this section, a complete solution for Problem 1 in Section 7.1.1 is given assuming no memory or time constraints. Then, a practical solution for Problem 2 is presented that will be the key to speeding up compression time.

7.2.1 THE BUBBLING ALGORITHM

In the digital case, when there is enough memory to work on the whole set $\Re^N$, i.e., the number $|\Re^N|$ of all vectors in $\Re^N$ is manageable, there is a perfect algorithm for generating a perfect instantaneous clustering, i.e., $\varepsilon(\Omega) = \sigma(\Omega) = 0$. It is called the *bubbling algorithm*.

The null speed $\sigma(\Omega) = 0$ implies that there are exactly $M=|\Omega|$ clusters with one vector in each cluster. So the main part of the algorithm is to define the map c.

It is assumed that the number of all possible vectors in $\Re^N$ is not too large for a computer to handle. A buffer of the size $|\Re^N|$ is allocated to mark the values of the map c for all vectors in $\Re^N$.

The algorithm begins by initializing the whole space with value M. Next, the value k is assigned to the only vector from the cluster Ω_k, for all $k = 0, 1, \cdots, M-1$. Then, *bubble* (expand) those values from the vectors with a mark less than M to the vectors with the mark M in one step, one direction, and one component at a time, until the whole space has been filled and there is not another vector in $\Re^N$ with the mark M.

BUBBLING ALGORITHM:

Given a finite set of vectors Ω in $\Re^N$, assume that the classification map c is described in a buffer indexed by vectors from $\Re^N$. Let the *ith unit vector* be denoted by

$$\mathbf{u}_i = (\overbrace{0,0,\cdots,0}^{i-1 \text{ zeros}},1,0,\cdots,0). \tag{7.2.1}$$

Step 0: For all $\mathbf{r} \in \Re^N$ set $c(\mathbf{r}) = M$.

Step 1: For any $\mathbf{o}_k \in \Omega_k$, $k = 0, 1, \cdots, M-1$, set $c(\mathbf{o}_k) = k$.

Step 2: Set the loop flag λ to 0. For all $\mathbf{r} \in \Re^N$, if $c(\mathbf{r}) = M$, set $\lambda = 1$; otherwise, for all $i = 0, 1, \cdots, n-1$, let $c(\mathbf{r} + \mathbf{u}_i) = c(\mathbf{r})$ if $c(\mathbf{r} + \mathbf{u}_i) = M$, and let $c(\mathbf{r} - \mathbf{u}_i) = c(\mathbf{r})$ if $c(\mathbf{r} - \mathbf{u}_i) = M$.

Step 3: If the loop flag λ is 1, go back to *Step 2*. Otherwise the algorithm is completed.

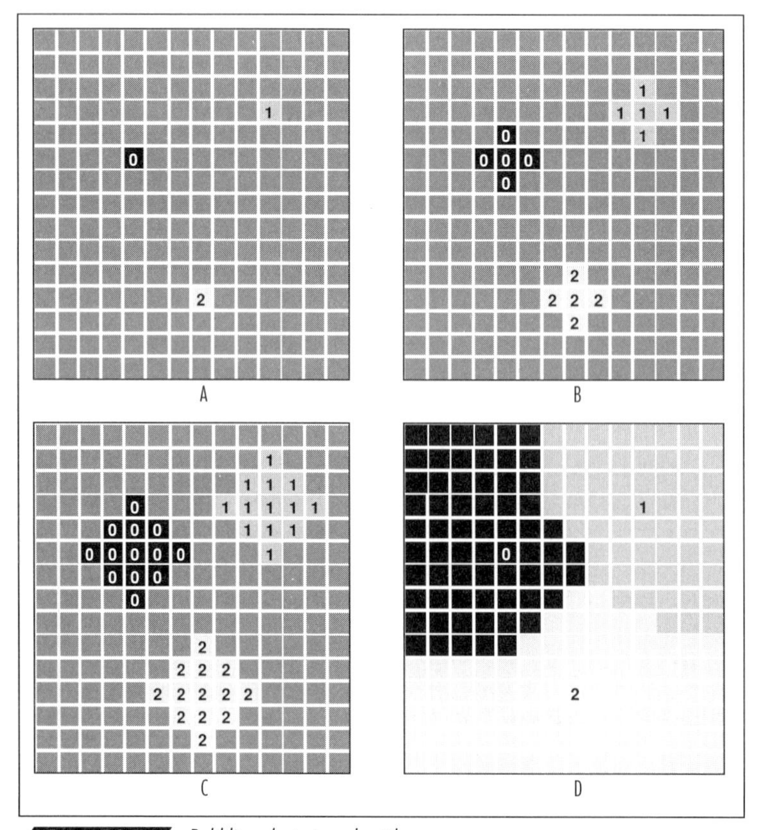

FIGURE 7.2.1 *Bubbling clustering algorithm*

Figure 7.2.1 gives an intuitive illustration of the algorithm. In that example, the set Ω of three reference vectors are marked 0, 1, and 2 in Figure 7.2.1*a*. Step 2 of the algorithm is repeated until there is no further change. Figures 7.2.1*b* and *c* show the first two iterations of the process, and Figure 7.2.1*d* displays the final classification map.

This algorithm is not practical in most cases; it can only be applied to very small image destination region pieces. In fact, it is useful only when the dimension is not too high, e.g., $n = 2$ or 3. A good use of this algorithm will be discussed in Chapter 11, which deals with the color mapping problem.

7.2.2 HECKBERT CLUSTERING

When there is not enough memory to store the buffer of marks on $\Re^N$, the classification map c needs to be described in an algorithmic manner. *The Heckbert clustering algorithm* splits the clusters using hyperplanes that are perpendicular to some coordinate direction. At the same time, the splitting can be used as a part of the algorithm in describing the map c.

Inductively, choose the largest current cluster, find the coordinate direction such that the cluster has the largest spread, split the cluster using a hyperplane perpendicular to that direction at the value that makes the generated sizes of the two clusters as close as possible. Repeat the process until the predetermined objective (speed, cluster size, or cluster number) is achieved.

In general, the method introduced in this section will not mathematically provide the best clustering (as the bubbling algorithm did). However, its effectiveness in dealing with massive points or vectors in the space proves to be an acceptable method for many applications.

HECKBERT ALGORITHM:

Given a finite set of vectors Ω in $\Re^N$ and an integer m to be the desired number of clusters, the *Heckbert algorithm* defines the classification map c by a sequence of $(n-1)$ splittings. Each splitting is defined by four parameters, (k, k', i, v), where the kth cluster is split by the ith hyperplane on value v into two clusters, the kth and the k'th clusters for $x_i < v$ and $x_i \geq v$ respectively. Briefly, the algorithm splits *from* the kth cluster *to* k'th cluster, *at* the ith variable, *on* value v. The criterion of determining the splittings are the following:

Step 0: Let $c_0^0 = \Omega$.

In the first splitting, $s = 1$:

Step 1-a: Choose the cluster $k_1 = 0$.

Step 1-b: Find the direction i_1 with one of the widest spreads, i.e.,

$$\max_{\mathbf{x},\mathbf{y}\in c_0^0}\left\{x_{i_1}-y_{i_1}\right\}=\max_{0\le i<N}\left\{\max_{\mathbf{x},\mathbf{y}\in c_0^0}\left\{x_i-y_i\right\}\right\}. \qquad (7.2.2)$$

Step 1-c: Find the middle split value v_1, so that

$$\left|\#\left\{\mathbf{x}\in c_0^0 \mid x_{i_1}<v_1\right\}-\frac{\#c_0^0}{2}\right|=\inf_{v\in\Re}\left|\#\left\{\mathbf{x}\in c_0^0 \mid x_{i_1}<v\right\}-\frac{\#c_0^0}{2}\right|.$$

Note that for any finite set S, $\#S$ denotes the *number of elements* in the set.

Step 1-d: Split the cluster into two:

$$c_0^1=\left\{\mathbf{x}\in c_0^0 \mid x_{i_1}<v_1\right\} \text{ and } c_1^1=\left\{\mathbf{x}\in c_0^0 \mid x_{i_1}\ge v_1\right\}. \qquad (7.2.3)$$

In the following splittings, $s=2, 3, \cdots, n-1$:

Step s-a: $s=2, 3, \cdots, n-1$. Choose one of the largest clusters k_s, i.e.,

$$\#c_{k_s}^{s-1}=\max_{0\le k<s}\#c_k^{s-1}. \qquad (7.2.4)$$

Step s-b: Find the direction i_s with one of the widest spreads, i.e.,

$$\max_{\mathbf{x},\mathbf{y}\in c_{k_s}^{s-1}}\left\{x_{i_s}-y_{i_s}\right\}=\max_{0\le i<N}\left\{\max_{\mathbf{x},\mathbf{y}\in c_{k_s}^{s-1}}\left\{x_i-y_i\right\}\right\}. \qquad (7.2.5)$$

Step s-c: Find the middle split value v_s, so that

$$\left|\#\left\{\mathbf{x}\in c_{k_s}^{s-1} \mid x_{i_s}<v_s\right\}-\frac{\#c_{k_s}^{s-1}}{2}\right|=\inf_{v\in\Re}\left|\#\left\{\mathbf{x}\in c_{k_s}^{s-1} \mid x_{i_s}<v\right\}-\frac{\#c_{k_s}^{s-1}}{2}\right|. \qquad (7.2.6)$$

Step s-d: Split the cluster $c_{k_s}^{s-1}$ into two:

$$c_{k_s}^{s}=\left\{\mathbf{x}\in c_{k_s}^{s-1} \mid x_{i_s}<v_s\right\} \text{ and } c_s^s=\left\{\mathbf{x}\in c_{k_s}^{s-1} \mid x_{i_s}\ge v_s\right\}. \qquad (7.2.7)$$

And set the others to the same as before, i.e.,

$$c_k^s = c_k^{s-1} \text{ , for } 0 \le k < s \text{ and } k \ne k_s. \tag{7.2.8}$$

Therefore, given a vector **x** from the set Ω, the classification map c is defined by the hyperplanes.

Figure 7.2.2*a* shows a simple example of a reference set of 13 vectors split into four clusters in three splittings. Figure 7.2.2*b* gives the most probable three splittings based on common sense. It tells how this method could be far apart from the optimal solution (still, it provides a practical solution to our problem).

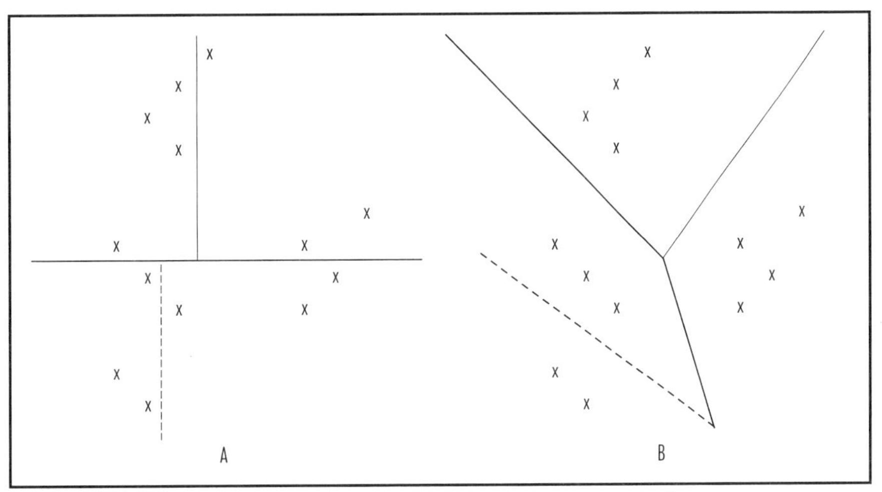

FIGURE 7.2.2 *Heckbert algorithm in a simple case: (A) Heckbert; (B) visual*

7.2.3 HECKBERT CLUSTERING IN C

Writing in C, the classification map can be described in a few lines in the routine `ClassificationMap( )`, which will be a part of the clustering class.

Here is the C++ class: `Clustering`. It performs the cluster splitting for a given set of vectors for a specified number of splittings, and it defines the classification map through a subroutine.

```
1   #define     N   16      // vector space dimension
2   #define     M   5000    // maximum number of clusters

3   class Clustering
4   {
5       public:
```

```
6           Clustering();
7           ~Clustering();

8           LONG        NumVector;  // number of reference vectors
9           VECTOR      *Vector;    // reference vectors
10          HECKBERT    Heck;       // Heckbert clustering structure

11          int     MakingClusters( );        // make clustering
12          LONG    ClassificationMap( VECTOR v );
                                          // identify cluster

13      private:

14          LONG    CluNum;  // number of Heckbert clusters
15          LONG    *CluList;    // vector reordering list
16          LONG    *CluSize;    // cluster sizes
17          LONG    *CluBegn;    // begining mark in CluList

18  };
```

VECTOR and HECKBERT have the following structure:

```
1   typedef struct Vector
2   {
3       SHORT       v[N];       // vector in N-dim vector space
4   } VECTOR;

5   typedef struct Simple_Heckbert_Clustering
6   {
7       SHORT       snum;       // number of splittings
8       SPLIT       s[M];       // splittings
9   } HECKBERT;

10  typedef struct Simple_Heckbert_Splitting
11  {
12      SHORT       at;         // splitting variable component
13      SHORT       on;         // splitting value
14      SHORT       fr;         // splitting cluster index
15      SHORT       to;         // new created cluster index
16  } SPLIT;
    // where the SPLIT structure should reads:
    // split from cluster [fr] to [to]
    //          at variable [at] on value [on]
```

The clustering structure–making routine is implemented—directly follow the algorithm. The clusters are described by `CluList`, a reordering list of numbers from `0` to `NumVector-1`, `CluNum`, the number of clusters, which is equal to `Heck.snum+1`, one more than splitting; `CluSize[i]`, the size of the *i*th cluster; and `CluBegn[i]`, where the *i*th cluster begins in the cluster reordering list. Therefore, for the reference vectors, `Vector[0]`, `Vector[1]`, ..., `Vector[NumVector-1]`, for any `i=0,1,...,CluNum-1`, the *i*th cluster consists of the vectors

```
    Vector[CluList[CluBegn[i]]],
    Vector[CluList[CluBegn[i]+1]], ... ,
    Vector[CluList[CluBegn[i]+CluSize[i]-1]].

1   #define MAX_PIXEL_VALUE     255

2   int Clustering::MakingClusters( )
3   {

4       LONG    i,j,k,n;
5       LONG    v,z;
6       LONG    *lst,*nxt;
7       VECTOR  max,min;
8       LONG    count[MAX_PIXEL_VALUE];

    //  start with one cluster
9       CluNum = Heck.snum + 1;
10      CluBegn[0] = 0;
11      CluSize[0] = NumVector;
12      for(i=0;i<NumVector;i++) CluList[i]=i;

    //  loop after number of splittings
13      for(n = 1; n < CluNum; ++n){

14  ANOTHER_ONE:

        //  find the index of one of the largest clusters
15          k = 0;
16          z = CluSize[0];
17          for(i = 1; i < n; i++){
18              if(CluSize[i]>z){
19                  z = CluSize[i];
20                  k = i;
21              }
22          }
23          Heck.s[n].fr = k;
24          Heck.s[n].to = n;

        //  break out if no cluster has more than 2 vectors
25          if(z<2){
26              do{
27                  Heck.s[n].at = 0;
28                  Heck.s[n].on = 0;
29                  Heck.s[n].to = k;
30              }while((++n)<Heck.snum);
31              break;
32          }

        //  get the splitting variable of the k-th cluster

            // 1) calculate max and min in each variable
33          lst = CluList + CluBegn[k];
34          for(i=0; i<N; i++) max[i]=min[i]=Vector[*lst].v[i];
```

```
35          for(j = CluSize[k]-1; j>0; j--){
36              ++lst;
37              for(i = 0; i<N; i++){
38                  if( max[i]<(v=Vector[*lst].v[i]) ) max[i] = v;
39                  else if( min[i] > v ) min[i] = v;
40              }
41          }

            // 2) find the variable of the largest dispension
42          z = max[0] - min[0]; i = 0;
43          for(j = N-1; j>0; j--){
44              if( max[j] - min[j] > z ){
45                  z = max[j] - min[j];
46                  i = j;
47              }
48          }
49          Heck.s[n].at = i;

            // 3) if all vectors are the same try another cluster
50          if(!z){
51              CluSize[k] = 1;
52              go to ANOTHER_ONE;
53          }

        //  get the splitting value
            // 1) count i-th vector component distribution
54          memset(count,0,(MAX_PIXEL_VALUE<<2));
55          lst = CluList + CluBegn[k];
56          for(j = CluSize[k]; j>0; j--){
57              ++count[Vector[*(lst++)].v[i]];
58          }

            // 2) split cluster into v[i]<=v & v[i]>v
59          j = ((CluSize[k]-1)>>1);  // half of the cluster size
60          v = min[i] - 1;
61          z = 0;
62          while( (z += count[++v]) < j );
63          if(((z-j)<<1)>count[v]) z -= count[v--];
64          Heck.s[n].on = v;

        //  sort the vector list to separate the cluster
            // 1) set new clusters:
65          CluSize[n] = CluSize[k] - z;
66          CluBegn[n] = CluBegn[k] + (CluSize[k] = z);
67          lst1 = (lst0 = RanLst + CluA[n]) + z;

            // 2) switch, so that the k-th cluster has v[i]<=v
            //    and the n-th cluster has v[i]>v
68          nxt = (lst = CluList + CluBegn[k] - 1) + z;
69          for(j = CluSize[k]; j >0; j--){
70          if(Vector[*(++lst)].v[i]>v){
71              while(Vector[*(++nxt)].v[i]>v);
72              z = *nxt;
73              *nxt = *lst;
74              *lst = z;
```

```
75            }}
76        }
77                return(0);
78    }
```

And the classification map is given by

```
1   int Clustering::ClassificationMap( VECTOR v )
2   {
3       int     c,k;

4       c = 0;
5       for(k=1;k<Heck.snum;k++){
6           if(c == Heck.s[k].fr
7           && v.v[Heck.s[k].at]>=Heck.s[k].on )
8               c = Heck.s[k].to;
9       }
10     return(c);
11  }
```

Indeed, from the above routine, the classification map c is well-defined on the whole vector space $\Re^N$. The set Ω is used only for defining the map c. This observation is the key of the compression scheme.

Finally, the C++ class is completed with the constructor and destructor:

```
1   voidClustering::Clustering( )
2   {
3       GetMemory(CluList, (NumVector<<2));
4       GetMemory(CluBegn, ((Heck.snum+1)<<2);
5       GetMemory(CluSize, ((Heck.snum+1)<<2);
6   }
```

and

```
1   voidClustering::~Clustering( )
2   {
3       FreeMemory(CluSize);
4       FreeMemory(CluBegn);
5       FreeMemory(CluList);
6   }
```

7.2.4 COMPRESSION USING CLUSTERING

Instead of the exhaustive global searching routine of *Codec 1.0* and *Codec 2.0*, use the new clustering searching algorithm described in the previous paragraph. The compression core algorithm diagram in Figure 5.1.6 will be refined to Figure 7.2.3.

Table 7.2.1 shows the result in comparison with the exhaustive searching in *Codec 1.0*. Visually, there are really no differences.

Notice that having more clusters does not mean that the algorithm is faster. It takes more time for a destination region vector to be determined.

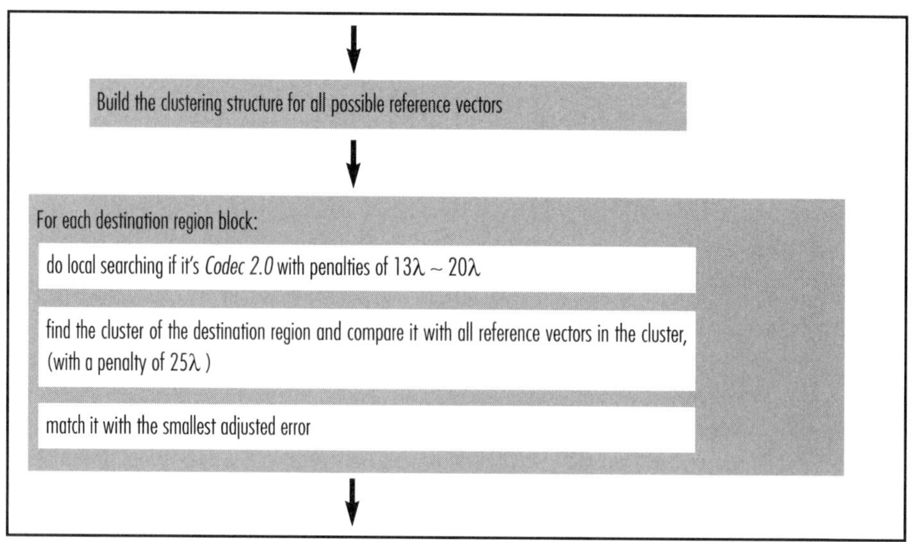

FIGURE 7.2.3 *Clustering searching compression core*

TABLE 7.2.1 *Speed vs quality using the* Heckbert algorithm

Cluster No.	Speed (sec.)	Cluster Speed	Lena		Golden Hill	
			rmse	*PSNR*	*rmse*	*PSNR*
1	**28,800**	100%	3.3453	37.64	4.1366	35.80
64	**450**	3.06%	3.4985	37.25	4.3481	35.37
128	**240**	1.97%	3.5287	37.18	4.3987	35.26
256	**130**	1.08%	3.5705	37.08	4.4532	35.16
512	**80**	0.65%	3.6216	36.95	4.5155	35.04
1024	**65**	0.35%	3.6732	36.83	4.5890	34.90
2048	**75**	0.20%	3.7419	36.67	4.6700	34.74
4096	**120**	0.13%	3.8203	36.49	4.7585	34.58

7.3 VARIATIONS OF HECKBERT CLUSTERING

This section will present some variations of the Heckbert clustering algorithm. They improve various aspects of the algorithm: the speed in determining a destination region, the efficiency of the classification map structure, the performance in using different image block variables, etc.

7.3.1 BALANCED TREE CLUSTERING

A rough description of the Heckbert algorithm given in the previous section is this: This algorithm starts from the cluster of all vectors and keeps cutting the largest cluster into two clusters of similar sizes until it reaches the desired cluster number.

Note that initially, all vectors form the only cluster, e.g., A. In the first step this cluster is cut into two, e.g., B and C. In the second step the one from B and C with larger size, e.g., B, will be cut into two new ones, e.g., D and E. Thus, at this moment there are in all three clusters: C, D, and E. In the next step, almost for sure, the one that will be cut will be C, since C has the largest size in general (because C has size similar to B, and D and E have sizes about half the size of B). As a result, in every step, instead of cutting the cluster having the largest size, all clusters will be cut in half. Such a step is also called a *level*. Such an algorithm is called the *balanced tree clustering algorithm*.

BALANCED TREE CLUSTERING ALGORITHM:

Given a finite set of vectors Ω in $\Re^N$, and an integer l denoting the desired level of splittings, there will thus be $n = 2^l$ clusters. The classification map c, the same as the *Heckbert algorithm*, is defined by a sequence of $(n-1)$ splittings of four parameters, (k, k', i, v), which is divided into l levels. The exact algorithm is the following:

Step 0: Let $c_0^0 = \Omega$.

In *Step s*, $s = 1, 2, \cdots, l$.

Step s-a: For each cluster k, $0 \le k < 2^{s-1}$.

Step s-b: Find the direction i_k^s with one of the widest dispersion, i.e.,

$$\max_{x,y \in c_k^{s-1}} \left\{ x_{i_k^s} - y_{i_k^s} \right\} = \max_{0 \le i < N} \left\{ \max_{x,y \in c_k^{s-1}} \left\{ x_i - y_i \right\} \right\}. \tag{7.3.1}$$

Step s-c: Find the middle split value v_k^s, so that,

$$\left| \#\left\{ x \in c_k^{s-1} \,\middle|\, x < v_k^s \right\} - \frac{\# c_k^{s-1}}{2} \right| = \inf_{v \in \Re} \left| \#\left\{ x \in c_k^{s-1} \,\middle|\, x < v \right\} - \frac{\# c_k^{s-1}}{2} \right|.$$

Step s-d: Split the cluster c_k^{s-1} into two:

$$c_k^s = \left\{ x \in c_k^{s-1} \,\middle|\, x_{i_k^s} < v_k^s \right\} \text{ and } c_{2^{s-1}+k}^s = \left\{ x \in c_k^{s-1} \,\middle|\, x_{i_k^s} \ge v_k^s \right\}. \tag{7.3.2}$$

The main advantage of this scheme is that a destination region can be classified quickly by using the balanced tree structure, since clusters are cut from 1 to 2, to 4, ..., to 2^s, etc. Using the cluster order given in Step *s-d*, the cluster splitting procedure forms a balanced tree as in Figure 7.3.1.

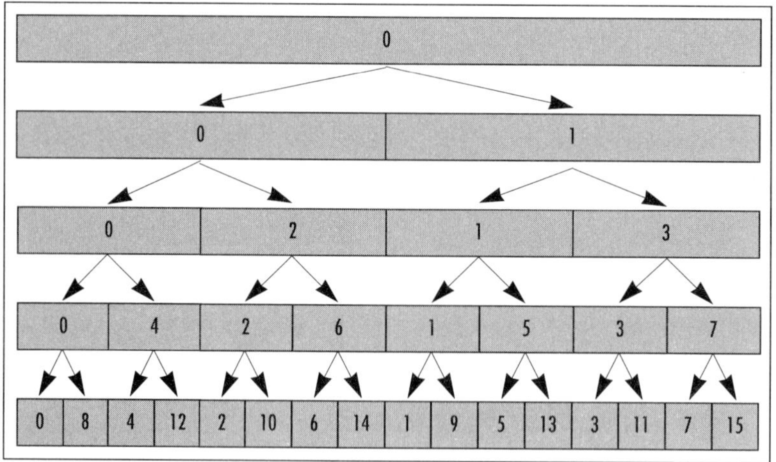

FIGURE 7.3.1 *Balanced cluster tree*

In the C++ class, the Heckbert splitting structure is redefined by adding one more entry: slevel, the *splitting level*, which is the number of levels that will be split.

```
1   typedef struct Balanced_Heckbert_Clustering
2   {
3       SHORT       slevel;     // number of splitting levels
4       SHORT       snum;       // number of splittings
5       SPLIT       s[M];       // splittings
6   } HECKBERT;
```

Thus, the number of clusters is 2 to the power of the splitting level, and the number of splittings is one less than that.

```
1       Heck.snum = (Clunum = (1<<Heck.slevel))-1;
```

As a convention, the array of splittings is divided into levels:

level 1: `s[1]`,

level 2: `s[2]`, `s[3]`,

level 3: `s[4]`, `s[5]`, `s[6]`, `s[7]`, and in general,

level l: `s[n+1]`, `s[n+2]`, `...` , `s[n+(n-1)]`, where `n=(1<<(l-1))`.

In this cluster order, for any vector $\mathbf{x} \in \Re^N$ the classification map c can be calculated in the following routine:

```
1   int Clustering::ClassificationMap( VECTOR v )
2   {
3       int     l,n,k;

4       l = 0; n = 1;
5       for(k=0;k<Heck.slevel;k++){
6           if(v.v[Heck.s[n+l].at]>=Heck.s[n+l].on) l += n;
```

```
7           n <<= 1;
8       }
9               return(OK);
10  }
```

Clearly, the classification speed has been reduced from the number `Heck.snum` to `Heck.slevel`, that is, from the number of clusters to its logarithm in base 2.

The cluster making procedure will be simpler as well. The following code shows the necessary modification of the `MakingClustering` routine.

```
2    int Clustering::MakingClusters( )
3    {
4        LONG    i,j,l,k,n;
...      // the same lines 5-8
     // cluster number is calculated from the splitting level
9        Heck.snum = (Clunum = (1<<Heck.slevel)) - 1;
...      // the same lines 10-12
     // loop of splitting becomes loop of level
13       for(l = 0; l < Heck.slevel; ++l){
     // lines 13-24: instead of finding the largest cluster,
     // split all
14       n = (1<<l);
15       for(k = 0; k < (1<<l); ++k,++n){
23           Heck.s[n].fr = k;
24           Heck.s[n].to = n;
     // lines 25-32: no split, if the cluster has only one vector
25           if(CluSize[k]<2) go to NOT_SPLIT;
...      // the same lines 33-49
     // lines 50-53: no split, if all vectors are identical
50           if(!z){
51               CluSize[k] = 1;
52               go to NOT_SPLIT;
53           }
...          // the same lines 54-75
76           go to NEXT_CLUSTER;
     // here is the not splitting case:
77   NOT_SPLIT:
78           CluSize[n] = CluSize[k] = 1;
79           CluBegn[n] = CluBegn[k];
80           Heck.s[n].at = 0;
81           Heck.s[n].on = 0;
82   NEXT_CLUSTER:
83           continue;
84       } // next k
85       } // next l
86      return(0);
87   }
```

TABLE 7.3.1 *Speed vs. quality using* Balanced Tree Algorithm

Cluster No.	Speed (sec.)	Cluster Speed	Lena		Golden Hill	
			rmse	*PSNR*	*rmse*	*PSNR*
1	**28,800**	100%	3.3453	37.64	4.1366	35.80
64	**450**	3.06%	3.4981	37.25	4.3481	35.37
128	**235**	1.97%	3.5297	37.18	4.3987	35.27
256	**125**	1.08%	3.5729	37.07	4.4541	35.16
512	**66**	0.65%	3.6235	36.95	4.5154	35.04
1024	**38**	0.35%	3.6766	36.82	4.5902	34.89
2048	**25**	0.20%	3.7274	36.70	4.6627	34.76
4096	**19**	0.13%	3.8001	36.54	4.7489	34.60

The performance comparison of using the balanced cluster tree is given in Table 7.3.1, in comparison with Table 7.2.1.

In the last case of 4096 clusters, 50% of the time is actually spent in making the cluster tree and the other 50% on searching the destination–reference matches.

7.3.2 ORTHONORMAL HIERARCHY VARIABLES FOR QUAD BLOCKS

In the compression scheme, the block region intensity is always set either explicitly to its mean value or implicitly to the brightness adjustment β-value. Thus, factoring this, there are only 15 actual independent variables in the 4×4 tiling case. A natural decomposition of 4×4 blocks, called the *S-transform decomposition* or the *Haar base hierarchy*, converts 16 pixel values into a set of 16 new variables, in which mean value is one of them. This decomposition is actually defined for any square block whose width is a power of 2.

Starting from 2×2, the Haar transformation is described below. Also, geometrically, it is illustrated in Figure 7.3.2.

$$\begin{aligned} y_{00} &= \frac{1}{2}\left(x_{00}+x_{01}+x_{10}+x_{11}\right), \\ y_{01} &= \frac{1}{2}\left(x_{00}-x_{01}+x_{10}-x_{11}\right), \\ y_{10} &= \frac{1}{2}\left(x_{00}+x_{01}-x_{10}-x_{11}\right), \\ y_{11} &= \frac{1}{2}\left(x_{00}-x_{01}-x_{10}+x_{11}\right). \end{aligned} \tag{7.3.3}$$

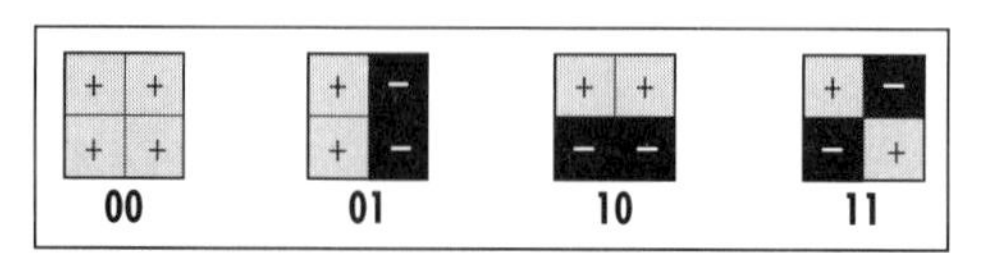

FIGURE 7.3.2 *2×2 Haar transformation*

As a consequence, the following equality is obtained:

$$\sum y_{ij}^2 = \sum x_{ij}^2 , \tag{7.3.4}$$

i.e., the L^2-distance errors measured by using the new variables are the same as those measured by using the old variables. The variable y_{00} is exactly the mean of the block.

In the 4×4 case, the Haar transformation is defined by the following formula:

$$y_{ijhk} = \frac{1}{4} \sum_{s=0}^{1} \sum_{r=0}^{1} \sum_{q=0}^{1} \sum_{p=0}^{1} (-1)^{i \cdot p} (-1)^{j \cdot q} (-1)^{h \cdot r} (-1)^{k \cdot s} x_{pqrs} , \tag{7.3.5}$$

for all $0 \le i, j, h, k \le 1$. Geometrically, it can be drawn as Figure 7.3.3. The C implementation of converting to new variables is straightforward.

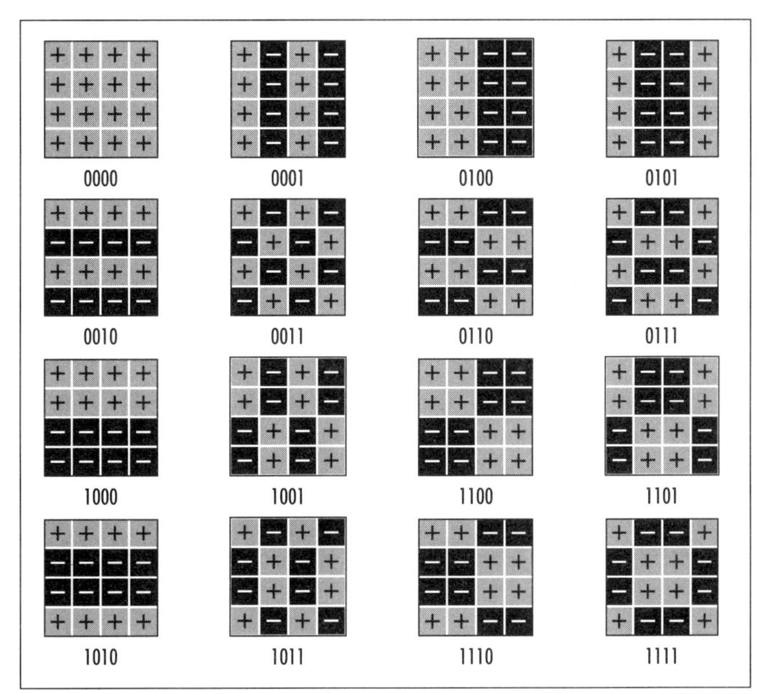

Figure 7.3.3 4×4 Haar transformation

```
1    #define  NxN 16 // number of pixels in the block

2    void GetNewVariables(
3        SHORT   *Var// for both i/o variable buffers
4    ){

5        SHORT        x0,x1,x2,x3;
6        SHORT        *v0,*v1,*v2,*v3;
```

```
7       SHORT       b,n;
8       SHORT       i,j;

9       b = 1; n = 4;
10      do{
11          v0 = &Var[0];
12          for(j=0;j<NxN;j+=n){
13              for(i=0;i<b;i++){
14                  v3 = (v2 = (v1 = v0 + b) + b) + b;
15                  x0 = *v0 + *v1;
16                  x1 = *v2 + *v3;
17                  x2 = *v0-*v1;
18                  x3 = *v2-*v3;
19                  *v0 = ( x0 + x1 )/2;
20                  *v1 = ( x0-x1 + (MAX_PIXEL_VALUE<<1) )/2;
21                  *v2 = ( x2 + x3 + (MAX_PIXEL_VALUE<<1) )/2;
22                  *v3 = ( x2-x3 + (MAX_PIXEL_VALUE<<1) )/2;
23                  v0++;
24              }
25              v0 += n-b;
26          }
27          b <<= 2;
28      }while((n<<=2) <= NxN);

29      return;
30  }
```

Apply the Heckbert clustering to the new variables. For the 4×4 case, factoring in both the speed and the performance quality, the results are slightly better than (but almost the same as) using the normalized pixel values (see Table 7.3.2). However, there is a major advantage in using Haar variables: *the match criterion can be weighted according to the importance of image information in different frequencies*. For example,

TABLE 7.3.2 *Speed vs quality using Haar base variables*

Cluster No.	Speed (sec.)	Cluster Speed	Lena		Golden Hill	
			rmse	*PSNR*	*rmse*	*PSNR*
64	**500**	3.06%	3.4629	37.34	4.3481	35.37
128	**260**	1.97%	3.5023	37.24	4.3643	35.33
256	**140**	1.08%	3.5488	37.13	4.4173	35.23
512	**75**	0.65%	3.5956	37.02	4.4725	35.12
1024	**40**	0.35%	3.6488	36.89	4.5503	34.97
2048	**28**	0.20%	3.7177	36.73	4.6156	34.85
4096	**21**	0.13%	3.7913	36.56	4.7004	34.69

1. In the large destination block cases (i.e., vectors having a large number of variable entries), by using the new variables the clustering process can be done based only on a subset of coarse variables.
2. For better visual quality, a weight table (similar to a JPEG DCT quantization table) can be assigned to the new variable set before matching comparison.

In practice, for the cases of 8×8, 16×16, etc., only the top 16 coarse variables will be used in the clustering classification, which is equivalent to considering the 16 variables obtained from resampling each of those regions to a 4×4 block using the average down-sampling method.

7.3.3 HIERARCHY VARIABLES FOR SQUARE BLOCKS

If the destination regions are square tiles of a width different from a power of 2, they can always be approximated by some resampling procedures. The resampling methods are filters weighted between simple subsampling and flat averaging.

Figure 7.3.4 subsamples a 5×5 to a 4×4, and Figure 7.3.5 double-samples a 7×7 to an 8×8. Yet, there is no reason to require the number of our clustering variables to be a power of two. The principle is to have a reasonable number—not too many that are redundant and not too few for capturing a good region classification—of variables that are geometrically meaningful.

Next, some recommendations on designing new variable sets for various square block tiles are given. Notationally, an $N\times N$ region vector is denoted by $v=\{v_{ji}\}_{1\le i,j\le N}$.

The 3×3 is a delicate case; one can use directly the nine variables or use the following two suggested variable sets:

$$u=\{v_{ji}-v_{22}\}_{1\le i,j\le N} \tag{7.3.6}$$

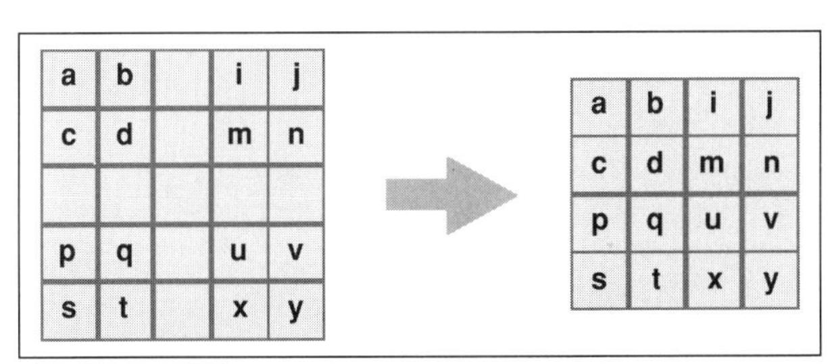

FIGURE 7.3.4 *Subsamples a 5×5 block to a 4×4 block*

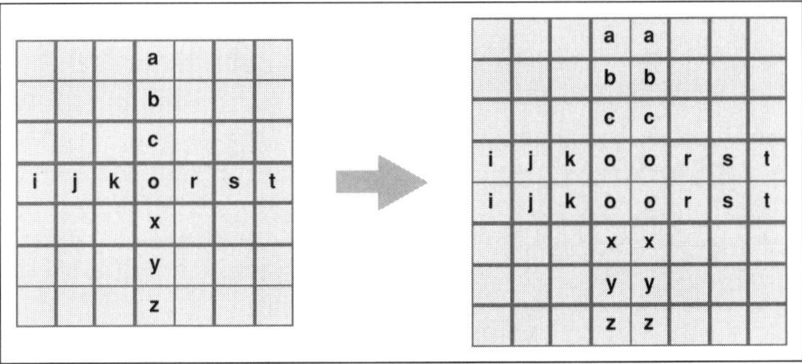

FIGURE 7.3.5 *Double-sampling a 7×7 block to a 8×8 block*

and

$$u = \begin{Bmatrix} v_{11} - v_{33}, v_{13} - v_{31}, v_{12} - \dfrac{v_{11} - v_{13}}{2}, v_{32} - \dfrac{v_{31} - v_{33}}{2} \\ v_{12} - v_{32}, v_{21} - v_{23}, v_{21} - \dfrac{v_{11} - v_{31}}{2}, v_{23} - \dfrac{v_{13} - v_{33}}{2} \end{Bmatrix}. \qquad (7.3.7)$$

The 5×5 can not only be subsampled to switch to a 4×4, but it also can be resampled to a 3×3 as the following:

$$u = \begin{Bmatrix} v_{11} - \dfrac{v_{12} + v_{21}}{2}, & \dfrac{v_{13} + v_{22} + v_{23} + v_{24}}{4}, & v_{15} - \dfrac{v_{14} + v_{25}}{2} \\ \dfrac{v_{31} + v_{22} + v_{32} + v_{42}}{4}, & v_{33} + \dfrac{v_{23} + v_{43} + v_{32} + v_{34}}{4}, & \dfrac{v_{35} + v_{24} + v_{34} + v_{44}}{4} \\ v_{51} - \dfrac{v_{41} + v_{52}}{2}, & \dfrac{v_{53} + v_{44} + v_{43} + v_{42}}{4}, & v_{55} - \dfrac{v_{54} + v_{45}}{2} \end{Bmatrix}. \qquad (7.3.8)$$

In general, by weighted averaging, an $N \times N$ block can be resampled to either a 3×3 or a 4×4. For example, a 6×6 block can be either averaged down to a 3×3 block or rescaled to a 4×4 block following the exact same procedure as shown in Figure 5.2.3 on how to resample 3×3 blocks to a 2×2 blocks.

7.3.4 CANONICAL SPATIAL FORMS

The next task is to reduce the searching loop of eight spatial symmetries by making them canonical. The method is called the *momentum classification method.*

Momentum classification is based on the average color intensity of four corners. Given a square tile

$$\{z_{ji}\}_{0 \le i,j < N}, \qquad (7.3.9)$$

four corner momentum values can be calculated as

$$a = \sum_{j=0}^{\left[\frac{N+1}{2}\right]} \sum_{i=0}^{\left[\frac{N+1}{2}\right]} z_{ji}, \quad b = \sum_{j=0}^{\left[\frac{N+1}{2}\right]} \sum_{i=\left[\frac{N}{2}\right]}^{N} z_{ji}, \quad c = \sum_{j=\left[\frac{N}{2}\right]}^{N} \sum_{i=0}^{\left[\frac{N+1}{2}\right]} z_{ji}, \quad d = \sum_{j=\left[\frac{N}{2}\right]}^{N} \sum_{i=\left[\frac{N}{2}\right]}^{N} z_{ji}. \qquad (7.3.10)$$

Based upon an isometric transform, the square tile always can be mapped to a tile with the property that

$$a - d \ge b - c \ge 0. \qquad (7.3.11)$$

Using the eight standard spatial forms listed in Chapter 2, Table 7.3.3 shows the actual isometric transforms used in all cases.

TABLE 7.3.3 *Momentum classification table*

$a < d$	$b < c$	$\vert a-d \vert < \vert b-c \vert$	isometry
0	0	0	0 *identity*
0	0	1	1 *x-flip*
1	1	1	2 *y-flip*
1	1	0	3 *180°-rotation*
0	1	0	4 *(x=y)-flip*
1	0	1	5 *270°-rotation*
0	1	1	6 *90°-rotation*
1	0	0	7 *(x+y=0)-flip*

TABLE 7.3.4 *Destination block and reference block spatial form index table*

R-region \ D-region	0	1	2	3	4	5	6	7
0 *identity*	0	1	2	3	4	6	5	7
1 *x-flip*	1	0	3	2	5	7	4	6
2 *y-flip*	2	3	0	1	6	4	7	5
3 *180°-rotation*	3	2	1	0	7	5	6	4
4 *(x=y)-flip*	4	6	5	7	0	1	2	3
5 *270°-rotation*	*5*	*7*	*4*	*6*	*1*	*0*	*3*	*2*
6 *90°-rotation*	6	4	7	5	2	3	0	1
7 *(x + y = 0)-flip*	7	5	6	4	3	2	1	0

After fixing both the destination region and the reference region, the resulting spatial form index can be identified in Table 7.3.4.

The performance is shown in Table 7.3.5, which is not very good but comparable with the speed gain. The reason for this being presented is for the idea. There are two types of classification that could be used to improve the compression speed. One is called the *quantity classification*, based on variable values as presented in earlier clustering schemes. The other one is called the *character classification*, based on the textures, edges, orientations, features, and other image-inherited properties.

In general, the character classifications could be very arduous, since those properties are difficult to identify and quantify. However, character classification could be very effective if a desirable one is found. Converting spatial forms into their canonical forms is one such approach. This algorithm is very effective to the block vectors that have clear tilting directions. Unfortunately, there are too many block vectors that do not belong to this category.

In Figure 7.3.6, all of the clustering methods are plotted in the same chart for a summarized comparison. At the moment, the most effective method is *balanced tree clustering using Haar variables*. So far, only the case of the *uniform 4×4 square image partition* has been analyzed. As exercises, one can apply this method to other fixed and mixed image partitions.

TABLE 7.3.5 *Speed vs quality using extra momentum classification*

Cluster No.	Speed (sec.)	Cluster Speed	Lena		Golden Hill	
			rmse	*PSNR*	*rmse*	*PSNR*
64	**104**	3.06%	3.6430	36.90	4.4975	35.07
128	**56**	1.97%	3.7152	36.73	4.5749	34.92
256	**34**	1.08%	3.8041	36.53	4.6620	34.76
512	**20**	0.65%	3.8930	36.33	4.7749	34.55
1024	**15**	0.35%	4.0107	36.07	4.9049	34.32
2048	**11**	0.20%	4.1669	35.73	5.0626	34.04
4096	**10**	0.13%	4.3394	35.38	5.2424	33.74

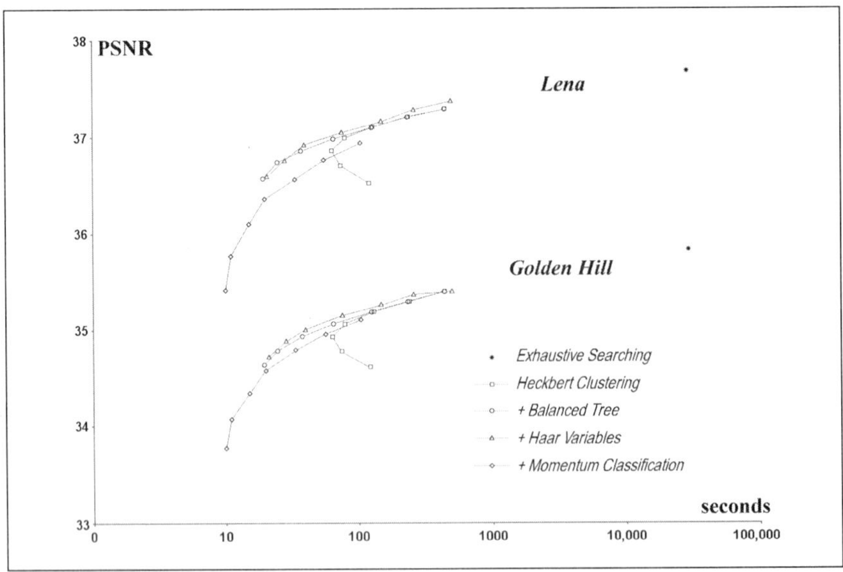

FIGURE 7.3.6 *Clustering searching performance log-cart*

7.3.5 OVERLAPPED BALANCED TREE CLUSTERING

The balanced tree clustering method can be further improved. In fact, the main problem of the Heckbert algorithm is possible misclassification. A better reference region vector could be missed for a destination region, simply because these two regions could be allocated on the different sides of a splitting hyperplane, though they are very close in distance. In the compression algorithm there is no reason to require that each reference region vector belong to only one cluster; it is convincing that the compression performance should improve by allowing the clusters to overlap with each other, for example, a fixed overlapping percentage p. Given a reference set of M vectors, it will therefore be split into two clusters of $\frac{1+p}{2}M$ vectors in the first level, a percentage p overlapping. In the second level, these two clusters split into four of $\left(\frac{1+p}{2}\right)^2 M$ vectors. And so on.

Note that in this algorithm the cluster size is fixed and controlled in every step.

OVERLAPPED CLUSTERING ALGORITHM:

Given a set of M vectors Ω in $\Re^N$ and an integer l to be the desired level of splittings, there will thus be $n = 2^l$ clusters. Let $0 \le p < 1$ be a given overlap percentage rate. Thus, all clusters will have the same size, $\left\lceil \left(\frac{1+p}{2}\right)^l M \right\rceil$. The classification map c, as in the *balanced tree clustering algorithm*, is defined by a sequence of $(n-1)$ splittings of four parameters (k, k', i, v), which is divided into l levels. Let $m(k)$ denote the integer $\left\lceil \left(\frac{1+p}{2}\right)^k M \right\rceil$, for any positive integer k. Here is the exact algorithm:

Step 0: Let $c_0^0 = \Omega$.

In *Step* s, $s = 1, 2, \cdots, l$.

Step s-a: $s = 1, 2, \cdots, l$, for each cluster k, $0 \le k < 2^{s-1}$.

Step s-b: Find the direction i_k^s with one of the widest dispersion, i.e.,

$$\max_{x,y \in c_k^{s-1}} \left\{ x_{i_k^s} - y_{i_k^s} \right\} = \max_{0 \le i < N} \left\{ \max_{x,y \in c_k^{s-1}} \left\{ x_i - y_i \right\} \right\}. \quad (7.3.12)$$

Step s-c: Order all the vectors in the cluster c_k^{s-1} according to their i_k^s th variable and find the middle split value v_k^s, so that

$$\left|\#\left\{ x \in c_k^{s-1} \,\middle|\, x < v_k^s \right\} - \frac{\# c_k^{s-1}}{2}\right| = \inf_{v \in \Re} \left|\#\left\{ x \in c_k^{s-1} \,\middle|\, x < v \right\} - \frac{\# c_k^{s-1}}{2}\right|. \quad (7.3.13)$$

Step s-d: Let the first $m(s)$ vectors form the cluster c_k^s, and the last $m(s)$ vectors form the cluster $c_{2^{s-1}+k}^s$, and use the splitting value v_k^s for determining future destination region vectors.

In the C++ class, everything remains the same as in the *balanced tree clustering algorithm*, except for a few changes: adding a public constant list $m(k) = \left\lceil \left(\frac{1+p}{2}\right)^k M \right\rceil$ in the class, enlarging the cluster list buffer in the cluster class constructor routine, and modifying some lines of the `MakingClustering` routine.

Therefore, the next line needs to be added to the `Clustering` class definition in the list of public variables:

```
#define MAX_LEVEL  32
LONG  M[MAX_LEVEL];  // cluster sizes in various level
```

The calculation of `M[MAX_LEVEL]`, and the memory allocation of `CluList` are handled in the class constructor.

```
1   #define P   25      // cluster overlap percentage in 100th

1   voidClustering::Clustering( )
2   {
        // get the cluster sizes for each level:
3       M[0] = NumVector*1000;
4       for(k=1;k<=Heck.slevel;k++) M[k] = M[k-1]*(100+P)/2;
5       for(k=0;k<=Heck.slevel;k++) M[k] /= 1000;

        // get the number of clusters
6       Heck.snum = (Clunum = (1<<Heck.slevel)) - 1;
7       GetMemory(CluList, ((M[Heck.slevel]*CluNum)<<2));
8       GetMemory(CluBegn, (CluNum<<2));
9       GetMemory(CluSize, (CluNum<<2));

        // initialize cluster beginnings
10      for(i=0;i<CluNum;i++){
            // get reverse binary number
11          j = 0;
12          for(k=0;k<Heck.slevel;k++){
13              if(i&(1<<k)) j |= (1<<(Heck.slevel-1-k));
14          }

15          CluBegn[i] = j*(CluSize[i] = M[Heck.slevel]);
16      }
17      return;
18  }
```

Here is the new `MakingClustering` routine.

```
1    int Clustering::MakingClusters( )
2    {
3        LONG    i,j,k,n;
4        LONG    v,z;
5        LONG    *lst,*nxt,*tmplist;
6        VECTOR  max,min;
7        LONG    count[MAX_PIXEL_VALUE];

     //  start with one cluster
8        for(i=0;i<NumVector;i++) CluList[i]=i;
9        GetMemory(tmplist, (NumVectors<<2));

     //  loop after number of splittings
10       for(l = 0; l < Heck.slevel; ++l){
11           n = (1<<l);   // current number of clusters

12           for(k = 0; k < n; ++k){
13               Heck.s[n+k].fr = k;
14               Heck.s[n+k].to = n+k;

             //  get the splitting variable of the k-th cluster

                 // 1) calculate max and min in each variable
15               lst = CluList + CluBegn[k];
16               for(i = 0; i < N; i++)
17                   max[i] = min[i] = Vector[*lst].v[i];
18               for(j = M[l]-1; j>0; j--){
19                   ++lst;
20                   for(i = 0; i<N; i++){
21                       if( max[i] < (v = Vector[*lst].v[i]))
22                           max[i] = v;
23                       else if( min[i] > v ) min[i] = v;
24                   }
25               }

                 // 2) find the variable of the largest dispension
26               z = max[0] - min[0]; i = 0;
27               for(j = N-1; j>0; j--){
28                   if( max[j] - min[j] > z ){
29                       z = max[j] - min[j];
30                       i = j;
31                   }
32               }
33               Heck.s[n+k].at = i;

             //  sorting the vector list:

                 // 1) count i-th vector component distribution
34               memset(count,0,(MAX_PIXEL_VALUE<<2));
35               lst = CluList + CluBegn[k];
36               for(j = M[s-1]; j>0; j--)
37                   ++count[Vector[*(lst++)].v[i]];

                 // 2) get the accumulated distribution
38               for(j = min[i]; j<=max[i]; j++)
39                   count[j] += count[j-1];

                 // 3) sort the list into new one
40               lst = CluList + CluBegn[k];
```

```
41              for(j = M[l]; j>0; j--){
42                  tmplist[count[Vector[*lst].v[i]-1]++]=*lst;
43                  lst++;
44              }

            //  create new clusters
45              lst = CluList + CluBegn[k];
46              nxt = tmplist;
47              for(j = M[l+1]; j>0; j--) *(lst++) = *(nxt++)
48              lst = CluList + CluBegn[n+k];
49              nxt = tmplist+M[s-1]-M[s];
50              for(j = M[l+1]; j>0; j--) *(lst++) = *(nxt++)

            //  get the splitting value
51              v = Vector[tmplist[M[s-1]>>1]].v[i];
52              Heck.s[n+k].on = (M[s-1]<(count[v-1]+count[v-2]))? v-1 : v;
53          }
54      }

55      FreeMemory(tmplist);
56      return(0);
57  }
```

Here are some interesting questions for ending the section.

Exercise: How does the overlapping balanced cluster tree perform? What is the right overlapping percentage? Are the percentage rates image-dependent?

7.4 SEED IMAGE CLUSTERING

In Chapter 5, compression, using a seed image, is presented as an alternative to global searching. However, in this case, the compression condition is slightly different from that of using the image itself as reference, because the reference region vectors will be used, not for just one single image, but for many images of a kind in every compression. As a result, this allows much more time to make a good clustering structure for faster and accurate searching. Furthermore, good clustering structure information can always be stored as a part of the seed image data file for future use.

7.4.1 CENTROID CLUSTERING

Most clustering techniques developed for the VQ algorithm can be easily transferred here. As an example, the implementation presented here uses 16 Haar variables, as shown in Figure 7.3.3.

CENTROID CLUSTERING ALGORITHM:

Step 1: The first step builds the initial set of cluster centroids. Since all reference region vectors are normalized to their means, the variable 0000 can be assumed to be zero all the time. Using either a 90° rotation or an

$(x + y = 0)$-flip, vertical textures can be converted to horizontal textures. This implies that the variables 0010, 1000, 1010, 0110, and 0111 can be assumed to have higher dispersions than the variables 0001, 0100, 0101, 1001, and 1011, respectively. For the diagonal case, the variables 0011, 1100, and 1111 are more significant. The remaining variables 0111 and 1110 are, lacking geometry the least important to the human visual system. Therefore, the core clusters will use the eight main variables: 0010, 1000, 1010, 0110, 0111, 0011, 1100, and 1111. Applying the vertical flip, the variable 0010 is assumed to be a positive number. And applying the horizontal flip, the variable 1100 is assumed to be positive as well. Now, split each variable, e.g., x_i, for some i, into three zones by two hyperplanes: $x_i = -H$ and $x_i = H$, for a given threshold H, e.g., $H = 16$. Thus, there are in total $2^2 3^6 = 2916$ initial clusters. The centroids of those clusters, eliminating the empty ones, will be the initial set of centroids.

Step 2: In the second step, those cluster centroids converge on a local optimal centroid set by an annealing procedure. That is, the clustering is redone by mapping each vector to its closest centroid and calculating the new cluster centroids again. Then, repeat this procedure until it *stabilizes*. (It does stabilize, because the sum of all distances between vectors and their corresponding centroids decreases in every repetition. This sum must converge to a constant in the discrete case.)

Step 3: Given any hyperplane, the *number of clustering violations* is defined to be the number of vectors that are split from their cluster centroids by a clustering hyperplane. The third step chooses a hyperplane that has the least violation number to be a Heckbert hyperplane, which will be used for the definition of the classification map.

Step 4: Now the set of reference vectors and the set of clustering centroids are split into two pairs of sets. For each pair of set reference vectors and set of clustering centroids, if there is more than one centroid, repeat the second step and the third step on the clusters. This procedure is continuously repeated until a complete set of Heckbert hyperplanes gives the complete clustering classification map.

Step 5: Finally, for any cluster that has too many reference vectors, repeat the following procedure: First find the variable with the largest dispersion, and then split it into to two equal sized clusters by a hyperplane perpendicular to the axis of that variable. Then calculate two centroids, and use them as the set of initial centroids to repeat the second step and the third step to obtain two new clusters.

This scheme may be improved to overlapped clustering as shown in the previous section. The exact C implementation is left as an exercise.

7.4.2 FIXED RADIUS DYNAMIC CLUSTERING

All the above clustering techniques contribute to good matches but often miss the best ones. A new method, called *fixed radius dynamic clustering*, will fix that. For each destination vector, a dynamic cluster is formed by reference vectors contained in the cube of a fixed diameter having the destination vector as the center. This fast searching algorithm will always return the best reference region within the given radius. But no match or an unsatisfactory match will be given if no matching vectors exist within the radius. In that case, often the destination region ends up with other alternative choices anyway, particularly for seed image searching.

DYNAMIC CLUSTERING ALGORITHM

Given is a set of M reference vectors $\left\{\mathbf{r}_j = \{r_{ji}\}_{i=1,2,\cdots,N}\right\}_{j=1,2,\cdots,M}$.

Step 1: In the first step, for each variable i, give a list of all reference vector indices, $\{l_k^i\}_{k=1,2,...,M}$, ordered in their ith variable values, which is actually a permutation of M indices. And also give an indicator list $\{n_v^i\}_v$ of the size of all possible variable values such that for each variable value v, the entry of the indicator list corresponding to the value v is the location of the reference vector list where all vectors in the front of this location have variable value less than v and all behind it have variable value greater than or equal to v. Consequently, the indicator list is a list of monotonically increasing numbers.

Step 2: Given a fixed radius ρ, for any destination vector $\mathbf{d} = \{d_i\}_{i=1,2,\cdots,N}$, the cluster of $\mathbf{d}$ is defined to be the set of all reference vectors that are contained in the cube $\prod_{i=1}^{N}[d_i - \rho, d_i + \rho]$, which is obtained using a marking process described below. Set the set of marks $\{m_k\}_{k=1,2,...,M}$, one for each reference vector, to zero. Then, for each variable, i, increase the marks of the reference vectors whose ith variable is in the interval $[d_i - \rho, d_i + \rho]$ by 1:

$$m_{l_k^i} += 1, \text{ for all } n_{d_i-\rho}^i \le k < n_{d_i+\rho+1}^i. \tag{7.4.1}$$

Therefore, the final cluster is

$$C(\mathbf{d}) = \left\{ \mathbf{r}_k \mid \text{if } m_k = N \right\}. \tag{7.4.2}$$

The disadvantage of this algorithm is that there is no control over the cluster size. In fact, if the radius is too large, a huge cluster becomes inevitable for the destination vectors that are nearly flat. Thus, in this particular case searching time is wasted when dealing with simple, flat regions. If the radius is too small, there will be too many empty clusters. Thus, some regions will have no codes. To avoid both extreme cases, the following modifications are necessary:

1. *Eliminate the flat vectors from the reference vector list.* Thus, the method is perfect for seed image application, since one should exclude the flat reference vectors from the total list of the indexed possible reference vectors for better performance. Indeed, to a flat destination block there should be no difficulty in finding a matching code.
2. *Switch the fixed constant radius* ρ *to variable-dependent and value-dependent fixed lower bound and upper bound functions:* $\rho_i^0(v)$ *and* $\rho_i^1(v)$. Thus the new cube, $\prod_{i=1}^{N}[d_i - \rho_i^0(d_i), d_i + \rho_i^1(d_i)]$, may have the destination vector not necessarily in the center. By finding the appropriate boundary functions, one should be able to generate clusters of the right size.

In fact, by adjusting the radius, the cluster sizes are controlled in a very precise manner.

7.4.3 SUPREMUM SEARCHING

How to obtain the exact cluster size by adjusting the radius?

Let S denote the target cluster size. In the special case of $S = 1$, the algorithm gives the best supremum metric matching. The *supremum metric* is also called the L^∞-metric.

For any destination vector $\mathbf{d} = \{d_i\}_{i=1,2,\cdots,N}$, sort out first the radius ρ, which is the smallest radius such that the size of the cluster of reference vectors in the cube of radius ρ centered at $\mathbf{d}$ is greater than or equal to the number S. For the reference vectors on the cube's boundary, calculate their distances from $\mathbf{d}$ using the image metric and throw the furthest ones out until the exact cluster size S is obtained.

Exercise: Implement such a clustering searching algorithm for $S = 64$.

PLATE 1

Red and Green Leaves

(*Source:* Original photo copyright © Ning Lu 1996; 1800x2700x24; 142 KB in FIF; Compression 100:1.)

PLATE 2-A

Lena; 512x512, Original.

(*Source:* Reproduced by special permission of Playboy Magazine© 1972 Playboy.)

PLATE 2-B

Mandrill Baboon; 512x512, Original.

(*Source:* Courtesy of ISO JPEG Standard Testing Image, scanned by IBA.)

PLATE 3-A

Plava Laguna; Original.

(*Source:* Courtesy of ISO JPEG Standard Testing Image, scanned by IBA.)

PLATE 3-B

Golden Hill; Original.

(*Source:* Courtesy of ISO JPEG Standard Testing Image, scanned by IBA.)

PLATE 4

Flowers – Decompressing and Iterating

(*Source:* Original photo copyright© Ning Lu 1993.)

(A) *Starting from an Arbitrary Image*

(B) *After First Iteration*

(C) *After Second Iteration*

(D) *After Third Iteration*

(E) *After Fourth Iteration*

(F) *After Sixteenth Iteration*

PLATE 5

Hawaiian Orchid – An Image Coded in Various Resolutions

(*Source:* Original photo copyright© Philip Greenspun, http://photo.net/philg.)

(A) *Compressed Using 16x16 Tiling*

(B) *Compressed Using 12x12 Tiling*

(C) *Compressed Using 8x8 Tiling*

(D) *Compressed Using 6x6 Tiling*

(E) *Compressed Using 4x4 Tiling*

(F) *Hawaiian Orchid (512x512)*

PLATE 6

A City Plaza, Fractal Zoom In

(*Source:* Original photo copyright© Ning Lu 1984.)

PLATE 7

Fractal Enhancement of a Chrysanthemum

(*Source:* Original photo copyright© Ning Lu 1993.)

(A) *256x256, Original; 64 dpi*

(B) *2048x2048; 521 dpi*

PLATE 8

Eiffel Tower

(*Source:* Original photo copyright© Ning Lu 1985; 1800x2700x24; 128 KB in FIF; Compression 111:1.)

PLATE 9

In the Cathedral of Pisa

(*Source:* Original photo copyright© Ning Lu 1985; 1800x2700x24; 163 KB in FIF; Compression 88:1.)

PLATE 10

Benches and Lamps – Choosing Good Color Tables

(*Source:* Original photo copyright© Ning Lu 1996.)

(A) *Full-Color Image*

(B) *Heckbert VQ – 128 Colors*

(C) *Heckbert VQ – 256 Colors*

(D) *Weighted Heckbert VQ – 64 Colors*

(E) *Weighted Heckbert VQ – 128 Colors*

(F) *Weighted Heckbert VQ – 256 Colors*

PLATE 11

San Jose Live – Color-Mapping in Uniformed Color Tables

(*Source:* Original photo copyright© Ning Lu 1995.)

(A) *Full-Color Image*

(B) *Uniformed in RGB 343 Colors, Dithered*

(C) *Uniformed in YUV 251 Colors, Not Dithered*

(D) *Uniformed in YUV 1024 Colors, Dithered*

(E) *Uniformed in YUV 512 Colors, Dithered*

(F) *Uniformed in YUV 251 Colors, Dithered*

PLATE 12

Face Painting – Color Decomposition

(*Source*: Original photo copyright© Ning Lu 1996.)

(A) *A Color Image*

(B) *The Chrominance Image: U+V (i.e., the Y values are set to 128)*

(C) *The Redless Image: Y+U (i.e., the V values are set to 128)*

(D) *The Chrominance Image: Y+V (i.e., the U values are set to 128)*

PLATE 13

A Street Corner with ATM – Color Image Compression

(*Source:* Original photo copyright© Ning Lu 1996.)

(A) *A Color Image, 1024x1024*

(B) *Compressed Using 4x4 Tiling for Y and Using 8x8 Tiling for U and V*

(C) *Compressed Using 4x4 Tiling for Y, U and V Are Linear of Y in 4x4 Tiling*

(D) *Compressed Using 4x4 Tiling for Commonly Addressed G, R, and B*

PLATE 14

Boston Harbor – Progressive Decompressing

(*Source:* Original photo copyright© Ning Lu 1992.)

80K

PLATE 15

Peachtree One Building – CFA Missing Color Interpolation

(*Source:* Original photo copyright© Ning Lu 1996.)

(A) *Original Full-Color Image, 640x320x24*

(B) *Hibbard Interpolation After Bayer CFA Masking*

(C) *New Interpolation After Bayer CFA Masking*

PLATE 16

A Temple in Hangzhou

(*Source:* Original photo copyright© Ning Lu 1994; 1800x2700x24; 324 KB in FIF; Compression 44:1.)

Months pass.... Islands of memory begin to rise above the river of his life. At first they are little uncharted islands, rocks just peeping above the surface of the waters. Round about them and behind in the twilight of the dawn stretches the great untroubled sheet of water; then new islands, touched to gold by the sun.

So from the abyss of the soul there emerge shapes definite, and scenes of a strange clarity. In the boundless day which dawns once more, ever the same, with its great monotonous beat, there begins to show forth the round of days, hand in hand, and their forms are smiling some, sad others. But ever the links of the chain are broken, and memories are linked together above weeks and months....

The River... the Bells... as long as he can remember—far back in the abysses of time, at every hour of his life—always their voices, familiar and resonant, have rung out.

—The Dawn

Romain Rolland

FRACTAL REALIZATION

8

Decompression, the reverse procedure of compression, is always considered a part of compression and representation. Quite often, an image is compressed only once and decompressed thousands of times thereafter. As a result, decompression speeds become critical when dealing with real-world applications. This chapter is devoted to this problem.

First, an extremely fast decompression algorithm, called the *pixel chaining algorithm*, is introduced. It applies to a fractal image compression system that uses *subsampling* as a rescaling method and *brightness adjustment* β*-values* as intensity parameters.

Then, another fast decompression scheme, the *pyramid decompression algorithm*, will be presented. In this case, the reference region addresses are required to be in even steps.

The Internet is fast becoming a mainstream medium, and *rapid decompression speeds* have different meanings there. The Internet requires not only a quick viewer that decompresses an image as soon as the image is completely downloaded, but also an instant viewer that displays the updated image using the up-to-the-moment partial data during downloading. This technique is called *progressive decompression* and is becoming a popular feature in on-line applications. One display choice of progressive decompression is described as follows: an image is first displayed as a thumbnail (a small version of the image picture used to label the image), and then the image data are gradually updated either to enhance the resolution by enlarging the displayed screen dimension or to improve the displayed image quality by adding additional image information.

Concerning this updating procedure, any smaller portion of the image can be zoomed to an arbitrary dimension—the same principle applies to fractal image enhancement.

Theoretically decompression is considered to be the *dual* of compression (for any compressor must have a decompressor to prove its value). When someone says, "*This image cannot be compressed*," this clearly means that the given compressor cannot transform the current image data into a smaller data file that could restore the image with satisfactory quality. But when someone says, "*This file cannot be decompressed*," this could mean two totally different scenarios: the compressor has screwed up the image—the packed data are corrupted—or the decompressor is incapable of handling the compressed data properly.

In a lossless coding system, a decompressor must recover the exact original data, so the comparison between a good decompressor or a bad one is based purely on running speeds, minimal memory requirements, and other mechanical performance ratings. In a lossy image compression procedure, a decompressed image is an approximation. As a consequence, if the degree of lossiness is known, and if extra data information can be assumed (e.g., all images are photo pictures), then a decompressed image may be rendered more realistically. Therefore, without changing the fractal codes, a decompressor with a good rendering system can outperform the standard decompression algorithm. Some of those rendering techniques, especially artifact removing methods, will be discussed in this chapter.

8.1 PIXEL CHAINING

There is a quick decompression technique that applies only to the subsampling compression and decompression system. It is called *pixel chaining.*

A subsampling compressor can be easily obtained by modifying an *averaging compressor* presented in the early chapters. Actually, all one needs to do is to switch the resampling function from averaging to subsampling—by changing one line of the codes in the image reference region initialization routine `Initialize_R_Region( )` in the C implementation of Section 3.4.2:

```
1    INT Encoder_100::Initialize_R_Region( )
...
11       *(px0++) = PTbl[(px1[0]<<2)]; //change for subsampling
...
```

A decompressor can be obtained from a similar modification. Despite this, a faster one (the so-called *pixel chaining scheme*)—developed by Iterated Systems in 1989—will be introduced. (See Figure 8.1.1)

8.1.1 REFERENCE PIXEL ORBIT

In the subsampling case, for each pixel in the image there is a unique pixel, called its *associated reference pixel*, given by the fractal transformation. If the reference pixel value is known, the corresponding destination pixel value can be derived in one step. By viewing this reference pixel as a destination pixel, another, new, reference pixel is associated to it. Continuing this procedure, a chain of pixels is built. This chain stops when it hits either a pixel already in the chain or a known pixel that has been computed from an earlier chain. If it stops at a known pixel, all pixel values of the chain will be generated by chasing the chain back. Otherwise, the last pixel in the chain is calculated by chasing it back enough reference pixels, with the end pixel value set to 128. For example, when the γ-value is set to $\gamma = 3/4$, then 20 pixels must be chased for the last pixel value to be accurate. The number 20 is deduced from the following calculation: Let the last pixel be denoted by d and its associated reference pixel by r_0. Their relation is given by the formula

$$d = \frac{3}{4} r_0 + \beta_0 \tag{8.1.1}$$

for some β-value β_0. Let the associated reference pixel of r_0 be denoted by r_1, the associated reference pixel of r_1 by r_2, ..., the associated reference pixel of r_n by r_{n+1}, ... Thus,

$$d = \frac{3}{4} r_0 + \beta_0 = \frac{3}{4}\left(\frac{3}{4} r_1 + \beta_1\right) + \beta_0 = \beta_0 + \frac{3}{4}\beta_1 + \cdots + \left(\frac{3}{4}\right)^n \beta_n + \left(\frac{3}{4}\right)^{n+1} r_n, \tag{8.1.2}$$

for some β-values from fractal codes. If replace r_n by 128, the error is estimated from

$$\left| d - \left(\beta_0 + \frac{3}{4}\beta_1 + \cdots + \left(\frac{3}{4}\right)^{n-1} \beta_{n-1} + \left(\frac{3}{4}\right)^{n} 128 \right) \right| \le \left(\frac{3}{4}\right)^{n} 128 \approx 0.4059, \quad (8.1.3)$$

for $n = 20$. That is, the decoding error falls into the truncation error. For an arbitrary γ-value, a similar calculation can be made.

Furthermore, in the example case of $\gamma = 3/4$, instead of storing the brightness adjustments, store the *scaled brightness adjustments* $k_n = \frac{3}{4}\beta_n$, for all n, equivalently. So the above formulas will become

$$d = \frac{3}{4}(r_0 + k_0) = \frac{3}{4}\left(\frac{3}{4}(r_1 + k_1) + k_0\right) = \frac{3}{4}\left(k_0 + \frac{3}{4}\left(k_1 + \cdots + \frac{3}{4}(k_n + r_n)\right)\right). \quad (8.1.4)$$

This will further improve the decompression implementation. In fact, many pairs of multiplication and addition operations have been switched into single-step table-look-ups.

Figure 8.1.1 shows two steps in a pixel chaining:

$$160 \leftarrow \frac{3}{4}100 + 85 \text{ and } 90 \leftarrow \frac{3}{4}160 - 30. \quad (8.1.5)$$

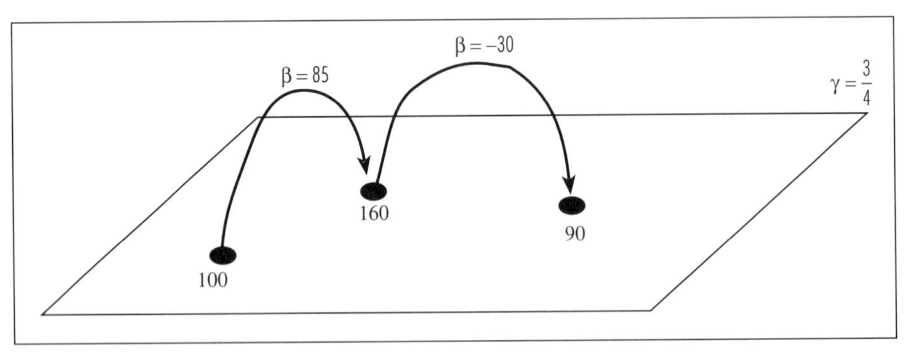

FIGURE 8.1.1 Pixel chaining

8.1.2 C IMPLEMENTATION

For those who like to disassemble music boxes or watches, here is the C code.

```
1   INT Decoder_300::Fractal_Decompression(
2   ){
3       LONG        i,j,n,z;
4       BYTE        *ppp;
```

```
5        SHORT       *sss;
6        FRACODE     *fc;
7        SHORT       *QTbl,qtable[512];
8        SHORT       **pq,**pqmap;
9        LONG        *xy,*xymap;
10       LONG        chain[1024];

     //  initiate pq division table
11       QTbl = qtable + 256;
12       for(n=0;n<256;n++){
13           QTbl[-n] = -( QTbl[n] = ((n<<2)+1)/3; )
14       }

     //  create a temporary working image for iteration
15       GetMemory(xymap,(z=image.w*image.h)*4);
16       GetMemory(pqmap,(z<<2));

     //  set standard indexed spatial forms
17       Initialize_S_Form( );

     //  create the reference map from the fractal codes
18       fc = fcode;
19       for(n=fcode_num,n>0;n--){
20           xy = xymap + (z = fc->dx + image.w*fc->dy);
21           pq = pqmap + z;
22           z = fc->rx + image.w*fc->ry;
23           sss = STbl[fc->s];
24           ppp = PTbl + QTbl[fc->q];
25           for(j=0;j<N;j++){
26               for(i=0;i<N;i++){
27                   *(xy++) = z + *(sss++);
28                   *(pq++) = ppp;
29               }
30               xy += image.w - N;
31               pq += image.w - N;
32           }
33       }

     //  mark every pixel to 254(undefined)
34       memset(image.px,254,(z=image.w*image.h));
35       while((z--)>0){
36           if(image.px[z] == 254){

         //  mark pixels in the chain to 1(active on chain)
37               image.px[chain[n=0]=z] = 1;
38               while( (image.px[chain[n+1] = xymap[chain[n]])
39                   == 254 )
40                   && (++n)<1000 )
41                   image.px[chain[n]] = 1;

         //  either the chain back to itself, or too long
42               if((image.px[chain[n]] == 1) ||
43                   (image.px[chain[n]] == 254) ){
44                   for(i=1;i<20;i++)
45                       chain[n+i] = xymap[chain[n+i-1]];
46                   x = 128;
47                   for(i=19;i>=0;i--) x = pqmap[chain[n+i]][x];
48               }
```

```
        //  or a known pixel:
49              else x = image.px[chain[n]];

        //  assign the value along the chain
50              for(i=n-1;i>=0;i--)
51              x = image.px[chain[i]] = pqmap[chain[i]][x];
52          }
53      }

54      FreeMemory(pqmap);
55      FreeMemory(xymap);

56      return(OK);
57  }
```

The gain in speed is significant. Consider the classic decompressor. Assume that there are 16 iterations, so each pixel needs 32 operations. In the pixel chaining algorithm each pixel is touched in an average of less than 2 operations. This algorithm is particularly useful for applications that require extremely high speeds and a relatively low compression performance.

Table 8.1.1 shows the performance difference between the average down-sampling method using the standard decompressor and subsampled down-sampling method using the pixel chaining scheme. The 4×4 block global even grid searching matches are used in both averaging and subsampling cases in the illustration. All sample images are grayscale in the resolution of 512×512. The images obtained by using an averaging method are always of higher quality than the images obtained using the subsampling down-sampling method. Visually, these two approaches are almost identical.

TABLE 8.1.1 *Decoding comparison*

	Lena	Golden Hill	Plava Laguna
rmse—averaging	3.3704	4.1520	5.0193
PSNR—averaging	37.58	35.77	34.12
rmse—subsampling	3.8899	4.8788	5.9869
PSNR—subsampling	36.33	34.36	32.59

The use of subsampling fractal image representation is used not only for gaining decompression speeds, but also for identifying image correlation in a new technology for image segmentation.

8.1.3 FRACTAL IMAGE SEGMENTATION

The essence of the fractal method is to complete the missing pixel data of an image piece using one of the similar "known" image pieces, which is called a *reference* image region. The phrase "*the most similar*" refers to the one image piece that has the best matching image piece in the partial information commonly known for both the completing image piece (i.e., destination region) and candidate reference image pieces. In fractal compression and representation a destination region is matched with a reference region by comparing the pixel values in the same sample grid by down-sampling the reference region. The fractal image segmentation is based purely on the same principle.

Start with a fractal image compression scheme that uses an image partition of destination regions and uses the subsampling method when matched to different scaling destination and reference regions. Given a coded image P, then, similar to the pixel chaining method discussed in the previous section, for any pixel x in the image, there is a unique associated reference pixel y such that there is some fractal transformation map between them. Let $\rho : P \longrightarrow P$ denote the map that carries each pixel x to its associated reference pixel y. Then the pixel chaining *reference orbit* of a pixel is the sequence

$$x, \rho(x), \rho^2(x), \cdots, \rho^k(x), \cdots. \tag{8.1.6}$$

Thus, the segmentation criterion becomes that the pixels in the same orbit will be in the same segmented portion of the image.

Consider the sequence of sets

$$P \supseteq \rho(P) \supseteq \rho^2(P) \supseteq \cdots \supseteq \rho^{k-1}(P) \supseteq \rho^k(P) \supseteq \cdots. \tag{8.1.7}$$

Let $Q = \bigcap_{k>0} \rho^k(P)$. Therefore, the segmentation problem of the image P is replaced by the segmentation problem of the set Q. Therefore, the key to a successful segmentation of the image P is that the set Q have a clear, nontrivial segmentation.

8.1.4 SEGMENTATION EXPERIMENTS

The detailed algorithm using fractal techniques to segment an image was first implemented by Ida and Sambonsugi of Toshiba R&D Center ([IS]). For simplicity, they used the 8×8 uniform square image partition, only the 2-to-1 trivial spatial form (i.e., no flip, no rotation), the fixed contrast factor $\gamma = 0.9$, the limited brightness adjustment values $\beta \in \{0, 1, 2, \ldots, 30\}$, and a local search matching within the square area $[-16, 16] \times [-16, 16]$.

The nontriviality of the set Q can be adjusted by the contrast factor γ-value and the brightness adjustment β-value range. Figure 8.1.2 shows how image regions are matched in a simple sample black and white picture. Figure 8.1.3 shows the sets $\rho^k(P)$ for different k values of the standard 512×512 grayscale *Lena* image. In a real implementation, the set Q is never actually obtained. Its approximation $\rho^k(P)$, for some large number k, is always used instead.

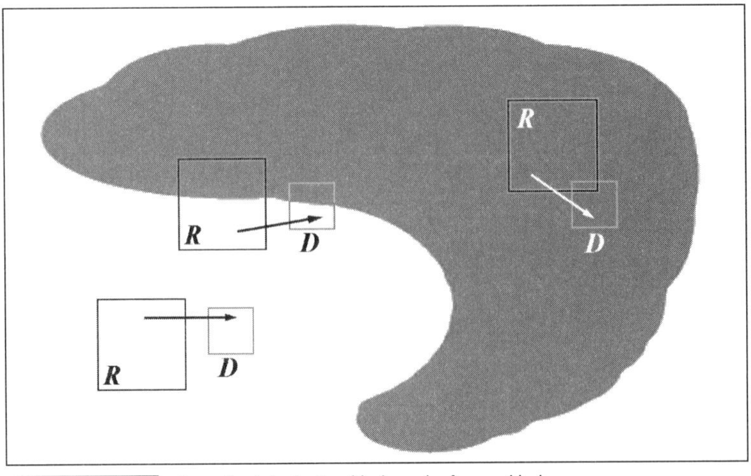

FIGURE 8.1.2 Example of destination blocks and reference blocks

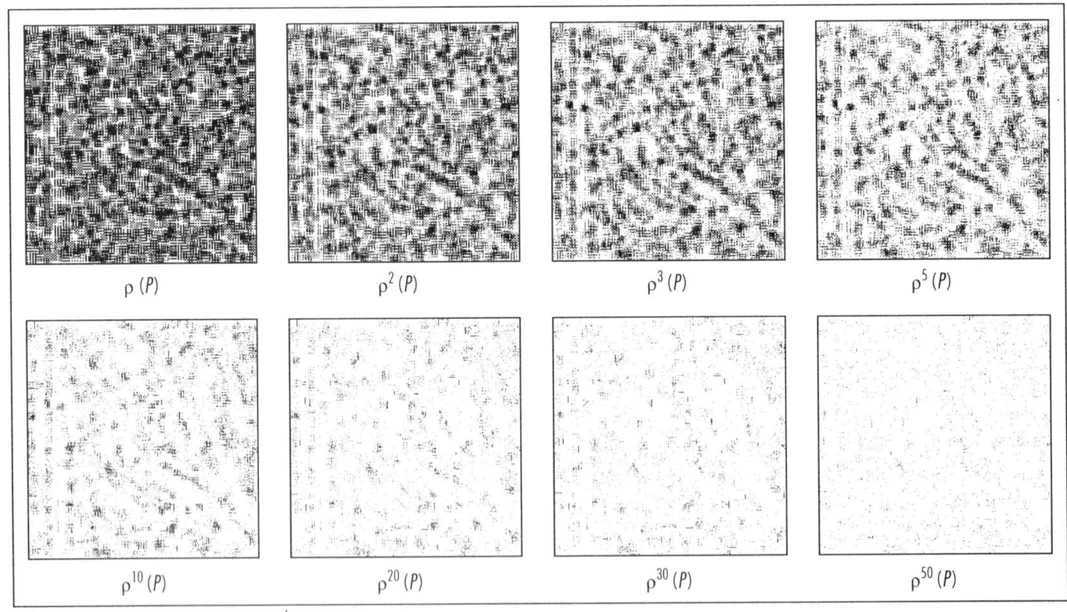

FIGURE 8.1.3 Image sets ρ^k (P), for some k

Figure 8.1.4*b* shows the result using $k=50$ and dividing the set ρ^{50} (P) into 32 clusters.

More experiments have been done with several parameters slightly modified: $\gamma=1$ and $\beta \in \{0, 1, 2, \ldots, 30\}$. The results are similar, as shown in Figures

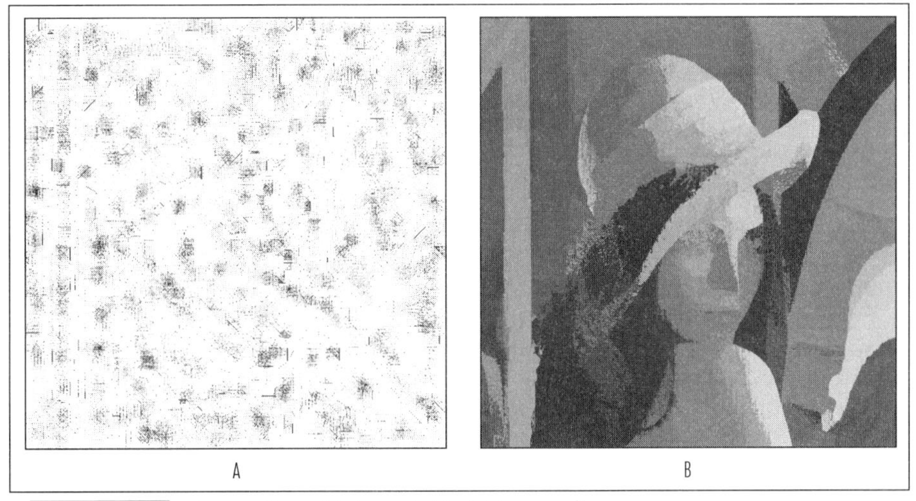

FIGURE 8.1.4 *Segmented* Lena: *(A) Segmented* Q; *(B) Segmented* P

FIGURE 8.1.5 *Segmented* Hawaiian Orchid *(Original photo courtesy of © Philip Greenspun, http://photo.net/philg; provided by Smadar Gefen)*

8.1.5 and 8.1.6. Intuitively, it is not difficult to understand that the image *Hawaiian Orchid* is more easily segmented than the image *Bobcat.*

This technique preserves the boundary for the area where the color intensity changes gradually. The special nature of this segmentation method is that the segmentation area boundaries are very fractal. How clear the segmentation boundaries are depends on how clearly the object boundaries are actually defined. As a result, the fuzziness, or the local fractal dimension, can be an

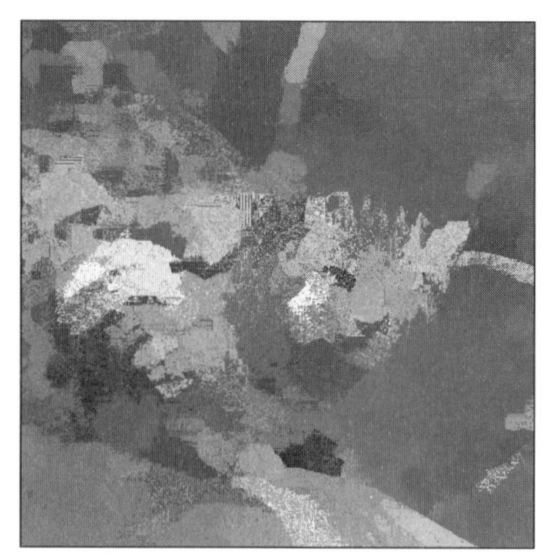

FIGURE 8.1.6 *Segmented Bobcat (Original photo courtesy of © Philip Greenspun, http://photo.net/philg; provided by Smadar Gefen)*

invariant that determines how clear the edges or boundaries are presented in an image. To some, the fractal region boundary may be considered a disadvantage for the lack of efficient methods in describing the exact region. Others may see it as a new feature, when the description memory is not quite as much of an issue.

More developmental research should move in this direction, and further research advances are expected.

8.2 THE PYRAMID ALGORITHM

When the reference region locations and the destination region locations have a common resolution grid factor (called the *address step*) greater than 1, a much smaller image can be decompressed at first, and then, only in the last iteration, is it blown up to the final resolution. This idea is another approach to the improvement of decompression speeds by imposing some restrictions on the types of fractal code that are used.

8.2.1 EVEN-ADDRESSED REFERENCE REGIONS

A reasonable restriction of a fractal image model is to assume that all destination regions are rectangles of an even resolution in both width and height, and all reference block location addresses use only the even grid. In other words, the fractal model can be viewed as a representation of a half-dimension quarter-size image, as shown in Figure 8.2.1.

For example, given a fractal model of $N \times N$ codes

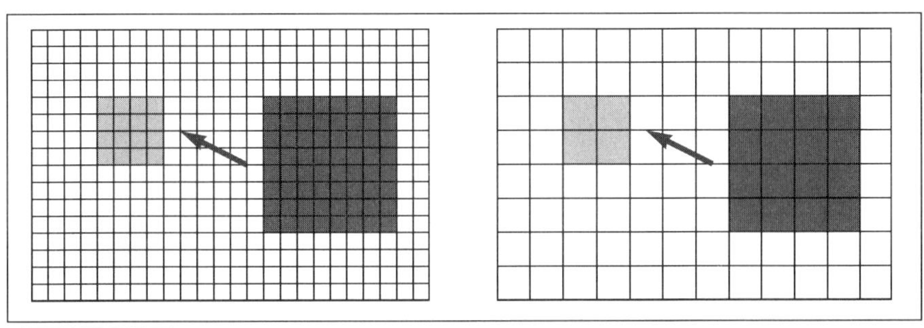

FIGURE 8.2.1 A fractal model for two different resolutions

$$\{((x_D, y_D),(x_R, y_R), \sigma, \tau, \gamma, \beta)\} \tag{8.2.1}$$

of a $w \times h$ image P, if all the numbers N, x_D, y_D, x_R, and y_R are even numbers, then the fractal model of $\frac{N}{2} \times \frac{N}{2}$ codes $\left\{((\frac{x_D}{2}, \frac{y_D}{2}),(\frac{x_R}{2}, \frac{y_R}{2}), \sigma, 2\tau, \gamma, \beta)\right\}$ represents some image Q of dimension $\frac{w}{2} \times \frac{h}{2}$. A typical straightforward verification will prove that: *The image Q is the half-dimension rescaled image of the image P, obtained by using the same resampling function used to match the reference regions to the derived regions.* That is, the image Q is the subsampling of the image P if the compression-decompression system uses subsampling, and the image Q is the average down-sampling of the image P if the compression-decompression system uses an averaging method.

Hence, first decompress the image Q, which takes only a quarter of the amount of calculation that for decompressing directly the image P. Then decompress image P directly from image Q, since all reference regions of P after resampling can be found exactly from image Q. As a result, if we consider that an iteration of Q takes a quarter of the time needed to iterate once the image P, i.e., 16 iterations of Q take the same amount of time as 4 iterations of P, then a 16-iteration decompressor is improved to a 5-iteration speed decompressor.The exact C-implementation will be left to the reader as an exercise.

8.2.2 RESOLUTION INDEPENDENCE

Taking the opposite point of view from that of the previous subsection, a fractal model can be viewed as an image model representing some image larger than its actual dimension.

For example, given a PIFS model of $N \times N$ codes

$$\{((x_D, y_D),(x_R, y_R), \sigma, \tau, \gamma, \beta)\} \tag{8.2.2}$$

25 dpi

50 dpi

100 dpi

FIGURE 8.2.2 Plava Laguna *(partial) fractally decoded in various resolutions (half, full, & double of the encoded resolution)*

of a $w \times h$ image P, the fractal model of $2N \times 2N$ codes $\left\{((2x_D, 2y_D), (2x_R, 2y_R), \sigma, \frac{\tau}{2}, \gamma, \beta)\right\}$ represents an image Q of dimension $2w \times 2h$. The image Q is called a *fractal zoom* of the image P.

Actually, the affine transformations (given in a fractal model) have nothing to do with image resolution. The above changes really correspond to a change of the spatial coordinate system by a scaling factor of 2. Figure 8.2.2 illustrates the same image portion decoded in different resolutions using the same fractal representation of the image *Plava Laguna*. This property (having no inherited resolution in its modeling) is called *resolution independence*. In the next section we will present it again, not as a consequence of fractal representation, but as a powerful tool in image prediction and enhancement. Although this model is resolution independent, the image resolution cannot be skipped in this discussion, since all captured digital images have some resolutions attached to them due to today's capturing devices.

In Chapter 15, a method of capturing a real-world image directly into its fractal format from its initial capturing will be presented. By then, the word "resolution" will have been pushed out of the concept "*digital imaging*" and will appear only in the display device.

8.3 FRACTAL BLOWUP

One of the most important applications using fractal image representation is *fractal image enhancement* (also called *fractal zoom* or *fractal blowup*, depending on whether one wants to print an image in a finer resolution grid or to view this image in a larger print).

This application takes advantage of one of the inherent fractal properties: the resolution-independent zoomability of fractal representation, which has been demonstrated in Chapter 3 for simple fractal objects and in the previous section of this chapter for real-world images.

Finding a fractal representation that produces a clear image (beyond the coded resolution) is a different task from that of finding an efficient fractal representation to represent a given image in its default resolution. It is preferable to preserve perceptual geometric structures between the destination regions and their corresponding reference regions in an image. Meanwhile, the image data file size is less important.

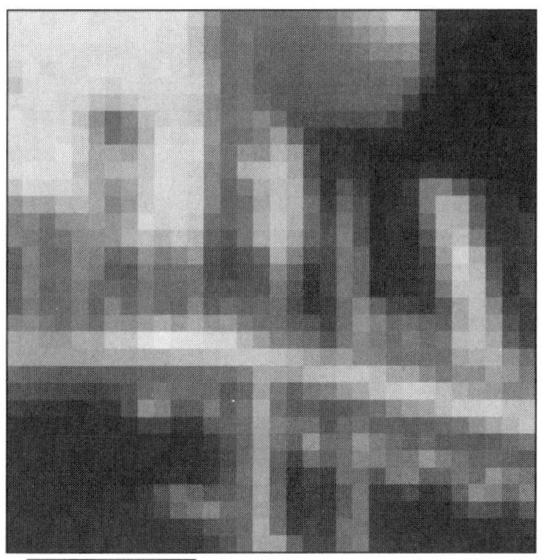

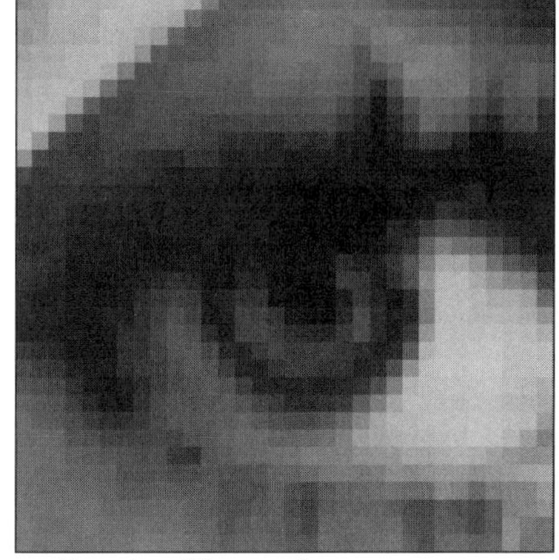

FIGURE 8.3.1 *Partial images pieces from* Plava Laguna *and* Lena *that will be zoomed*

8.3.1 FRACTAL IMAGE ENHANCEMENT

When compression is not an issue, it is natural to assume that all destination regions have the same square shape, e.g., an $n \times n$ block. In order to insure that the geometry is correct, the match searching may need to be compared using larger destination regions (e.g., $m \times m$ blocks, for $m \geq n$). It is tricky to choose the best block widths n and m. Small m's often produce good matches but poor geometric relations. Reciprocally, larger m's preserve the geometric patterns yet have less of a chance of finding good matches. So what are the right numbers to choose? For the best result (if speed is not an issue), one should start from some huge block covering and then try the smaller ones gradually until the matching errors are less than some predetermined threshold numbers.

However, the illustration uses a fixed square block dimension $n = 2$ and $m = 4$. The match searching is limited locally in a 16×16 area only among the 2:1 ratio not-rotated and not-flipped square blocks, and only a few γ-values: 1/2, 5/7, 3/4, 7/8, and 1 are used. Figures 8.3.2 and 8.3.3 show the performance of this algorithm by adding *zoomed image variation control* and *original mean correction.*

It is interesting to compare this with the direct zoom from a compressed fractal representation in Figure 8.2.2.

What is the performance difference between the direct zoom (i.e., output image in the desired resolution using the zoomed fractal codes) and the step-by-step zoom (i.e., repeatedly twice zoom the image to the desired resolution)? One should expect that both generate the same quality, except that the step-by-step zoom is much slower. However, as shown in Figure 8.3.4, the step-by-step zoom actually gives a slightly better image. In fact, by doing step-by-step zoom, finer reference block-grids are allowed in the zooming process. In most applications, the direct zoom method is preferred for its faster computational speeds.

The differences between one-step zoom and step-by-step zoom are:

- The fractal codes are reused in the one-step zoom, so it is faster.
- More reference candidates become available in the step-by-step zoom, so better matches can be found.

Taking the advantages from both of these methods, a new zoom algorithm optimized for both speed and quality, called the *one-step-plus zoom*, has been designed. The initial zoom step of this algorithm is the same as the first steps in both previously discussed zoom methods. In every step thereafter, the algorithm renews every fractal code by comparing the current code to the eight neighborhood codes (which are obtained by modifying the code reference location in all directions by 1 pixel grid). The best match among these 9 codes will be the renewed code for zooming. The complete implementation is assigned to the reader as an exercise.

Color Plates 6 and 7 show fractal enhancement and zooming.

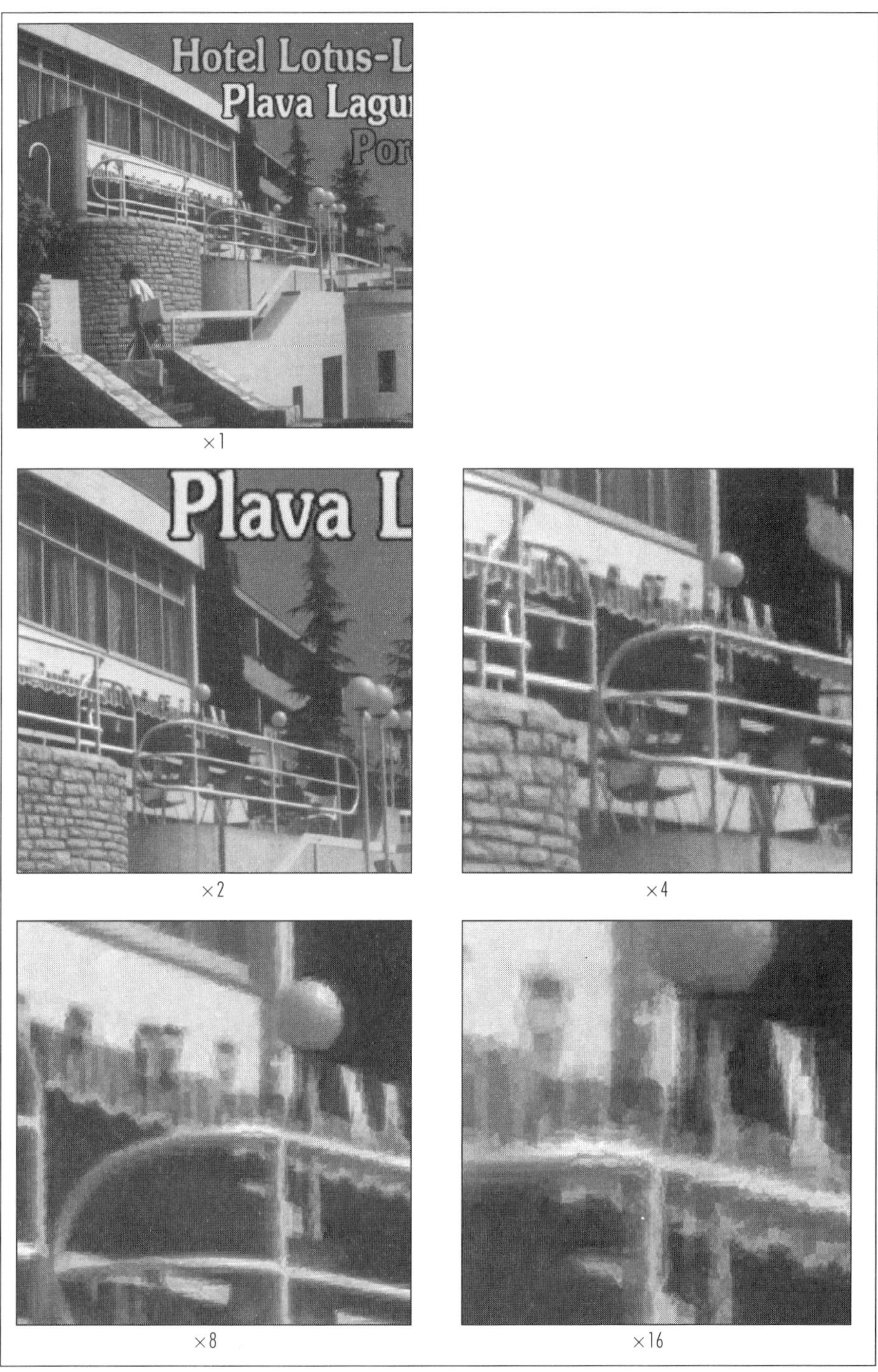

FIGURE 8.3.2 *Fractal blowup of* Plava Laguna *(provided by Echeyde Cubillo)*

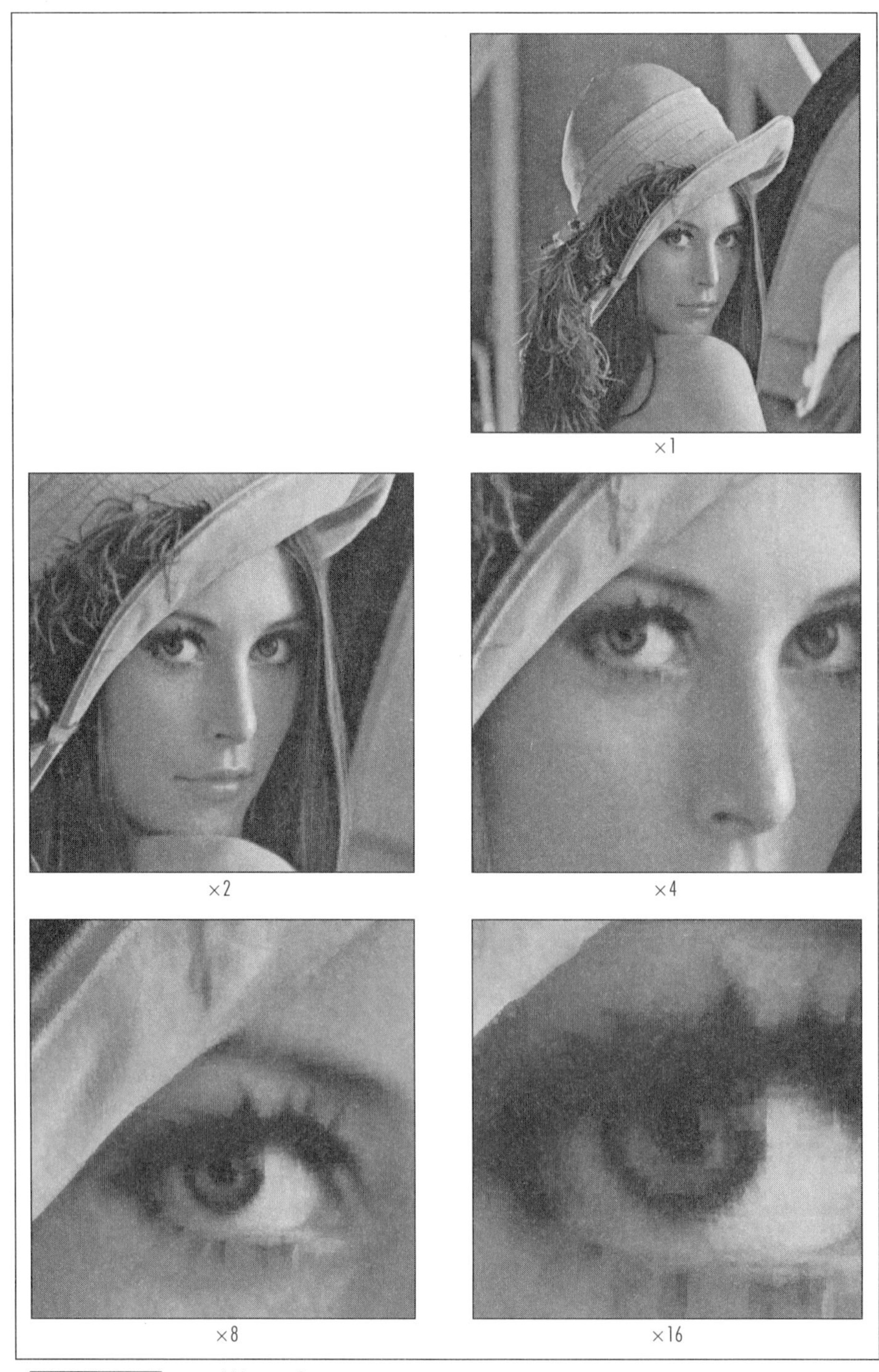

FIGURE 8.3.3 *Fractal blowup of* Lena

Step-by-step zoom (40 sec./MB for × 4)

One-step zoom (8 sec./MB for × 4)

FIGURE 8.3.4 *Step-by-step zoom vs. one-step zoom (provided by Echeyde Cubillo)*

8.3.2 Artifact prevention

Various types of artifacts can be found in the fractal blown-up images due to the imperfect fractal matching procedure. In general, this problem can be shifted to the encoder by stating that it is encoder's job to ensure that the images are clear. However, there are some common problems that are easier for the decoder to handle. For example, because of the blockiness artifact, as will be shown in the next section, the decompressor needs to improve image visual quality dramatically by adding a block-boundary filter.

One of the other artifacts that is detectable is called the *wrinkle artifact* (shown in Figure 8.3.5). In a compression routine, a destination region is usually matched with a down-sampled reference region. The down-sampling is calculated most often by taking its average. A high-dispersion and high-variation reference region could become a low-dispersion and low-variation region after down-sampling. This is exactly what happens here: a smooth destination region was matched by a high frequency region. To avoid this, the destination regions are forced not to be compared with reference regions with higher dispersions. However, this may cause a reduction in representation performance. Another choice is to add a conditional testing routine in the decompression side. For a destination region that has a higher blown-up dispersion, the pixel values are truncated back until the blowup region has smaller than the coarse image dispersion. As shown in Figure 8.3.5, this method works fairly well.

Artifact prevention is a task that involves both encoder and decoder. Quite often, this task is application dependent. After all, image interpolation is a totally different task from image compression.

FIGURE 8.3.5 *The wrinkle artifact of fractal blowup: (A) with artifact; (B) improved*

8.4 DECOMPRESSION TECHNIQUES

The goal of a decompressor, besides decoding truthfully the compressed image representation data, is to make the recovered image look visually appealing. By acknowledging the inherent deficiency of a specific compression system and by knowing the exact application usage, a good decompressor could exceed in decoded image quality above-average decompressors by using some extra initiatives.

For different applications a decompressor should be optimized differently. The key factors of a decompression system include, in general, the following: decompression speed, decompression quality, decompression feature, and decompression system adaptability. Lately, due to the fever of the Internet, progressive decompression becomes a necessity, a brief discussion of which will be given in this section.

To improve the actual image appearance, some modifications to the decoded image are inevitable. Altering the decoded image is surely going against all odds. However, there are cases that have been proven to be effective. For example, a blocky-artifact-removing filter not only improves the visual quality of the image , but also increases the signal-to-noise ratio in many cases.

8.4.1 BLOCK EDGE RENDERING

The blocky artifact is the most annoying problem in any block oriented coding scheme. The compression scheme may use a better image partition or covering to avoid this artifact (still, the square image partition is the most convenient and economic use of image partitioning). In the decompression end it is desirable to erase those block boundaries by using a filter. Nonetheless, true edges are often seen near some block boundaries. How can a smooth filter run without destroying the true edges? *The conditional block boundary blurring* is a commonly used technique.

The idea is to blur pixels near a block edge only if the dispersion of the pixel values near this edge are smaller than a predetermined, well-tuned threshold. This subsection will present a one-dimensional scheme. It will apply to a two-dimensional picture by applying this scheme first to every row and then to every pixel column.

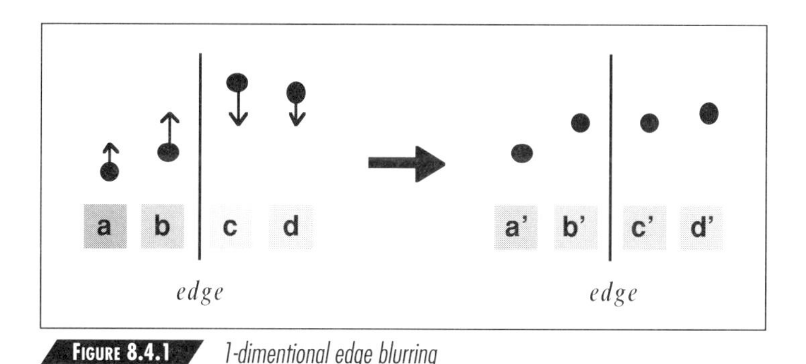

FIGURE 8.4.1 1-dimentional edge blurring

Figure 8.4.1 shows an edge blurring of depth 2, that is, 2 pixels in each side of the block edge that will be blurred.

In some of Iterated Systems's research reports, the blurring algorithm is described by the following formulas:

$$a' = a - \frac{2}{5}\varepsilon\,,\ b' = b - \frac{3}{5}\varepsilon\,,\ c' = c + \frac{3}{5}\varepsilon\,,\ d' = d + \frac{2}{5}\varepsilon\,, \tag{8.4.1}$$

where $\varepsilon = \dfrac{a+b-c-d}{4}$. The blurring will not be processed if $|\varepsilon| > T$, for some threshold T, to avoid blurring a true edge. Figure 8.4.2 shows the result of this algorithm, where $T = 8$ is used.

In Fisher and Menlove's work [F], a similar strategy is used. Their blurring algorithm is characterized by the formulas

$$a' = \frac{3a+2b+c}{6},\ b' = \frac{2b+c}{3},\ c' = \frac{b+2c}{3},\ d' = \frac{b+2c+3d}{6}. \tag{8.4.2}$$

Certainly, one can generalize the above blurring techniques to another depth. In general, a longer depth suits a larger block. Keeping the depth within a quarter of the block tile width is optimal.

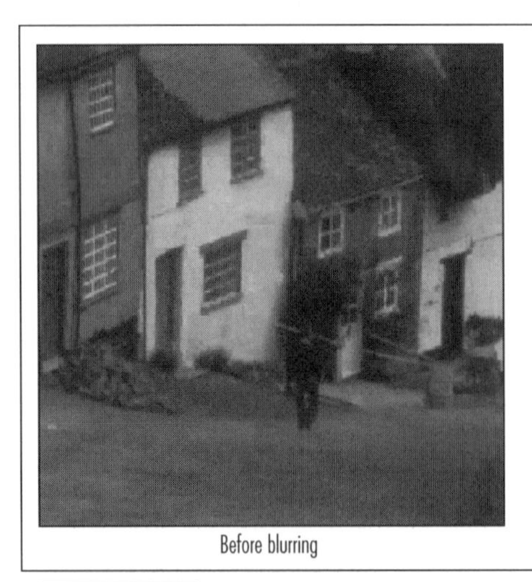

Before blurring

After blurring

FIGURE 8.4.2 *The edge-blurring postprocessing of* Golden Hill *(partial, coded in 6×6)*

8.4.2 COLOR RENDERING

Often, the two chrominance components of an image are compressed at a lower resolution than the one luminance component. In this case, chrominance screen interpolation becomes important. Some interpolation techniques will be presented in Chapter 11 and Chapter 15.

In most compression systems, because the chrominance images are compressed separately from the luminance image, a typical artifact occurs when an edge in an image is mis-aligned slightly in these images. In this case, a false color line could appear along the edge. However, this false color line is not difficult to detect and remove by using some standard image processing methods.

There are a couple of other color imaging features: *color mapping* and *color image dithering*. Both involve the task of image display using few colors. Color mapping locates the best color table of a fixed color number for a given image. Color image dithering displays the best image within the given color table. Both of them can be used separately or together.

A detailed discussion of color image representation will be covered in Chapter 11.

8.4.3 PROGRESSIVE DECOMPRESSION

As the Internet, Intranets, and the World Wide Web increase in popularity, now and in the future, software will in turn become more *net-oriented*. On the net, decompression speed is measured differently from that associated with a conventional end-user product. Though image downloading is slow, a user still expects to see the image when he clicks on the downloading button. Thus, a decompressor on the net needs to be able to decompress images using partially received data as soon as possible. The time it takes to view an image from the up-to-the-moment partially received data while transmitting and downloading are in process becomes more critical. In many cases, a user may decide to skip or even kill the transmission before an image is completely transmitted, or continue the process for every detail. The technique of providing this interactivity involves *progressive data decompression.*

Good progressiveness consists of two important factors: *good data organization* and *effective progressive access*. Data need to be organized such that the most significant information is sent first. Reading and accessing software needs to take advantage of every received data bit in a rapid and effective manner to maximize its performance.

In this case, the image data should be compressed and organized progressively, from lossy to lossless. Imagine that while the image data file is downloading a user stops the transmission when a clear enough decoded image is shown on the screen, and the up-to-the-moment received image data standing on itself is actually a complete file of a lossy representation of the image. Ideally, this lossy image data file would give a good compressed representation of the image with respect to its file size.

Thus, different from the pure compression task, in a progressive compression task a high-distortion-rate and high-compression-ratio representation data file is required to be the exact first part of a low-distortion-rate and low-compression-ratio data file. So no fractal codes are allowed to be switched, and only additional information can be added.

These additional data pieces could contain the extra precision of reference region locations in a higher-resolution grid or some enhanced terms of the intensity transformations of fractal codes or more residual information of the image in some higher resolution.

In Chapter 9 a multiresolution structure that fits this requirement will be introduced, and in Chapter 14 a progressive sample file format will be defined.

Once upon a time there dwelt two monks in a temple high in a mountain. One day, the old monk told a story to the young one. He said:

Once upon a time there dwelt two monks in a temple high in a mountain. One day, the old monk told a story to the young one. He said:

Once upon a time there dwelt two monks in a temple high in a mountain. One day, the old monk told a story to the young one. He said:

Once upon a time there dwelt two monks in a temple high in a mountain. One day, the old monk told a story to the young one. He said:

Once upon a time there dwelt two monks in a temple high in a mountain. One day, the old monk told a story to the young one. He said:

Once upon a time there dwelt two monks in a temple high in a mountain. One day, the old monk told a story to the young one. He said:

Once upon a time there dwelt two monks in a temple high in a mountain. One day, the old monk told a story to the young one. He said:

—Once Upon a Time

Anonymous

9 Multiresolution Decomposition

All compression techniques that include audio, still images, and video are based on one key element: *predicting the current data from the past data.* By using existing knowledge and up-to-the-moment received data, a probabilistic prediction of the next datum can be characterized by a small set of parameters. Then, possibly, one can add some correction terms to describe the difference between the real datum and the predicted one. Thus, when an accurate prediction is given, the code for the datum will simply be a signal saying that the datum is the same as the prediction. Then a decoder can obtain the real datum by repeating the same prediction process. Therefore, compression is achieved when accurate predictions are frequently given.

The interpolation technique established in the previous section gives an accurate prediction from a coarse resolution image to its finer resolution. As a natural extension, a new compressional approach (called a *multiresolution image compression scheme*) seems promising. This chapter will begin with such a compression algorithm using fractal interpolation. Then, another major image compression technique will be introduced: *image subband coding using wavelet and other reconstructive filters*, which many have developed in parallel, with the same philosophy behind it. Merging these techniques to create a promising future image data structure, *a progressive, resolution-independent, compressed, and practical image representation* is presented.

9.1 FRACTAL MULTIRESOLUTION COMPRESSION

Given an original image, first down-sample it using the averaging method. For example, a 768×512 image is down-sampled to an image of dimension 384×256. The inverse procedure is up-sampling, also called *interpolation*. An interpolation is a prediction: predicting the image pixel values in a higher resolution. The *correction data* are the differences between the original data and corresponding predicted data. Thus, instead of storing the 768×512 image data, one can store equivalently the 384×256 image data and the packed correction data. If the correction data is packed very tightly, a good compression is achieved. In other words, a good prediction is presented.

The above procedure can be repeated to the down-sampled image. Each down-sampling step is called a *level*. For example, the 5-level down-sampling of a 768×512 image has an image dimension 24×16. This repeating procedure of compression is called a *multiresolution compression*. Figure 9.1.1 illustrates a 3-level multiresolution procedure, where the arrow A refers to the average down-sampling operation and the arrow P is a prediction interpolation.

The *fractal multiresolution compression algorithm* presented here is abbreviated from a prototype designed by John Elton and John Muller in 1993 at Iterated Systems. They were motivated by the work of Mallat [M2] in wavelet compression.

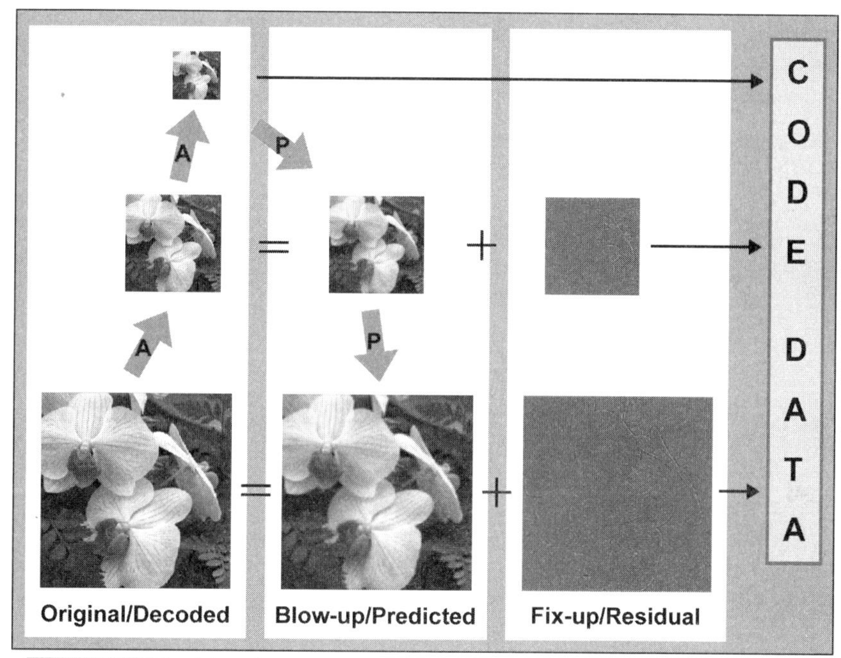

FIGURE 9.1.1 *Multiresolution compression*

The algorithm starts by down-sampling an original image to a thumbnail coarse image, which is stored in coded pixel values with a possible quantization. Then, this coarse image is up-sampled to a finer resolution: two times in both x and y directions, using Withers's *mean-preserving interpolation*, which expands each pixel P into four pixels, P_N, P_E, P_S, and P_W, symmetrically by the formulas

$$\begin{aligned}P_N &= P + \varepsilon \cdot (A + B - C - D + E - F - G + H),\\ P_E &= P + \varepsilon \cdot (C + D - E - F + G - H - A + B),\\ P_S &= P + \varepsilon \cdot (E + F - G - H + A - B - C + D),\\ P_W &= P + \varepsilon \cdot (G + H - A - B + C - D - E + F),\end{aligned} \tag{9.1.1}$$

as illustrated in Figure 9.1.2, where the number ε is a parameter with the default value 1/8, called the *expansion parameter*.

Now compare the up-sampled image with the same-resolution image that is down-sampled from the original image, block by block, in some fixed square partition—add a *fractal code* or an *escape code* to the blocks whose error reductions are greater than a predetermined quality distortion rate.

Fractal codes: Either spatial 2-to-1 contractive maps or spatial 1-to-1 copying maps, these codes are equivalent to maps that choose reference regions from either the coarse image or the up-to-the-moment decoded portion of the current

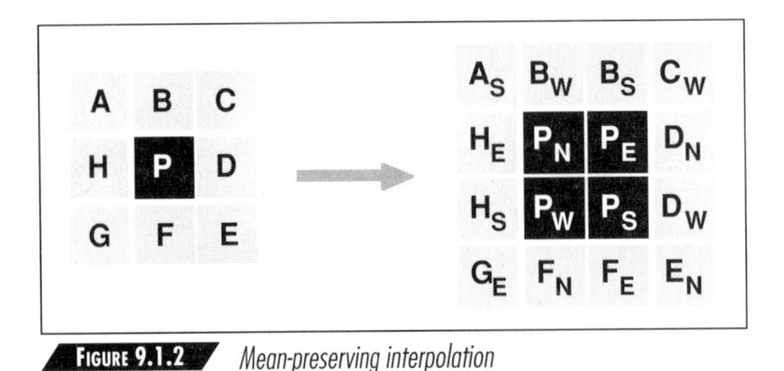

FIGURE 9.1.2 *Mean-preserving interpolation*

decoded image. In the implementation, two block layers are considered: 4×4 and 2×2. The reference blocks are indexed locally in a spiral manner.

Escape codes: Formed by three numeric correction components to 2×2 blocks, *horizontal correction*, *vertical correction*, and *diagonal correction*, these codes are exactly the three high-frequency variables of the Haar–Walsh–Hadamard decomposition:

$$
\begin{aligned}
E_h &= P_N - P_E - P_S + P_W, \\
E_v &= P_N + P_E - P_S - P_W, \\
E_d &= P_N - P_E + P_S - P_E.
\end{aligned}
\tag{9.1.2}
$$

In the extreme case of no fractal codes, this compression system becomes a variation of the Haar–Walsh–Hadamard-based subband wavelet coding (see later sections in the chapter). In fact, they are the same when the expansion parameter is set to zero.

For each block, the criterion for determining a code is purely by selecting the largest ratio of image quality improvement over packed code size increase.

When all blocks have been compared, one level of compression has been accomplished. Then the fine screen will be turned into the coarse screen, and the next resolution will be interpolated and coded in the same manner. This process will be repeated until the original image resolution is reached.

The exact implementation is left to the reader as an exercise. It has been proven that this algorithm is in the same range as any current state-of-the-art compression techniques.

In the above algorithm the coarse image up-sampling did not use the fractal interpolation because the speed of this process is quite slow—which is unbearable, not to an encoder, but definitely to a decoder in most applications, since the decoder must repeat the zooming procedure in exactly the same manner as the encoder. However, the compression performance improves significantly if fractal interpolation is used. The fact that today computer processing speed doubles every eighteen months (not considering possible special hardware assistance) suggests that ten years from now a present one-minute decoder will perform well within one second. At that time this method may well be the most practical one.

Using a coarse image to predict its refinement in a finer resolution, as fractal theory states, is based on the self-similarities of the image in different scales. In the language of frequency domain, this is really because of the self-similarities across various frequencies. This is exactly how wavelet subband image coding works. Therefore, in the rest of this chapter, this image compression technology will be introduced not as a comparison, but as a part of the same unity. Yes, these two approaches are seemingly so different, yet these explanations will show that many of the principles in both methods are similar.

9.2 FROM SPACE–TIME TO FREQUENCY SPECTRUM

Any theory can be traced back to some great moment. The wavelet subband coding is an offspring of the *Fourier series.*

9.2.1 THE FOURIER TRANSFORM

In 1888, Fourier discovered that [F4]

> *A periodic function f(t) with period T, i.e.,* $f(t+T)=f(t)$, *can be expressed as a linear combination of complex exponentials with frequencies* $\{n\omega\}_{n=0,1,2,\cdots}$, *where* $\omega=2\pi/T$. *In other words,*
>
> $$f(t)=\sum_{k=-\infty}^{k=\infty} F[k]\cdot e^{ik\omega t}, \tag{9.2.1}$$
>
> *where* $i=\sqrt{-1}$. *Furthermore,*
>
> $$F[k]=\frac{1}{T}\int_{-T/2}^{T/2} f(t)\cdot e^{-ik\omega t}dt. \tag{9.2.2}$$

The above expression for $f(t)$ is also called the *Fourier decomposition* of the function $f(t)$. Each component of this decomposition, e.g., $F[k]\cdot e^{ik\omega t}$, is a *simple* periodic term fractional period T/k. The coefficient T/k describes the contribution of the periodic component $e^{ik\omega t}$ of period T/k to the function.

This expression is generalized to arbitrary functions by letting T go to infinity. Thus, for any continuous integrable function f defined on $\Re^n$, the *Fourier transform* of f is defined by

$$F(\omega)=\int_{\Re^n} f(t)\cdot e^{-i\omega\cdot t}dt=<e^{i\omega\cdot t},f(t)>,\ \text{for any } \omega\in\Re^n. \tag{9.2.3}$$

Consequently, the inverse Fourier transform of F has the following formula:

$$f(t) = \frac{1}{2\pi} \int_{\Re^n} F(\omega) \cdot e^{i\omega \cdot t} d\omega = < \frac{1}{2\pi} e^{-i\omega \cdot t}, F(\omega) >, \text{ for any } t \in \Re^n. \quad (9.2.4)$$

The function F, which is the Fourier transform of the continuous integrable function f, is integrable but not necessarily continuous. When we extend the function space to a space where the Fourier transform could be completely defined, the *distribution space* is obtained, the exact image space discussed in Chapter 2. *The space in frequency domains is really the dual space of the apparent world we are seeing.* And the Fourier transform is a bridge between them.

In practice, most applications have been written mainly for the study of 1-dimensional audio or radio signals. As a result, the Fourier transform has been considered as a converter between the *time domain* of an analog signal and the *frequency domain* of its spectrum. Having seen the commonality between the spatial and temporal signals, it is not difficult to recognize the Fourier transform as a wider connection between the real-world signal domain and its dual frequency domain. Consequently, the expression *time domain* has been changed to *space-time domain.*

In a precise definition, for any distribution f, its *Fourier transform* $\hat{f}$ is given by the following formula:

$$< \hat{f}, \varphi > \;=\; < \int_{\Re^n} \varphi(\omega) e^{-i\omega \cdot t} d\omega, f >, \text{ for all } \varphi \in \mathcal{K}(\Re^n). \quad (9.2.5)$$

Conventionally, the Fourier transform is denoted by the addition of the hat "^". As an exercise, the inverse formula can be easily generated from (9.2.4).

Assuming that most readers are not familiar with distribution theory, the notations of functions will be used instead of distributions. Nonetheless, the distribution is an important concept, one that is more accurate for modeling real-world images. Most theories of the function space can be translated into the distribution space equivalently by following all concepts and notations carefully. This will be a good exercise for the advanced reader.

The spaces of the space-time domain and its frequency domain are both full of algebraic structures. The conversion of the Fourier transform carries most of those algebraic operations naturally and elegantly. Quite often, the easier forms are really in the frequency domain, for it is far more intrinsic. For example,

1. The Fourier transform preserves the *linearity* of vector spaces, i.e.,

$$h = \alpha f + \beta g \;\Leftrightarrow\; \hat{h} = \alpha \hat{f} + \beta \hat{g}. \quad (9.2.6)$$

2. The Fourier transform is *symmetric*, or *dual*, i.e.,

$$h = \hat{f} \quad \Leftrightarrow \quad \hat{h}(\omega) = 2\pi f(-\omega), \ \forall \omega. \tag{9.2.7}$$

3. A *shifting* in the space-time domain is equivalent to a phase factor in the frequency domain, i.e.,

$$h(t) = f(t-a), \ \forall t \quad \Leftrightarrow \quad \hat{h}(\omega) = e^{-i a \cdot \omega} \hat{f}(\omega), \ \forall \omega. \tag{9.2.8}$$

4. A *scaling* in the space-time domain results in the inverse scaling in the frequency domain, i.e.,

$$h(t) = f(at), \ \forall t \quad \Leftrightarrow \quad \hat{h}(\omega) = \frac{1}{|a|} f\left(\frac{\omega}{a}\right), \ \forall \omega, \tag{9.2.9}$$

where conventionally, $\hat{f}, \hat{g}$, and $\hat{h}$ are assumed to be the Fourier transforms of the distributions f, g, and h, respectively.

There are few other operations, such as differentiation, integration, and convolution, that can be carried out in the frequency domain as simple multiplication operations. For example, in the 1-dimensional case:

5. *Differentiation* and *integration* lead to multiplication and division, i.e.,

$$h(t) = \frac{\partial^n f(t)}{\partial t^n}, \ \forall t \quad \Leftrightarrow \quad \hat{h}(\omega) = (i\omega)^n \hat{f}(\omega), \ \forall \omega, \tag{9.2.10}$$

$$h(t) = \int_{-\infty}^{t} f(\tau)\, d\tau, \ \forall t \quad \Leftrightarrow \quad \hat{h}(\omega) = \frac{1}{i\omega} \hat{f}(\omega), \ \forall \omega. \tag{9.2.11}$$

6. The *convolution* of two functions and their expression in the Fourier transform are given by

$$h(t) = \int_{-\infty}^{\infty} f(\tau) g(t-\tau) d\tau = f(t) * g(t), \ \forall t \quad \Leftrightarrow \quad \hat{h}(\omega) = \hat{f}(\omega) \cdot \hat{g}(\omega), \ \forall \omega. \tag{9.2.12}$$

In the digital case, the differentiation, integration, and convolution operations become subtraction, addition, and mask filtration. The Fourier transform converts between the digital space-time domain and the discrete frequency domain. For example, for a digital image P of dimension $N \times N$, its discrete Fourier transform is defined to be

$$\hat{P}[s,t] = \mathrm{Re}\left(\alpha \cdot \sum_{k=0}^{N-1} \sum_{h=0}^{N-1} P[h,k]\, e^{-i\frac{2\pi sh}{N}} e^{-i\frac{2\pi tk}{N}} \right) = \alpha \cdot \sum_{k=0}^{N-1} \sum_{h=0}^{N-1} P[h,k] \cos \frac{2\pi(sh+tk)}{N}, \tag{9.2.13}$$

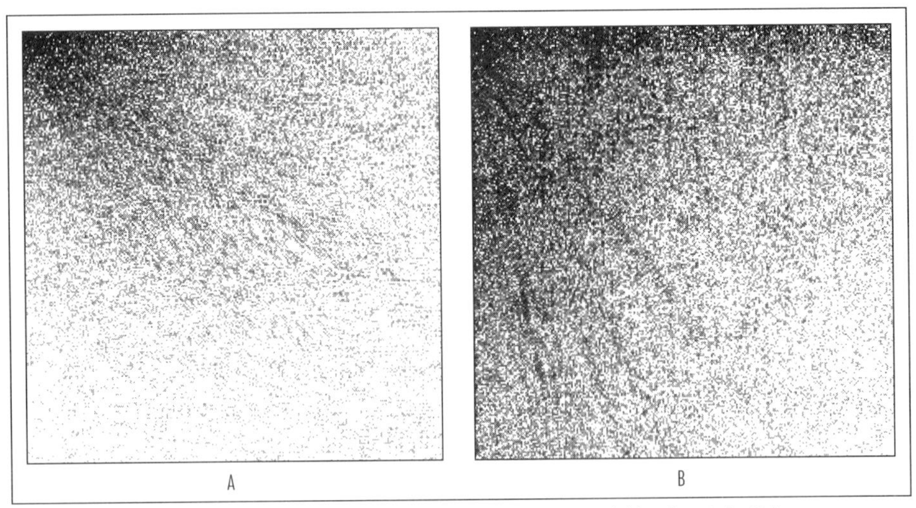

FIGURE 9.2.1 *Images in the Fourier transform frequency domain (provided by Zhiwu Lu): (A)* Lena *(256×256); (B)* Plava Laguna *(256×256)*

FIGURE 9.2.2 *Images recovered using a quarter (128×128) of frequency domain data (provided by Zhiwu Lu): (A)* Lena *(256×256); (B)* Plava Laguna *(256×256)*

where α is a normalization factor. Figure 9.2.1 shows the images *Lena* and *Plava Laguna* in their frequency domains. It is interesting to see that the images have high amplitude in low frequencies, and near-zero amplitude in high frequencies. If three-fourths of the frequency image data is set to zero (except the top left corner) the recovered images, shown in Figure 9.2.2, look perceptually very good. Certainly, there are few spots where reduction in image quality is visible, caused by the missing high-frequency data. Thus, instead of keeping all high-frequency information, it would suffice to keep some localized high-frequency information just for those few spots as a small amount of additional data.

This observation really inspires and outlines a new approach to digital image compression. There are two major steps in this approach: to identify a transformation that is *perfectly reversible*, has *packed energy*, and is *easy to compute*; and to build a structure to store additional localized higher-frequency-domain data for designated spots.

9.2.2 HAAR SUBBAND DECOMPOSITION

Could there be a better transform than the Fourier transform for the digital application? The Walsh–Hadamard and DCT transforms discussed in Chapter 6 are clearly good candidates.

We revisit the Walsh–Hadamard transform in a new variable order. For example, when $N = 8$,

$$H_8 = \frac{1}{\sqrt{8}}\begin{pmatrix} 1 & 1 & 1 & 1 & 1 & 1 & 1 & 1 \\ 1 & -1 & 1 & -1 & 1 & -1 & 1 & -1 \\ 1 & 1 & -1 & -1 & 1 & 1 & -1 & -1 \\ 1 & -1 & -1 & 1 & 1 & -1 & -1 & 1 \\ 1 & 1 & 1 & 1 & -1 & -1 & -1 & -1 \\ 1 & -1 & 1 & -1 & -1 & 1 & -1 & 1 \\ 1 & 1 & -1 & -1 & -1 & -1 & 1 & 1 \\ 1 & -1 & -1 & 1 & -1 & 1 & 1 & -1 \end{pmatrix}. \qquad (9.2.14)$$

Thus in general, for $N = 2^n$ the *Walsh–Hadamard transform* of a given digital image P of dimension $N \times N$ has the following expression:

$$\tilde{P}[p,q] = H_N P H_N[p,q] = \frac{1}{N} \cdot \sum_{k=0}^{N} \sum_{h=0}^{N} (-1)^{<\bar{p},h>} (-1)^{<\bar{q},k>} P[h,k], \quad (9.2.15)$$

where $<\bar{p}, h>$ and $<\bar{q}, k>$ are notations defined by the power of 2 expressions shown below. Assuming that

$$p = p_0 + 2\,p_1 + \cdots + 2^{n-1}\,p_{n-1} \text{ and } q = q_0 + 2\,q_1 + \cdots + 2^{n-1}\,q_{n-1}, \quad (9.2.15)$$

for some $p_0, p_1, \cdots, p_{n-1}, q_0, q_1, \cdots, q_{n-1} \in \{0,1\}$, then

$$\bar{p} = p_{n-1} + 2\,p_{n-2} + \cdots + 2^{n-1}\,p_0, \quad (9.2.16)$$

$$< p,q > = p_0 q_0 + p_1\,q_1 + \cdots + p_{n-1} q_{n-1}. \quad (9.2.17)$$

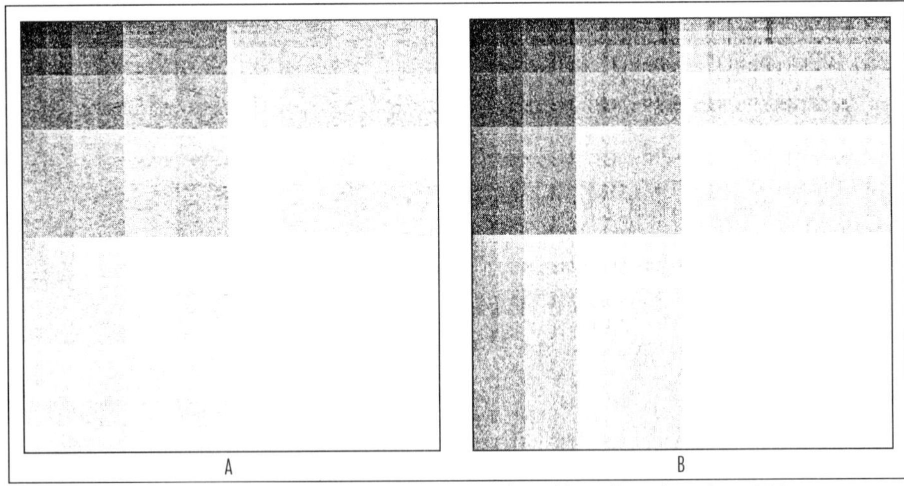

FIGURE 9.2.3 *Images in their frequency domain given by the Walsh–Hadamard transform: (A)* Lena; *(B)* Plava Laguna Hotel

Figure 9.2.3 shows that the Walsh–Hademard transform has a similar energy distribution in the frequency domain as that of the Fourier transformation. The nice thing about this transform is that it can be very easily computed.

For simplicity, from now on the image P will be assumed to have a dimension of $2^n \times 2^n$. Surely, an image can always be embedded into a larger support whose dimension has the required form.

It is not difficult to see that the transform is symmetrically defined in both variables and that it can be decomposed into a commutative pair of horizontal and vertical transforms. In fact, if we define

$$H(P)[p,q] = \frac{1}{\sqrt{N}} \sum_{h=0}^{N} (-1)^{<\bar{p},h>} P[h,q],$$

$$V(P)[p,q] = \frac{1}{\sqrt{N}} \sum_{k=0}^{N} (-1)^{<\bar{q},k>} P[p,k] \tag{9.2.18}$$

to be the horizontal and vertical transforms, respectively, it is easy to verify that

$$\tilde{P}[p,q] = V(H(P))[p,q] = H(V(P))[p,q]. \tag{9.2.19}$$

Therefore, in this case, only the 1-dimensional transform needs to be studied:

$$\tilde{P}[p] = \frac{1}{\sqrt{N}} \sum_{k=0}^{N} (-1)^{<\bar{p},k>} P[k]. \tag{9.2.20}$$

Next, a new look at this transform will be presented with additional insights. This transform was first implemented for fast calculation.

Consider the following pair of sampling filters $h=(h_0, h_1)$, the so-called *Haar filter*, as defined below:

$$h_0(x)=\begin{cases}\frac{1}{\sqrt{2}} & \text{for } -\frac{3}{2}\le x<-\frac{1}{2};\\ \frac{1}{\sqrt{2}} & \text{for } -\frac{1}{2}\le x<\frac{1}{2};\\ 0 & \text{otherwise.}\end{cases}\qquad h_1(x)=\begin{cases}\frac{-1}{\sqrt{2}} & \text{for } -\frac{3}{2}\le x<-\frac{1}{2};\\ \frac{1}{\sqrt{2}} & \text{for } -\frac{1}{2}\le x<\frac{1}{2};\\ 0 & \text{otherwise.}\end{cases}\qquad (9.2.21)$$

For a given digital image data string P of length 2^n, the sampling filtered strings are defined by convolution:

$$(h_0 * P)[k]=\frac{P[k]+P[k+1]}{\sqrt{2}} \text{ and } (h_1 * P)[k]=\frac{P[k]-P[k+1]}{\sqrt{2}}. \qquad (9.2.22)$$

Hence, digitally the filter is completely defined by the *kernels* of the filters: $h_0=\sqrt{2}/2\,(1,1,0)$ and $h_1=\sqrt{2}/2\,(-1,1,0)$ with a normalization factor. In the kernel expressions, zeros are added so that $h_0[0]$ and $h_1[0]$ are in the centers.

Doing a 2-to-1 down-sampling of the filtered strings:

$$P_L[k]=(h_0 * P)[2k],\quad P_H[k]=(h_1 * P)[2k], \qquad (9.2.23)$$

for all $k=0, 1, \cdots, 2^{n-1}-1$, respectively, a string of length 2^n has been converted into two strings each of length 2^{n-1}. The string P_L is called the *low pass filtered string* of P, or simply the *low band* of P; and the string P_H is called the *high pass filtered string* of P, or similarly, the *high band* of P.

Joining these two strings together yields

$$h(P)[k]=P_L[k] \text{ and } h(P)[2^{n-1}+k]=P_H[k], \qquad (9.2.24)$$

for all $k=0, 1, \cdots, 2^{n-1}-1$, respectively. The string $h(P)$ is called the *subband decomposition* of the two-channel filter $h=(h_0, h_1)$ of the image string P. The subband decomposition of the Haar filter is called *Haar decomposition.*

Repeat the same procedure for each of these bands and there will be twice as many bands of half the length (which is equivalent to decomposing the string $h(P)$ again in our current example). When the band length is finally equal to 1, the string obtained is exactly the Walsh–Hadamard transform of P.

THEOREM:
$h^n(P)=\tilde{P}$, i.e., the Walsh–Hadamard transform is equivalent to Haar decomposition.

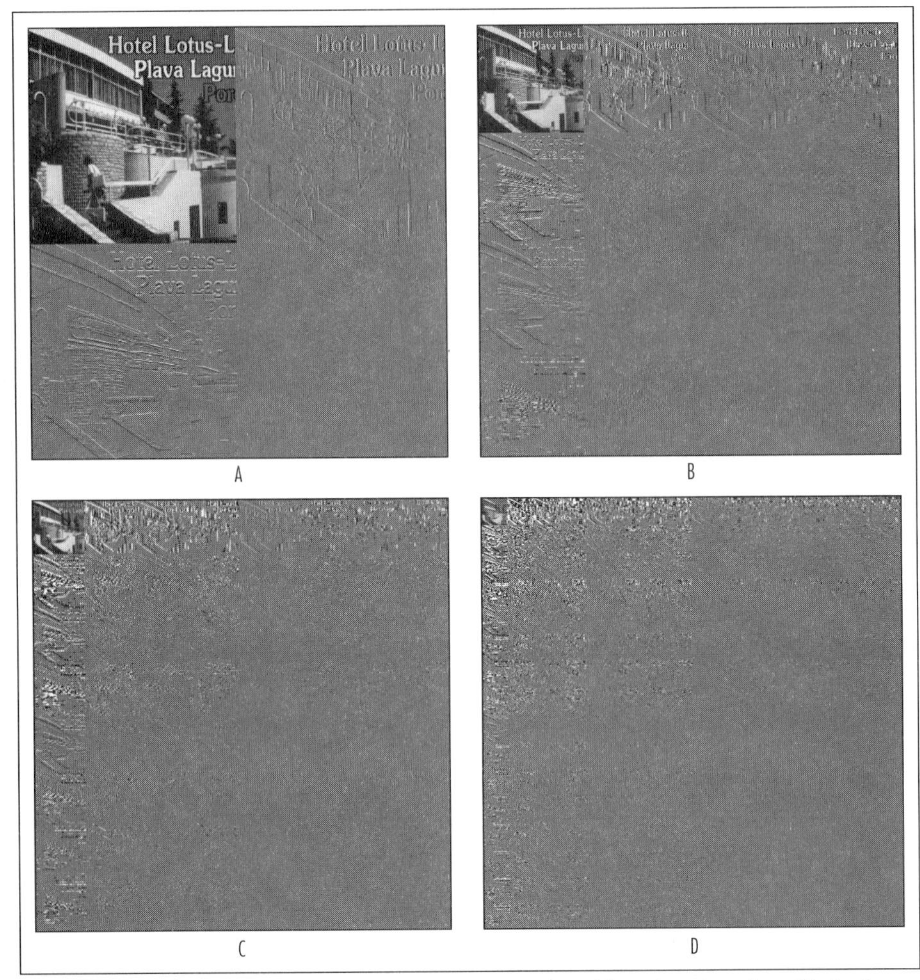

FIGURE 9.2.4 Subband decomposition of Plava Laguna using the Walsh–Hadamard transform: (A) after 1 step; (B) after 2 steps; (C) after 3 steps; (D) after 4 steps

In general, applying the same two-channel filter $h = (h_0, h_1)$ to a digital image, by filtering all rows horizontally first and then all columns vertically, the subband decomposed image will have four subbands: *low–low*, *high–high*, *vertically low–horizontally high*, and *vertically high–horizontally low*. The theorem remains true.

The proof is straightforward and will be left as an exercise. Figure 9.2.4 illustrates such a process.

Instead of decomposing every band in every step, the decomposition can stop if no further energy concentration occurs. By so doing the computation

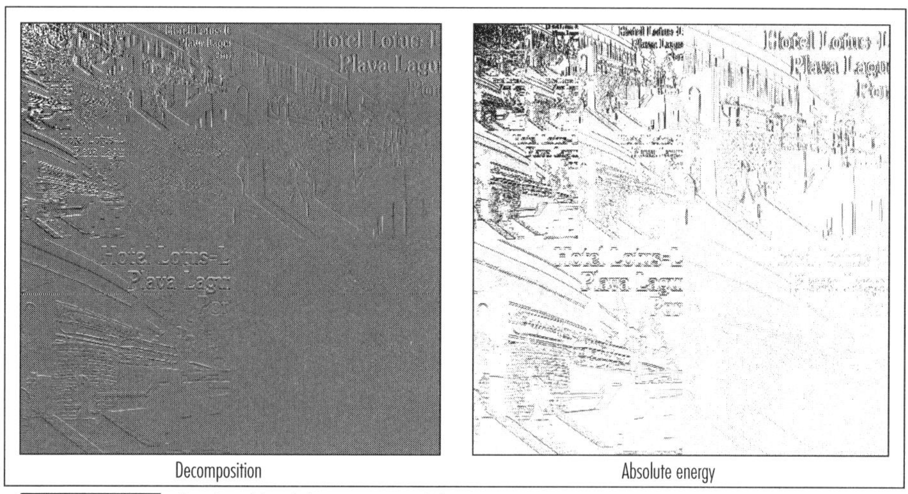

FIGURE 9.2.5 *Dyadic subband decomposition of* Plava Laguna

can be tremendously reduced. In the example of the Walsh–Hadamard transform (in every step), most of the energy is expected to be passed to the low band, and no significant energy will remain for the high band. Figure 9.2.4 also shows the energy redistribution of the image *Plava Laguna*. Similar results are generated for all testing images. In conclusion, only the low band needs to be decomposed.

As illustrated in Figure 9.2.5, the decomposition that decomposes only the *straight–low* band is called a *dyadic subband decomposition* of the two-channel filter $h=(h_0, h_1)$. In comparison to the complete subband decomposition, the computation drops from $2^n n$ entry calculations to 2^{n+1} for a 1-dimensional string and from $4^n n$ entry calculations to $4^{n+1}/3$ for a 2-dimensional image.

The other advantage of the dyadic subband decomposition is that the higher decomposition bands stop being filtered in the earlier steps, and thus the operations involved remain quite local—which can be used exactly as the localized high frequencies as the additional data when these frequencies have significant non-zero values. Therefore, the second part of the approach is processed effortlessly as well.

9.2.3 BIORTHOGONAL QUADRATURE FILTER

Given any *two-channel quadrature filter* (QF), $h=(h_0, h_1)$, which is a pair of finitely supported sequences centered around zero, and given any image data string P, the decomposition described in the previous subsection can be performed as a combination of convolution and decimation as formulated thus:

$$P_L[k] = (h_0 * P)[2k] = \sum_p h_0[p]P[2k-p],$$
$$P_H[k] = (h_1 * P)[2k] = \sum_p h_1[p]P[2k-p]. \qquad (9.2.25)$$

And $h(P)=(P_L, P_H)$, where h_0 is called the *low-pass*, and h_1 is called the *high-pass*. Because the high-pass should reduce the energy of a data string, applying it to the constant string (just as in the discussion of AC components in Chapter 6) the sum of all entries should be zero. Consequently, the following *normalization condition* will be assumed for all QFs:

$$\sum_k h_0[k] = \sqrt{2}, \text{ and } \sum_k h_1[k] = 0 . \qquad (9.2.26)$$

Furthermore, because the decomposition is useless unless the initial data can be recovered, the two-channel QFs—whose inverse processes can be established and the original data obtained—are the only filters we are interested in.

Given an image data string P, instead of viewing the string $h(P)$ as two joint bands P_L and P_H, from two subbands, an up-sampling two-channel filter $g=(g_0, g_1)$, is defined by

$$g(P_L, P_H)[k] = \sum_p g_0[k+2p]P_L[p] + \sum_p g_1[k+2p]P_H[p], \qquad (9.2.27)$$

as shown in Figure 9.2.6. A two-channel QF $h=(h_0, h_1)$ is said to be *reversible*, or *perfectly reconstructive* (*PR*), if there is an inverse up-sampling two-channel filter $g=(g_0, g_1)$ such that

$$g(h(P)) = P, \text{ for any image } P. \qquad (9.2.28)$$

For example, the Haar decomposition $h_0 = \sqrt{2}/2\,(1,1,0)$ and $h_1 = \sqrt{2}/2\,(-1,1,0)$ is PR. The inverse is given by $g_0 = \sqrt{2}/2\,(0,1,1)$ and

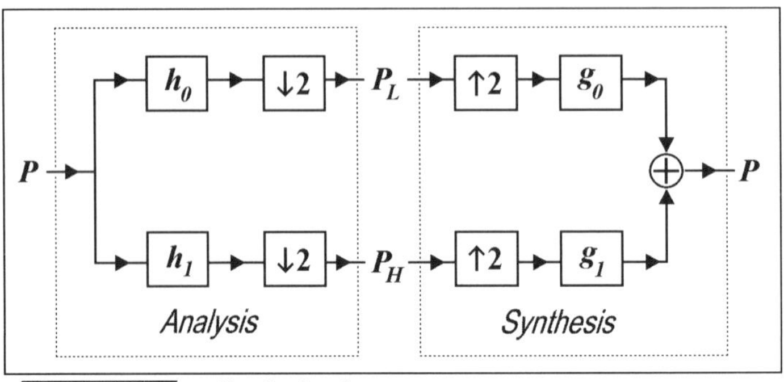

FIGURE 9.2.6 Subband coding diagram

$g_1 = \sqrt{2}/2\,(0,1,-1)$, or $g = \bar{h}$ in this case, where the *top-bar* notation is defined by $\bar{h}[k] = h[-k]$, for all k.

Here is another simple example: $h_0 = (1)$ and $h_1 = (0, 1, -1)$ define a PR quadrature filter, whose inverse is given by $h_0 = (1, 1, 0)$ and $g_1 = (-1, 0, 0)$.

In the compression system adapted by the RICOH group [ZASB], the quadrature filter is defined by

$$h_0 = \frac{\sqrt{2}}{2}(0,1,1) \text{ and } h_1 = \frac{\sqrt{2}}{16}(0,-1,-1,8,-8,1,1). \qquad (9.2.29)$$

It is easy to prove that

$$g_0 = \frac{\sqrt{2}}{2}(-1,1,0) \text{ and } g_1 = \frac{\sqrt{2}}{16}(-1,1,8,8,1,-1,0) \qquad (9.2.30)$$

form its inverse.

What is the condition for a two-channel QF to be PR?

Two two-channel filters $h = (h_0, h_1)$ and $g = (g_0, g_1)$ are said to be *biorthogonal* if

$$(\textit{Normalization}) \quad \sum_p h_0[p] = \sum_p g_0[p] = \sqrt{2}, \quad \sum_p h_1[p] = \sum_p g_1[p] = 0, \qquad (9.2.31)$$

$$(\textit{Duality}) \quad \sum_p g_0[p]h_0[2k-p] = \sum_p g_1[p]h_1[2k-p] = \delta[k], \qquad (9.2.32)$$

$$(\textit{Independence}) \quad \sum_p g_0[p]h_1[2k-p] = \sum_p g_1[p]h_0[2k-p] = 0, \qquad (9.2.33)$$

for all integer k.

THEOREM OF BQF:

A normalized perfect reconstructive two-channel quadrature filter $h = (h_0, h_1)$ is biorthogonal.

Proof: A useful way to denote a finite sequence is by a polynomial. A data string P will uniquely define the polynomial

$$P(z) = \sum_k P[k] \cdot z^{-k}, \qquad (9.2.34)$$

and vice versa. Similarly, the filters h_0 and h_1 can be defined in the same way as well:

$$\begin{aligned} H_0(z) &= \sum_k h_0[k] \cdot z^{-k}, \\ H_1(z) &= \sum_k h_1[k] \cdot z^{-k}. \end{aligned} \tag{9.2.35}$$

This polynomial expression is called a *Z-transform*. The nice thing about this notation is that the convolution of $h_i * P$ is given by the multiplication polynomial $H_i(z)P(z)$, for $i = 0$ and 1.

The low-pass filtering sampling of even terms is equivalent to nullifying the odd powers of z:

$$P_L(z^2) = \frac{1}{2}\left(H_0(z)P(z) + H_0(-z)P(-z)\right). \tag{9.2.36}$$

And similarly,

$$P_H(z^2) = \frac{1}{2}\left(H_1(z)P(z) + H_1(-z)P(-z)\right). \tag{9.2.37}$$

And the polynomial for $g(h(P))$ is

$$\begin{aligned} g(h(P))(z) &= G_0(z)P_L(z^2) + G_1(z)P_H(z^2) \\ &= \frac{1}{2}\left[G_0(z)H_0(z) + G_1(z)H_1(z)\right]P(z) \\ &\quad + \frac{1}{2}\left[G_0(z)H_0(-z) + G_1(z)H_1(-z)\right]P(-z) \\ &= T(z)P(z) + S(z)P(-z). \end{aligned} \tag{9.2.38}$$

Hence, for the filter $g = (g_0, g_1)$ to be the inverse is equivalent to

$$\begin{aligned} S(z) &= 0 \\ T(z) &= 1. \end{aligned} \tag{9.2.39}$$

The condition $S(z) = 0$, called the *alias-free condition*, implies that

$$\begin{aligned} G_0(z) &= F(z)H_1(-z) \\ G_1(z) &= -F(z)H_0(-z), \end{aligned} \tag{9.2.40}$$

for a scaling factor $F(z)$. From $T(z) = 1$, this scaling polynomial must be

$$F(z) = \frac{2}{H_0(z)H_1(-z) - H_0(-z)H_1(z)}. \tag{9.2.41}$$

Thus,

$$F(z) = -F(-z). \tag{9.2.42}$$

Consequently, $T(z) = 1$ is equivalent to

$$T(z) = \frac{1}{2}\big(G_0(z)H_0(z) + G_0(-z)H_0(-z)\big) = 1, \tag{9.2.43}$$

which is called the *duality condition*. This implies that the polynomial $Q(z) = G_0(z)H_0(z)$ must have no nontrivial even terms, i.e., it must have the form

$$Q(z) = 1 + \sum_p q_{2p+1} z^{-(2p+1)}, \tag{9.2.45}$$

which is equivalent to the convolution expression

$$\sum_p g_0[p]h_0[2k - p] = \delta[k], \tag{9.2.46}$$

which is the first half of (9.2.32). Substituting a different pair, G_1 and H_1, we have the other half of (9.2.32).

From (9.2.40) and (9.2.42) it is obvious that the polynomials

$$\begin{aligned} G_0(z)H_1(z) &= F(z)H_1(-z)H_1(z), \\ G_1(z)H_0(z) &= -F(z)H_0(-z)H_0(z) \end{aligned} \tag{9.2.47}$$

are both odd polynomials, which implies formula (9.2.33). In conclusion, a normalized PR QF must be biorthogonal. ♦

A two-channel QF $h = (h_0, h_1)$ is said to be *orthonormal* if $h = (h_0, h_1)$ and $\bar{h} = (\bar{h}_0, \bar{h}_1)$ are biorthogonal, where $\bar{h}[k] = h[-k]$, for all k.

THEOREM OF OQF:

If an orthonormal two-channel quadrature filter $h = (h_0, h_1)$ is perfect reconstructive, then

$$h_1[k - 2n - 1] = (-1)^k h_0[-k], \tag{9.2.48}$$

for all k, and for some integer n. With a shift, the following is always true:

$$h_1[1 - k] = (-1)^k h_0[k]. \tag{9.2.49}$$

Proof: Since $g = \bar{h}$ implies that

$$G_0(z) = H_1(\frac{1}{z}) \text{ and } G_1(z) = H_0(\frac{1}{z}), \tag{9.2.50}$$

substitute them into the first equation of (9.2.40) with $z \leftarrow u$ and into the second equation with $z \leftarrow -\frac{1}{u}$. Then multiply both sides of these two equations to obtain

$$F(u)F(-\frac{1}{u}) = -1, \tag{9.2.51}$$

which holds only if $F(z)$ is an odd monomial, i.e., $F(z) = z^{2n+1}$. Again use formula (9.2.40), and the first part of the theorem is done.

Shifting the high-pass left or right by 2, the PR property remains unchanged, while $F(z)$gets a factor of $z^{\pm 2}$, respectively. ♦

As a consequence, the support of an orthonormal QF whose low-pass and high-pass have the same finite support must contain an even number of points—which is actually true in most practical applications.

9.2.4 MRA AND WAVELETS

Another way to construct a subband decomposition filter is to use wavelets.

A *multiresolution analysis* (*MRA*) of $\mathbf{L}^2(\Re)$ is a chain of subspaces $\{\mathbf{V}_j\}_{-\infty<j<\infty}$ satisfying the following conditions:

1. *Containment:* $\mathbf{V}_j \subset \mathbf{V}_{j-1} \subset \mathbf{L}^2(\Re)$, for all j.
2. *Intersection:* the intersection $\bigcap_j \mathbf{V}_j = 0$.
3. *Span:* the closure of the union $\overline{\bigcup_j \mathbf{V}_j} = \mathbf{L}^2(\Re)$.
4. *Dilation:* $\theta(2t) \in \mathbf{V}_{j-1} \quad \Leftrightarrow \quad \theta(t) \in \mathbf{V}_j$.
5. *Generator:* There is a function $\phi \in \mathbf{V}_0$ with compact support such that the collection of its integer translates $\{\phi^k(t) = \phi(t-k)\}_{-\infty<t<\infty}$ is a *Riesz base* of the space $\mathbf{V}_0$, i.e., it is linearly independent and spans the space $\mathbf{V}_0$.

Note that it is unnecessary to require the function ϕ to have compact support in the above definition. However, only compact cases are of interest in all applications in this book, and by adding this condition some of the later discussions become simpler.

From conditions (4) and (5), the collection of scaled translates $\left\{\phi_j^k(t)=\phi(2^{-j}t-k)\right\}_{-\infty<t<\infty}$ is a Riesz base for the space $\mathbf{V}_j$, for all j.

Clearly, the function $\phi\in\mathbf{V}_0$, called the *scaling function* of the MRA, completely determines the MRA. One may assume without loss that $\|\phi\|=1$. Hence, an MRA is said to be *orthogonal* if $\left\{\phi^k(t)=\phi(t-k)\right\}_{-\infty<t<\infty}$ is an orthonormal base. That is,

$$<\phi^k,\phi^h>=\delta(k-h),\ \text{for all}\ -\infty<k,h<\infty. \tag{9.2.52}$$

This definition implies that $\phi\in\mathbf{V}_0\subset\mathbf{V}_{-1}=\text{span}\left\{\phi_{-1}^k\right\}$, which yields the following expression:

$$\phi(t)=\sqrt{2}\sum_{-\infty<k<\infty}h[k]\phi(2t-k), \tag{9.2.53}$$

for some square-summable sequence $\left\{h[k]\right\}_{-\infty<k<\infty}$.

Define

$$\psi(t)=\sqrt{2}\sum_{-\infty<k<\infty}(-1)^k h[1-k]\phi(2t-k), \tag{9.2.54}$$

which is called the *mother wavelet.*

If the MRA associated to ϕ is orthogonal, then clearly, the scaling function and the mother wavelet are orthogonal; that is, $<\phi,\psi>=0$.

Exercise: If the MRA is orthogonal, the collection of wavelets $\left\{\psi^k(t)=\psi(t-k)\right\}_{-\infty<t<\infty}$ is an orthonormal basis of some space $\mathbf{W}_0=\text{span}\left\{\psi^k\right\}\subset\mathbf{V}_{-1}$, Moreover,

$$\mathbf{V}_{-1}=\mathbf{V}_0\oplus\mathbf{W}_0. \tag{9.2.55}$$

Repeat this exercise with the dilation equation, and it is obvious that the whole function space $\mathbf{L}^2(\Re)$ has a *wavelet decomposition*

$$\mathbf{L}^2(\Re)=\bigoplus_{k=-\infty}^{\infty}\mathbf{W}_k \tag{9.2.56}$$

given by the wavelet ψ.

In general, given an orthogonal MRA (ϕ,ψ), a PR quadrature filter (h_0,h_1) is naturally defined as

$$h_0[k] = h[k] \quad \text{and} \quad h_1[k] = (-1)^k h[1-k], \tag{9.2.57}$$

for all k, where $\{h[k]\}_{-\infty<k<\infty}$ is the sequence from (9.2.53).

9.2.5 FRACTAL INTERPOLATION OF QUADRATURE FILTERS

The previous paragraph shows how to get a PR orthonormal QF from an orthogonal MRA. The inverse is true as well. The best way to construct the scaling function from a quadrature filter is to use an iterated function system (IFS) interpolation. As a result, the graph of any orthogonal wavelet scaling function is a fractal.

Given an orthogonal QF (h_0,h_1), from (9.2.49) we can assume the equations in (9.2.57). Thus, if an operator H is defined from expression (9.2.53) as

$$H(\theta)(t) = \sqrt{2} \sum_{-\infty<k<\infty} h[k]\theta(2t-k), \tag{9.2.58}$$

then the scaling function ϕ is a normalized fixed point of the operator H, i.e., $H(\phi) = \phi$, and $\|\phi\| = 1$.

On the other hand, if ϕ is a normalized fixed point of the operator H associated to an orthogonal QF, then clearly an orthogonal MRA (ϕ,ψ) can be defined by repeating the process in the previous subsection.

As discussed in Wickerhauser's book [W2] and Reissell's note [R], if we apply the Fourier transform to both sides of equation (9.2.53), then we obtain

$$\hat{\phi}(\xi) = \frac{\sqrt{2}}{2} \hat{h}(\frac{\xi}{2}) \hat{\phi}(\frac{\xi}{2}), \tag{9.2.59}$$

where $\hat{h}(\xi) = \sum_k h[k] e^{-2\pi i k\xi}$ is as defined for the discrete case.

And if we continue applying the same procedure, we obtain

$$\hat{\phi}(\xi) = 2^{-\frac{L}{2}} \hat{h}(\frac{\xi}{2})\hat{h}(\frac{\xi}{4}) \cdots \hat{h}(\frac{\xi}{2^L}) \hat{\phi}(\frac{\xi}{2^L}). \tag{9.2.60}$$

The normalization conditions of the scaling function and the QF implies that $\hat{\phi}(0) = 1$ and $\hat{h}(0) = \sqrt{2}$. Thus,

$$\hat{\phi}(\xi) = \prod_{k=1}^{\infty} \frac{\sqrt{2}}{2} \hat{h}(\frac{\xi}{2^k}). \tag{9.2.61}$$

By using expression (9.2.53), the values of the scaling function ϕ at all dyadic points $k/2^n$, where $n > 0$ and k is an arbitrary integer, can be computed if the values of ϕ at integers are known. Thus, the graph of ϕ can be completely drawn iteratively. For this reason, expression (9.2.53) is also called the *refinement equation*.

There are two standard methods to calculate the values of ϕ at integers: the *cascade iterative algorithm*, and the *eigenvector calculation*.

The cascade algorithm is purely fractal. We set the initial string:

$$f_0 = \left\{ f_0[k] = \delta(k) \right\}_{-\infty<k<\infty} = \left\{ \cdots, 0, 0, 0, 1, 0, 0, 0, \cdots \right\}. \tag{9.2.62}$$

Then we iterate:

$$f_0, f_1 = H(f_0), \cdots, f_{i+1} = H(f_i), \cdots \quad \to \quad \left\{ \phi[k] \right\}_{-\infty<k<\infty}. \tag{9.2.63}$$

The other method is algebraic. If the filter length is N, the integer values of ϕ, $\{\phi[k]\}_{-\infty<k<\infty}$, form the eigenvector corresponding to the eigenvalue 1 of the $(N-1)\times(N-1)$ matrix $\{\sqrt{2}\ h[j-2i]\}_{0<i,j<N}$. This can be seen easily by substituting integers 0, 1, $\cdots$ for t into the refinement equation (9.2.53).

As an example, the Daubechies D4 scaling function is illustrated in Figure 9.2.7. A more detailed IFS construction of the wavelet scaling function can be found in Geronimo's lecture [GL]. As a direct construction of IFS described in Chapter 2, Daubechies D4 scaling function has the support $[0, 3]$ and can be described as three IFS pieces—one for each unit interval:

$$W_i = W_{[y_i, z_i, y_{i+1}]}, \tag{9.2.64}$$

for $i=0, 1, 2$, where

$$y = \left(\phi(0), \phi(1), \phi(2), \phi(3) \right) = \left(0, 1, \sqrt{3} - 2, 0 \right), \tag{9.2.65}$$

$$z = \left(\phi(\tfrac{1}{2}), \phi(\tfrac{3}{2}), \phi(\tfrac{5}{2}) \right) = \left(\frac{\sqrt{3}+1}{4}, 0, \frac{3\sqrt{3}-5}{4} \right). \tag{9.2.66}$$

Each IFS $W_i = W_{[y_i, z_i, y_{i+1}]}$ consists of two transformations:

$$w_i^0 \begin{pmatrix} x \\ y \end{pmatrix} = \frac{1}{4} \begin{pmatrix} 2 & 0 \\ 4(z_i - y_i) + (1+\sqrt{3})(y_i - y_{i+1}) & 1+\sqrt{3} \end{pmatrix} \begin{pmatrix} x \\ y \end{pmatrix} + \begin{pmatrix} 0 \\ (3-\sqrt{3})y_i \end{pmatrix}, \tag{9.2.67}$$

$$w_i^1 \begin{pmatrix} x \\ y \end{pmatrix} = \frac{1}{4} \begin{pmatrix} 2 & 0 \\ 4(y_{i+1} - z_i) + (1-\sqrt{3})(y_i - y_{i+1}) & 1-\sqrt{3} \end{pmatrix} \begin{pmatrix} x \\ y \end{pmatrix} + \begin{pmatrix} 0 \\ 4z_i - (1-\sqrt{3})y_i \end{pmatrix}, \tag{9.2.68}$$

for $i=0, 1, 2$. Figure 9.2.7 was plotted using this set of three IFSs.

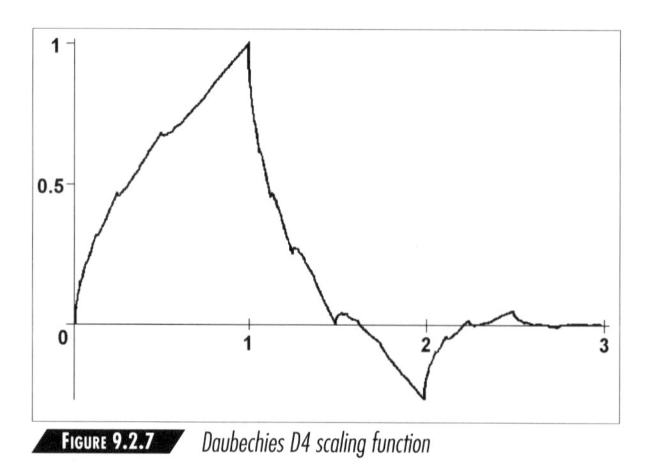

FIGURE 9.2.7 Daubechies D4 scaling function

9.2.6 SAMPLE PR QUADRATURE FILTERS

The following list of well-known PR QFs is simple enough for image coding. Most of them can be found in both Wickerhauser's book [W3] and Akansu and Haddad's book ([AH]). As we already knew, all the orthogonal filters are equivalent to orthogonal wavelets.

In all orthogonal QFs, only the low-pass expression is given. The high-pass can be deduced from expression (9.2.57). We will let `sr(k)` denote the *square root* of k.

1. *Daubechies orthonormal filters*
   For a given length, Daubechies filters maximize the smoothness of the associated scaling function.

Filter	Formula	Number
D2	`sr(2)`	0.7071067812
Haar	`sr(2)`	0.7071067812
	`((1+sr(3))/8)*sr(2)`	0.4829629131
D4	`((3+sr(3))/8)*sr(2)`	0.8365163037
	`((3-sr(3))/8)*sr(2)`	0.2241438680
	`((1-sr(3))/8)*sr(2)`	−0.1294095226
	`A = sq(2)*[sr(sr(40)+5)`	
	`+ sq(10)+1]/32`	0.3326705530
	`sr(2)/8+A*2-B`	0.8068915093
D6	`sr(2)*3/8-B*2`	0.4598775021
	`sr(2)*3/8-A*2`	−0.1350110200
	`sr(2)/8+B*2-A`	−0.0854412738
	`B = 3/(A*256)`	0.0352262919

2. *Coifman orthonormal filters*
   Coifman filters are designed so that both the scaling function and the mother wavelet will have vanishing moment.

Filter	Formula	Number
C6	((sr(15)- 3)/32)*sr(2)	0.0385807777
	((1-sr(15))/32)*sr(2)	−0.1269791254
	((3-sr(15))/16)*sr(2)	−0.0771615550
	((sr(15)+ 3)/16)*sr(2)	0.6074916414
	((sr(15)+13)/32)*sr(2)	0.7456875589
	((9-sr(15))/32)*sr(2)	0.2265842652

3. *More 6-tap orthonormal filters*
   There are more orthogonal filters that have been studied.

0.312656005	0.348319026	0.398655794	0.442766931
0.754045521	0.758774508	0.792728512	0.805049213
0.543768338	0.510327483	0.420459801	0.352529377
−0.108851490	−0.121232755	−0.141949922	−0.146445561
−0.149317562	−0.151539728	−0.112008814	−0.088189527
0.061912751	0.069565029	0.056328191	0.048503129

4. *Biorthogonal integer quadrature filters*
   In implementations, integer filters are often used. The following is a list of some of them.

	Filter entries									Normalization
Low-pass	1	2	1							2*sr(2)
High-pass	−1	2	−1							2*sr(2)
Low-pass	1	3	3	1						4*sr(2)
High-pass	−1	3	−3	−1						4*sr(2)
Low-pass	2	6	3	−1						5*sr(2)
High-pass	1	3	−6	2						5*sr(2)
Low-pass	−1	2	6	2	−1					4*sr(2)
High-pass	−1	−2	6	−2	−1					4*sr(2)
Low-pass	−1	1	8	8	1	−1				8*sr(2)
High-pass	−1	−1	8	−8	1	1				8*sr(2)
Low-pass	3	−9	−7	45	45	−7	−9	3		32*sr(2)
High-pass	−3	−9	7	45	−45	−7	9	3		32*sr(2)
Low-pass	3	−6	−16	38	90	38	−16	−6	3	64*sr(2)
High-pass	3	6	−16	−38	90	−38	−16	6	3	64*sr(2)

5. *Other PR filters*
There are some other perfect reconstructive quadrature filters that have been used in high-performance image coding. The following 9-tap filter is a good example.

Forward Low-pass	Forward High-pass	Inverse Low-pass	Inverse High-pass
0.037829	0	0.037829	0
−0.023849	0.064539	−0.064539	0.023849
−0.110624	−0.040690	−0.110624	−0.040690
0.377403	−0.418092	0.418092	−0.377403
0.852699	0.788485	0.852699	0.788485
0.377403	−0.418092	0.418092	−0.377403
−0.110624	−0.040690	−0.110624	−0.040690
−0.023849	0.064539	−0.064539	0.023849
0.037829	0	0.037829	0

9.3 WAVELET IMAGE COMPRESSION

Subband decomposition, using a local wavelet quadrature filter, transforms images from their real viewing space to an equivalent new spectrum variable space. The local image redundancies then become apparent for extraction. In this section the exact algorithm for extracting and coding will be described. Meanwhile, the natural relationship between this scheme and the fractal image compression will be explored.

9.3.1 ZERO TREE OF WAVELET CODING

A wavelet subband compression algorithm consists of two major components: the image data decorrelating transform and the data symbol entropy coding. In addition to these two steps, a data symbol processing step is often inserted between them. It regroups data symbols based on some quantization so that it achieves the desired lossy compression.

The previous section shows how to design a wavelet image data decorrelating transform using a quadrature filter dyadic decomposition. For example, applying the Haar filter to the sample image piece B in (6.3.17) as shown in Chapter 6,

$$B=\begin{pmatrix}139&144&149&153&155&155&155&155\\144&151&153&156&159&156&156&156\\150&155&160&163&158&156&156&156\\159&161&162&160&160&159&159&159\\159&160&161&162&162&155&155&155\\161&161&161&161&160&157&157&157\\162&162&161&163&162&157&157&157\\162&162&161&161&163&158&158&158\end{pmatrix}. \tag{9.3.1}$$

Thus,

$$H(B)=\begin{pmatrix}1259.6&0.1&-13.2&1.5&-6.0&-3.5&1.5&0\\-17.4&-12.9&-0.5&5.0&-3.5&-0.5&1.5&0\\-20.2&4.0&-3.2&0&-0.5&-0.5&5.0&0\\-2.0&-3.0&-1.5&0&0&-1.0&5.0&0\\-6.0&-3.5&-2.5&-1.0&1.0&-0.5&-1.5&0\\-7.5&0.5&-2.5&-3.0&-1.5&-2.5&0.5&0\\-1.5&0.5&0&-2.0&-0.5&-0.5&2.0&0\\0&1.0&1.0&-1.0&0&-1.0&0&0\end{pmatrix}. \tag{9.3.2}$$

The next step is quantization. If we set a threshold, e.g., 9, the block becomes

$$Q(H(B))=9\cdot\begin{pmatrix}140&0&-1&0&-1&0&0&0\\-2&-1&0&1&0&0&0&0\\-2&0&0&0&0&0&1&0\\0&0&0&0&0&0&1&0\\-1&0&0&0&0&0&0&0\\-1&0&0&0&0&0&0&0\\0&0&0&0&0&0&0&0\\0&0&0&0&0&0&0&0\end{pmatrix}. \tag{9.3.3}$$

Thus, the recovered one will be

$$B'=\begin{pmatrix}138&147&152&152&156&156&156&156\\147&156&152&152&156&156&156&156\\152&152&161&161&156&156&156&156\\161&161&161&161&156&156&156&156\\161&161&161&161&165&156&156&156\\161&161&161&161&165&156&156&156\\161&161&161&161&165&156&156&156\\161&161&161&161&165&156&156&156\end{pmatrix}. \tag{9.3.4}$$

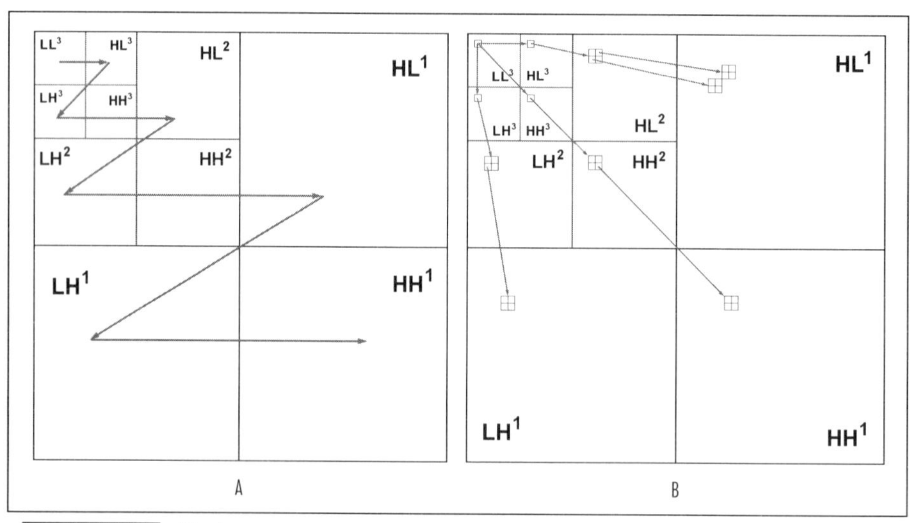

FIGURE 9.3.1 *Wavelet zero tree*

Finally, formula (9.3.3) needs to be coded. What is the most efficient order in which to code these entries? Shapiro provided an answer by coding them in a tree order, which he called an *embedded zero tree* [S2], and which, (as shown in the next subsection) is exactly a refinement of the quadtree previously discussed in Chapter 4.

In a dyadic subband decomposition, let **R** denote all pixel coordinates of the root low–low image. These coordinates are defined to be the *roots* of the *zero tree*. The root image **R** is also called the *highest layer* image.

Let w and h denote the width and height of the root image **R**. As shown in Figure 9.3.1*a*, for any coordinates of the form (i,j) of **R**, there are three *offspring*:

$$\mathbf{O}(i,j) = \{(i+w,j),(i,j+h),(i+w,j+h)\}, \qquad (9.3.5)$$

one from each of the high–low, low–high and high–high images of the last decomposition step. For any nonroot coordinates, i.e., $(i,j) \notin \mathbf{R}$, there are exactly four offspring:

$$\mathbf{O}(i,j) = \{(2i,2j),(2i,2j+1),(2i+1,2j),(2i+1,2j+1)\}, \qquad (9.3.6)$$

except for the coordinates of the form (i,j) such that the coordinate $(2i,2j)$ is off the image dimension, which are called *leaves* of the tree. Let **L** denote the set of leaves.

A *descendant* is an offspring, or an offspring of an offspring, etc. Mathematically, the descendant set $\mathbf{D}(i,j)$ is completely defined by the following two conditions:

$$\mathbf{O}(i,j) \subseteq \mathbf{D}(i,j) \text{ and } (k,h) \in \mathbf{D}(i,j) \setminus \mathbf{L} \Rightarrow \mathbf{O}(k,h) \subseteq \mathbf{D}(i,j). \qquad (9.3.7)$$

Similarly to the DCT coding in Chapter 6, the entries of the root image **R**, called the DC coefficients of the wavelet decomposition, are usually entropy coded by subtracting each entry from its proceeding one.

Again, most AC coefficients are expected to be zero. Mark a coordinate (i, j) to be EOB (*end of branch*) if all coefficients of its descendants are zero. AC coefficients that are not descendants of some EOB coordinates are coded in the scan order as shown in Figure 9.3.1*b*. The EOB symbols are coded immediately after the codes of their corresponding EOB coordinates.

The process of coding the sample image piece is illustrated in Figure 9.3.2. The AC coefficient symbol string of the sample image piece will be

```
0,-2,-1E; -1,0E,0E,1, -2,0E,0E,0E;

-1E,0E,0E,0E, 1E,0E,1E,0E, -1E,-1E,0E,0E.
```

The Haar decomposition will be chosen in our discussion for its simplicity. As an exercise, the reader could write a complete wavelet compression system using other sophisticated quadrature filters.

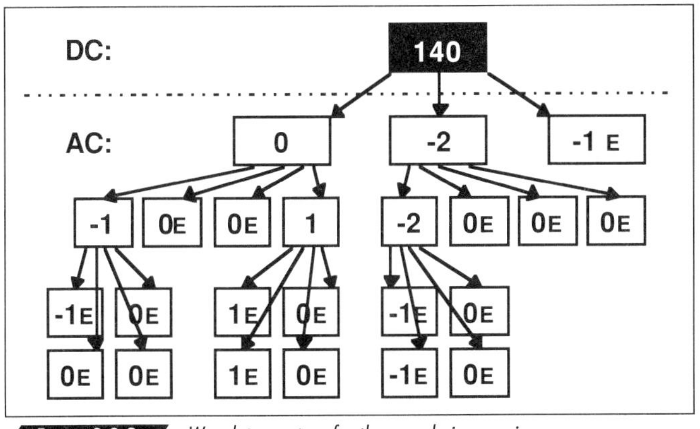

FIGURE 9.3.2 *Wavelet zero tree for the sample image piece*

9.3.2 GEOMETRY OF THE WAVELET COMPRESSION ALGORITHM

What advantage does a wavelet compression system have that other compression techniques do not? The essence of wavelets is variable manipulation. As a result, the gains are reflected in two geometric aspects.

The first gain is in decorrelating the *vertical*, *horizontal*, and *diagonal* frequencies of an image locally. Indeed, most image pieces have definite textures in one of the vertical, horizontal, or diagonal directions. Actually, when an image piece significantly involves all directions, it would be considered a noisy piece, where the higher distortion would be acceptable.

In each wavelet decomposition step, an image piece is decomposed into four subimages: the *low–low*, the *high–low*, the *low–high*, and the *high–high*

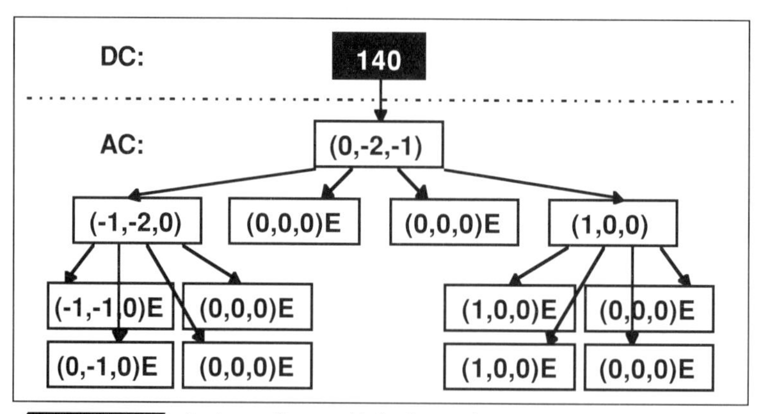

FIGURE 9.3.3 *Quadtree in Haar variable for the sample image piece*

images, which geometrically are exactly the *mean–like*, the *vertical frequency*, the *horizontal frequency*, and the *diagonal frequency* images. Thus, the zero tree is exactly the quadtree decomposed into three components: the *vertical*, the *horizontal*, and the *diagonal* components. Figure 9.3.3 shows the quadtree of the same sample image piece. Comparing this quadtree to the zero tree in Figure 9.3.2, the reason to treat these three components separately seems to be clear.

Therefore, each node in a quadtree is a square block. This square block corresponds to a triplet of three values (i.e., three nodes) in the zero-tree. Comparing Figure 9.3.2 and Figure 9.3.3, it is clear that these three zero-tree nodes must be in the same layer, rooted to the same pixel in the root image, and one value from each descendant set of the three offspring of this root pixel.

The second geometric gain derives from the delicate choice of wavelet. As discussed before, the Haar–Walsh–Hadamard decomposition is equivalent to the square image partition; its lossy image representations always give blocky images, which are visually intolerable. Choosing a wavelet with a long tail is equivalent to taking neighborhood pixel values into consideration.

Geometrically, a long-tapped wavelet is equivalent to an overlapping weighted image covering as described in Chapter 4. Algebraically, this wavelet is very well formulated for forward and backward calculation.

9.3.3 FRACTAL WAVELET COMPRESSION

Referring to Chapter 4, the classic fractal image compression was presented in quadtree structure (like the illustrations in Figure 9.3.3). If the reference blocks are required to be a subset of possible destination blocks, then a fractal spatially 2-to-1 block match is identical to a match of a pair of branches in the tree.

A branch is formed by all descendants of a node. In a Haar-based wavelet decomposition, a branch corresponds to a block described in Haar-based variables

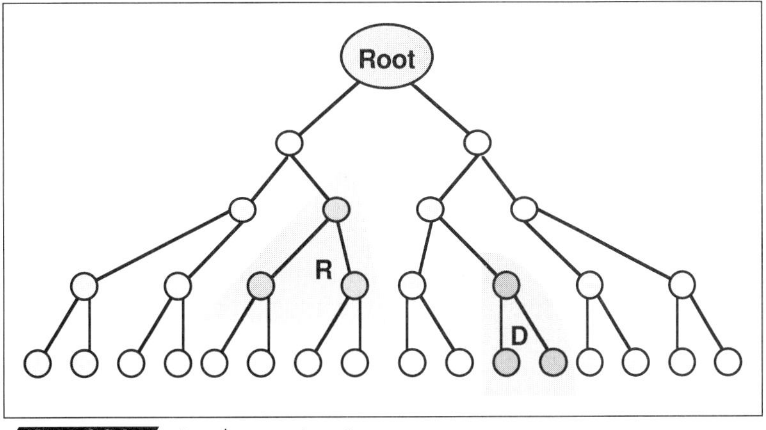

FIGURE 9.3.4 *Fractal compression using a tree*

(as discussed in Chapter 7). In this context, the destination-reference matching in classic fractal compression can be described as the matching of branches in the tree. As illustrated in Figure 9.3.4, a destination block is a tree branch, and a 2-to-1 spatial contractive reference block (if it is aligned to destination block boundaries) of the destination block is a branch starting from a node in one layer higher than the destination branch node.

Realizing this relation between a fractal compression scheme and wavelet compression scheme, Davis [D2] combined both techniques with a new algorithm. It preserves both geometric properties, discussed in the previous subsection, given by the wavelet method and allows the self-crossing reference offered by fractals as well.

Once both wavelet and fractal techniques are understood, the description of the algorithm is simple:

> *Begin from a zero tree of some wavelet decomposition. Consider one layer each time from top down, and skip the top two layers. For each node in the layer, view it as a potential destination node, and compare its descendant branch with all descendant branches of nodes in the layer that is one step higher, which are the potential reference branches. If the best match is found to be good enough, the destination branch is chopped off and is replaced with a reference to the best matching node.*

In Davis's paper, the performance of this scheme has been reported to be very similar to the zero-tree wavelet scheme. In his implementation, the reference branches are required to be a part of the zero tree; thus, the reference pool size is too small for the fractal part to find good matches. The value of this scheme is not in its performance but concerns the relation between the two main compression techniques shown.

How can we enlarge the reference pool without breaking the zero-tree structure?

9.3.4 FRACTAL NODES IN A WAVELET TREE

There is no reason to limit the reference of a destination branch to be a branch in the zero tree. Viewing the zero tree as a vector tree of the quadtree, a fractal code is allowed to be added at any node of the quadtree if it gives a better image representation (subject to a given distortion rate). By doing that, a unified system is established that gives a complete fractal system on one hand and a complete wavelet system on the other.

The algorithm can be precisely described in the following steps:

1. *Run a wavelet subband coding to the desired image quality.*
2. *Add a fractal code to each subband tree branch end.*
3. *Improve the fractal representation by merging fractal codes.*
4. *Slice the image data based on data resolution priorities.*
5. *Entropy pack the residual image as the last image slice if lossless compression is required.*

The advantages of hybridizing the wavelet and the fractal is not just for pure compression performance. In fact, there are two important features that are strongly demanded in today's digital imaging market: *progressiveness*, which is requested by the tremendous market of the World Wide Web, and *resolution independence*, which is a key requirement for desktop publishing.

As a matter of fact, there are technologies that can achieve either feature well but not both, until now. The wavelet subband coding technology has a natural progressive structure that has been proven by Said and Pearlman [SP]. And the fractal image representation gives a powerful resolution-independent image representation that serves for multi-usage. Our mixture gives an elegant answer to it: subband coding with fractal nodes.

From thumbnails to poster prints, the current market solution is to use a multi-resolution binding file, as in the case of Photo CD, which is not only clumsy but also unnecessary. Now the fractal wavelet imaging technique represents progressive resolutionless images once and for all. It pushes the resolution to infinity, far beyond what the subband wavelet method can achieve, and maintains a complete fractal description. As research moves in this direction, more advanced results will be discovered.

Interestingly enough, by now the links between fractal technology and other compression methods have been shown: VQ in Chapter 5, JPEG DCT in Chapter 6, and wavelet subband coding in this chapter. Among all of them, in practice, fractal and VQ are closer to each other because of their similar searching and matching techniques—the DCT transform and wavelet subband are more alike, since they both deal with the frequency domain and relate to the same piece of classical mathematics: the Fourier transform.

9.3.5. MORE FRACTAL WAVELET HYBRIDS

A few more interesting fractal wavelet hybrid schemes have been introduced. They reduce fractal searching complexity and improve coded image quality.

Cesbron and Malassenet [CM] use wavelet transforms (e.g., the Haar transform) to simplify fractal compression. Their algorithm first transforms an image into four quarter-sized subband images using a wavelet transform, and then compresses the low–low image using the standard fractal method and compresses the other three subband images by matching reference blocks from the low–low image. The art of implementing this type of algorithm is in how to balance between the code bit rates of the four decomposed subband images and how to share the image partitioning information across those subband images. Cesbron and Malassenet reported good performance in their paper.

Another interesting approach was patented by Berger ([B4]) of OLiVR. His idea is to compress textural residual images using fractal techniques. An image is decomposed into two orthogonal entities: the *coarse* part, which can use any preferred compression method (e.g., wavelet or JPEG DCT), and the *fine* part, which uses the classic fractal compression method. The extreme cases of this system are either totally nonfractal or purely fractal.

Given an original image O, as shown in Figure 9.3.5, it is decomposed into *prediction component* and *residual component* in two different resolution levels: *course level* and *fine level*. Thus, there are four images—*coarse prediction CP*, *coarse residual CR*, *fine prediction FP*, and *fine residual FR*—satisfying the following conditions:

$$\mathrm{O} = \mathrm{CP} + \mathrm{CR} = \mathrm{FP} + \mathrm{FR}. \tag{9.3.7}$$

We fractally code the fine residual image FR using the coarse residual image CR as the reference image. The compressed image code data are formed from the compressed description of the fine prediction image FP and the fractal codes of FR.

For example, we can choose FP to be a highly compressed JPEG or wavelet image, CP to be a bilinear interpolated image from mean values of 8×8 blocks or a coarse image of FP generated using some predetermined smoothing method. The images CR and FR are the residuals of the image O from images CP and FP respectively.

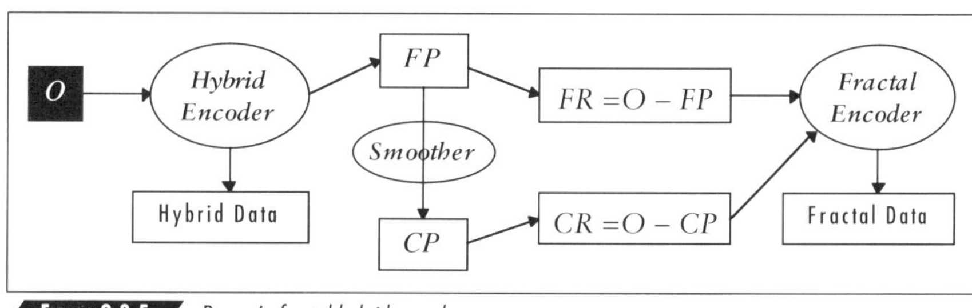

FIGURE 9.3.5 Berger's *fractal hybrid encoder*

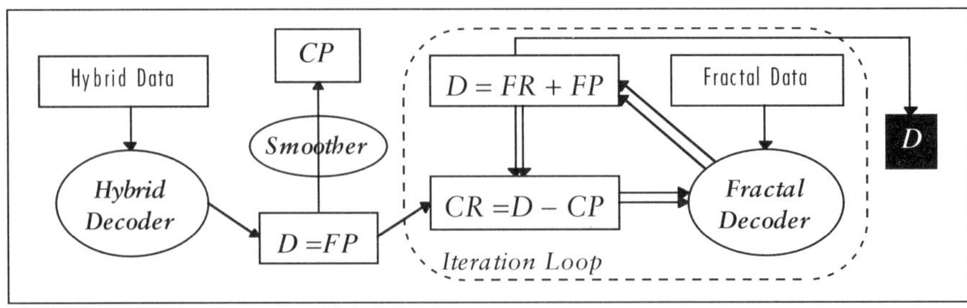

FIGURE 9.3.6 Berger's *fractal hybrid decoder*

The decoding algorithm is iterative as shown in Figure 9.3.6. The fine prediction *FP* is constructed first, and then the coarse prediction *CP* is generated using the same method as the encoder. Set the initial decoded image *D* to be *FP*. Then we iterate the following process:

- Create the coarse residual by differentiating images: $CR = D - CP$.
- Generate the fine residual *FR* from the fractal codes by referencing *CR*.
- Calculate the decoded image: $D = FP + FR$.

All the above ideas need further investigation. You may start your own new project from these.

How goes my comedy? Oh yes, it has moved forward a little. Still the same old story about my Ghosts, friendly Ghosts, brutal, mean, joyous, stupid, unbelievably stupid, kind, hot, warm, cold, inane, anxious Ghosts.

— Images

Ingmar Bergman

Images in Motion

10

The motion picture has evolved through a hundred-year period and has drastically altered our society—from the way people view themselves and others to the way they live. The stages have progressed through silent films, the advent of sound, the introduction of color, and finally, the brilliant use of "*computer technology*" to produce fantasies that could not have been realized through conventional cinematic techniques (see Figure 10.1.1). The 1960s changed the way American culture saw itself—for the first time television sets were in many homes. It was not until the mid-1970s that any motion picture company would have even considered using computerized special effects on a large scale. Movies continued to become more sophisticated, until a revolution in film technique occurred. *Star Wars* can be considered to have been the first super-special-effects motion picture. Though in 1977 computer technology was limited, millions were awestruck by the outer-space scenes—no one had ever experienced such visualization. As a consequence, movies became extremely expensive to produce, and in the early 1990s a figure of more than fifty million dollars to produce even more impressive computer techniques was not uncommon. *Jurassic Park* was the "crown jewel" for the computer scientists and computer artists. Dinosaurs looked strikingly real next to human characters throughout this movie. It is

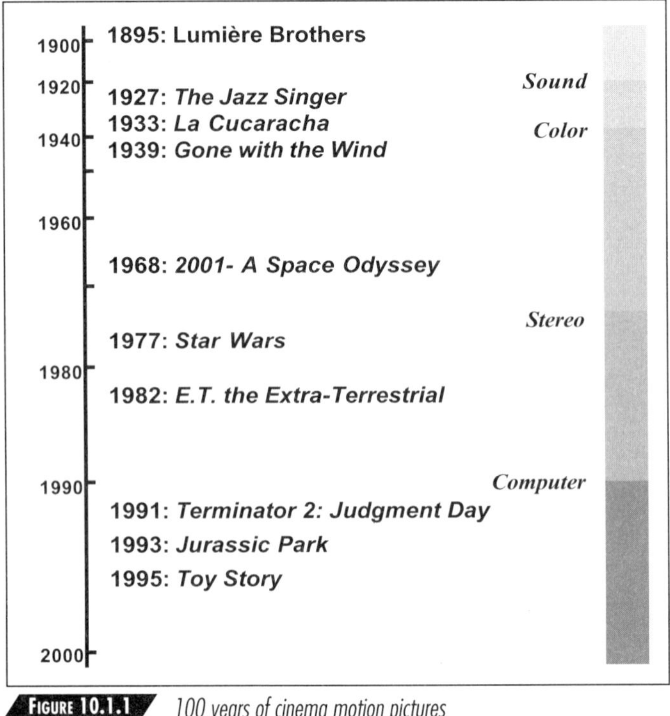

FIGURE 10.1.1 100 years of cinema motion pictures

not difficult to imagine that in the near future, say in the next twenty years, individuals will have access to such sophisticated computer equipment as to create their own full-length, fully digitized motion pictures. The characters, the scenery, the sound will be composed completely on the computer. Think what an innovation this would be. In fact, the first fully computer-animated feature film, *Toy Story*, was released in the mid-1990s. This 79-minute computer animation, formed by 114,240 frames of computer animation and 1635 completed shots (scene changes), used more than 400 models and over 800,000 machine hours to render the final elements.

Imaging having previously been discussed where still pictures are concerned, this chapter will be devoted to video motion pictures (images in motion). A video clip is really a set of sequential still frames. Therefore, the same techniques used for still pictures can be applied to video. Furthermore, in the time axis, the adjacent frames (reflecting a small amount of time shifting) look fairly similar. For this reason video frames are highly redundant and can be used for a much higher compression in comparison to still images. At the same time, as video is digitally transferred, the storage and transmission become critical where video is concerned. Besides the extra technological advantage and increasing market demand, there is also a new challenge: faster compression speeds. Take as an example 15 to 30 frames per second in a video clip—if each picture is compressed in one minute (this is quite similar to the time for a still image), it will take one to two weeks to compress a ten-minute clip. Therefore, by taking advantage of temporal redundancies, the compression system must be faster to serve real-world applications.

10.1 TEMPORAL COMPRESSION

Temporal redundancy is unique to video compression. This redundancy comes from the fractal similarity on the time axis. Thus, theoretically, an appropriate approach is to compress the time direction just as the other spatial directions with a different error distortion weight factor—which is exactly the *3-dimensional fractal image compression algorithm*. Unfortunately, this seems to be unrealistic and ineffective in today's computer applications because it takes a colossal amount of memory and an exponential amount of coding time. A small 128×128 image takes 32K bytes in pixel form, but a few seconds of a video clip of the same tiny 128×128 screen may easily take $128 \times 128 \times 128 = 4$M bytes. Still, this algorithm will be introduced as the first compression method by assuming that sooner or later its advantages will overcome its deficiencies because of rapid technological growth. However, today's practical solution cannot afford to view the time direction as a variable. At any coding moment no more than one frame is in coding status and only a few adjacent frames can be used as helping references to reduce information redundancy. A practical variation is to use *block motion compensation*. This is equivalent to a 2-dimensional fractal compression—by using previous frames as reference images and by using the fractal codes of those previous frames as predictors of the fractal codes of the current frame.

10.1.1 3-D FRACTAL COMPRESSOR

Both the computer's accessible memory and the computational clock rates continue to reach unbelievable speeds. Twenty years ago the largest PC RAM was set to a limit of 640K bytes with an ingenuous 20-bit structure—this limited number was considered astronomical for that time. Today an average personal computer now sells with of 16M bytes or more of RAM (more than 25 times larger than the 640K limit and more than 2000 times larger than the PC of the time). Similarly, in the last twenty years the computational clock rate has been cut in half every eighteen months. With this kind of momentum, in another twenty years 3-D fractal compression could well be the state-of-the-art solution. Of course, many new technological innovations must be made in the process that will reshape today's limited view. For example, in the very near future, memory chips could be replaced by a new-generation 3-dimensional accessible memory cube that will definitely diminish or evolve the popular concept of the "memory disk."

The algorithm of a 3-dimensional fractal compressor can be viewed as a straightforward generalization of the 2-dimensional case. For example, from the very beginning, an *event* is defined to be a distribution on a 3-dimensional cube of an image on a 2-dimensional rectangle, and a *video clip* is a digitization of an event using some 3-dimensional filter—as a picture is of an image. Then, no distinction is made between a video clip and its event, for the same reason that there is no difference between a picture and its image. The 2-dimensional Dirac function is a light source in an image, and the 3-dimensional Dirac function reflects a flash light source in a video clip.

An image partition of square tiles can be generated to a video-clip partition of square cubes. An image segmentation based on objects and their occupied spaces turns into a video-clip segmentation based on objects and their occupancy in both temporal and spatial spaces. A quadtree hierarchy of the image partition presented in Chapter 4 will be replaced by an octaltree hierarchy of a video-clip partition—since each cube can be split into exactly eight cubes of half width.

As a consequence, a basic fractal video-clip compression algorithm can be deduced easily and directly from the 2-dimensional sample coder presented in Chapter 4 if there are no constraints on memory usage and coding speeds. In this system a video-clip fractal code is defined by

$$(dx, dy, dt, rx, ry, rt, \gamma, \beta, \sigma),$$

where (dx, dy, dt) is the destination cube origin location, (rx, ry, rt) is the reference cube origin location, γ is the contrast γ-scaling factor, β is the brightness

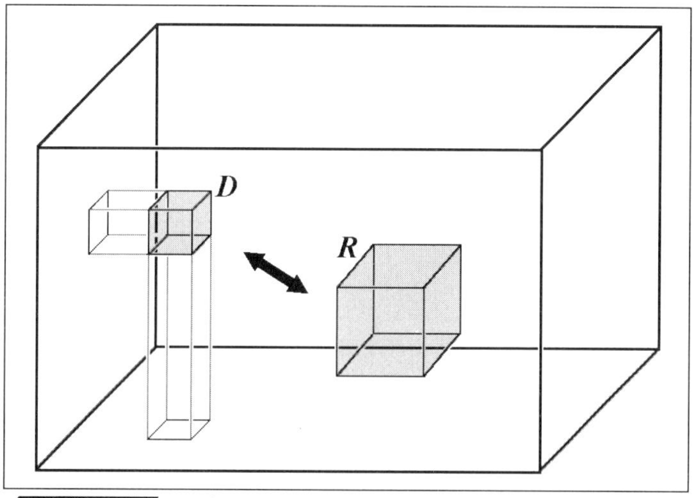

FIGURE 10.1.2. 3-D fractal video representation

β-tuning factor, and σ is the index of a cube-symmetric deformation, as illustrated in Figure 10.1.2.

The main advantage of using this representation is *resolution independence*. Spatially, the same video signal data can be decoded for monitors and TV sets with different output resolutions. And temporally, this same video signal data can generate a different frame rate.

Similarly to the image enhancement and interpolation tool given in Chapter 8, this resolution independence property can also be used as a powerful frame enhancement and interpolation tool to insert additional frames between the original coded image frames.

Figure 10.1.3 shows a few sections of an MPEG standard testing video-clip $352 \times 288 \times 300$ cube: *Paris*. By comparing the set of $t = \text{constant}$ (Figure 10.1.3*a*, *b*, and *c*) with the sets of $x = \text{constant}$ (Figure 10.1.3*d*, *e*, and *f*) and $y = \text{constant}$ (Figure 10.1.3*g*, *h*, and *i*), it is not difficult to see that the video clip has more redundancies in the temporal direction than in the spatial direction (the resolution in the temporal direction is measured in frame rate per second, which is different from the spatial resolution unit). Apparently, there are differences that should be taken into account when dealing with the time direction with respect to the spatial directions.

Notice that in *g* and *h* the actual movement of the woman's right hand is clearly recorded. From *i* one can tell that the woman in the picture had much more movement than the man next to her.

When the 3-D fractal compression scheme is practically and realistically acceptable, then there will be enough motivation to push this research area to develop many refinements for higher performance. Meanwhile, no significant progress should be expected.

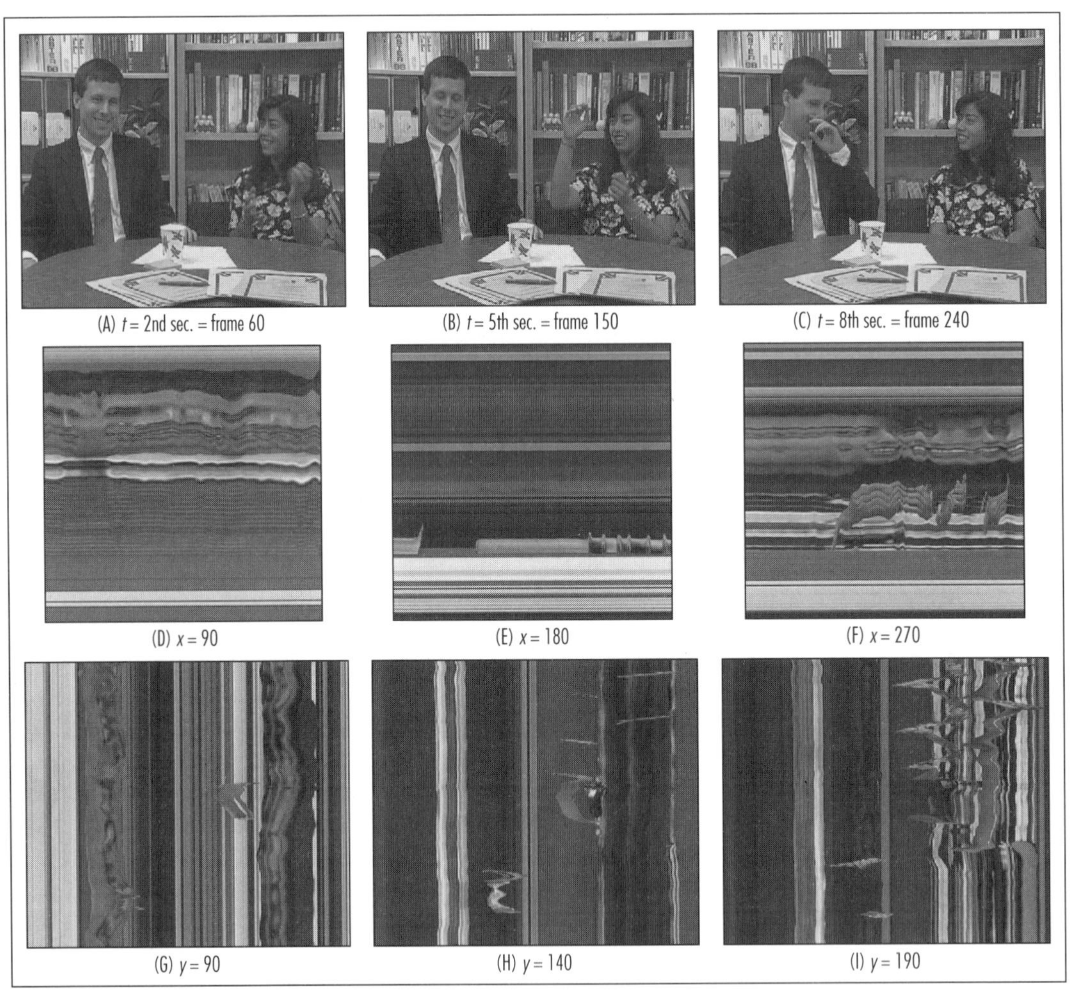

FIGURE 10.1.3 *Sections from Paris—an MPEG standard testing video clip*

10.1.2 MOTION COMPENSATION

Today it takes too much memory to process a large number of frames at the same time. The alternative to coding the redundancy between frames is to code the current frame and then compare it with the previous one, which is identical to the 2-dimensional fractal still image compression system presented in early chapters if the reference image is specified to be the previous frame.

In two consecutive frames of the same scene, it is clear that very often most regions of the picture on both frames look almost identical—with possibly a slight spatial shift because of camera movement. In this case, when the later frame is referenced from the previous one, the fractal codes should be configured to the following setting: *a 1-to-1 spatial ratio is always given between destination and reference,* i.e., no spatial contraction or expansion; *all references addressed locally,* e.g., a $(2r+1)\times(2r+1)$ square region of radius *r:*

$$\{-r, -r+1, \ldots -1, 0, 1, 2, \ldots, r\}\times\{-r, -r+1, \ldots -1, 0, 1, 2, \ldots, r\};$$

the contrast γ*-value is set to 1,* i.e., no contrast changes; *the brightness* β*-value is set to* 0, i.e., no brightness adjustment; and *the spatial deformation index is set to* 0, i.e., no rotation or flip. These fractal codes describe exactly the *motion compensation* used in various video coding systems.

In practice, higher performance has been achieved by allowing a nonzero brightness adjustment (cf. [CC-R]). In fact, a few quantized values around zero should be good enough for visual improvement.

10.1.3 MOTION VECTOR DIFFERENCE

The motion codes, consisting mainly of the reference relative locations, are also called *motion vectors,* which are considered as the *zero-order redundancy* in the temporal direction. In consecutive frames of a video clip (often they have similar motion vectors), physical motion in that video clip is accomplished continuously through a sequence of frames. This interframe motion vector correlation is viewed as the *first-order motion redundancy* and can be characterized as the *motion velocity.*

Depending on the coding complexity and the video-clip class, the motion velocities can be coded explicitly or implicitly. In the explicit case, multiframe codes are allowed, which is equivalent to saying, for example,

> *For the next 10 frames, the* 8×8 *block at (128, 208) of the ith frame is referenced from (120, 200) of the* $(i-1)$*st frame.*

And in the implicit case, the continuity of frames is never broken. The coding of velocity is hidden as a part of conditional entropy coding. For example,

> *Because the* 8×8 *block at (128, 208) of the previous frame is referenced from (120, 200) of its preceding frame, the same block at this frame will have a high chance of being referenced from some area near (120, 200).*

In practice, the motion velocity is not coded directly, but a similar parameter is used, called the *motion vector difference.* Believing the motion vector of the current region to be similar to the motion vector of its neighborhood, a *predictor* is built based on neighboring regions that have already been coded. The motion vector difference will be the difference between the current motion vector and this predictor.

In the case of square block tiling, the predictor could be constructed from the neighborhood motion vectors *MVa*, *MVb*, and *MVc*—from the left, above, and above right—as shown in Figure 10.1.4*a*. Used in video coding standards,

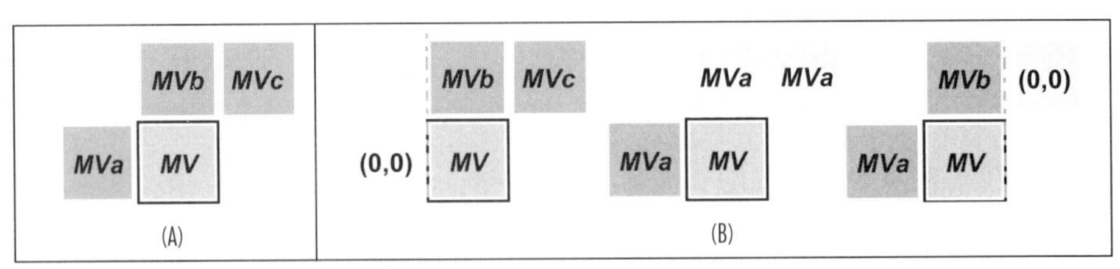

FIGURE 10.1.4 *Motion vector predictor*

each component of the predictor is defined to be the median value of the corresponding component values of these three motion vectors.

In the cases when a candidate predictor is not available, a substitute is illustrated in Figure 10.1.4*b*.

10.2 FRACTAL VIDEO COMPRESSION

Video clips and motion pictures are naturally divided into segments according to scene changes. Each segment, beginning with an initial frame, is often coded by itself as a still image; hence, it is called an *intracoded frame*, or *I-frame*. Each frame thereafter can be coded mainly using the motion codes by referencing its preceding frame, which is called a *P-frame*, as a *predicted frame* from its predecessor, as shown in Figure 10.2.1.

In practice, for the same compressed video clip to be used in various applications at various frame rates, more frames are allowed and can be added between any two of the I-frames and P-frames. Thus, the I-frames and P-frames are also called *coarse frames*, and these additional frames are called *bidirectional frames*,

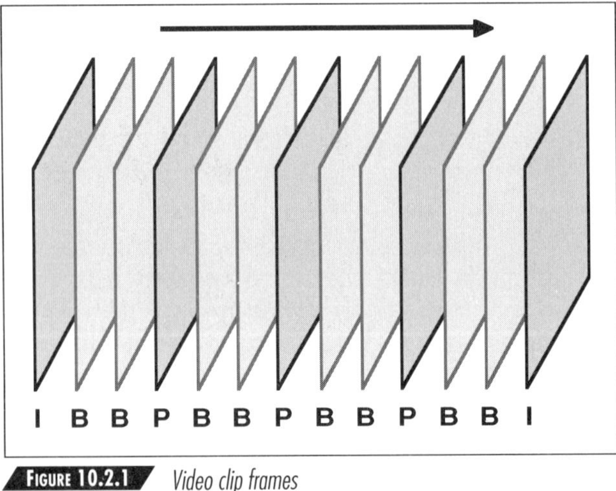

FIGURE 10.2.1 *Video clip frames*

or *B-frames*. Each B-frame is coded using the prediction from both coarse frames immediately before and after it.

A B-frame is often inserted between two P-frames. In that case, for the same block region, if two similar motion vectors are given in each of the P-frames, then an interpolated motion vector for the B-frame at the same block region is always expected. B-frames usually have a very high compression ratio, for they are coded by referencing both coarse frames: before and after.

10.2.1 FRACTAL VIDEO COMPRESSION

In a *2-dimensional fractal video compression system*, the I-frames are compressed using still compression techniques. For the necessity of fast compression (keeping in mind the fact that I-frames form only a small part of the total compressed data in most current implementations), a simplified still compressor may well serve this need. For example, based on our experience, a two-level quadtree compressor with one local and one global code setting of 2-to-1 spatially contractive fractal codes on each level will be good enough in most applications. Thus, each I-frame will be coded by 2-to-1 local and global self-referencing fractal codes.

Notice that when we decode such an I-frame, a hidden frame has actually been created in each iteration. For example, if an I-frame F is created by 10 iterations from some initial image F_0 with self-referenced fractal codes, we set 10 consecutive P-frames having the same set of codes that are identical to the I-frame fractal codes but referenced to the preceding frame instead of itself. Then, starting from the same initial image F_0, by the end the tenth P-frame is clearly the same I-frame obtained in the first procedure. As a result, *a fractal represented I-frame can be replaced by a sequence of P-frames if a time delay is allowed.* (See Figure 10.2.2)

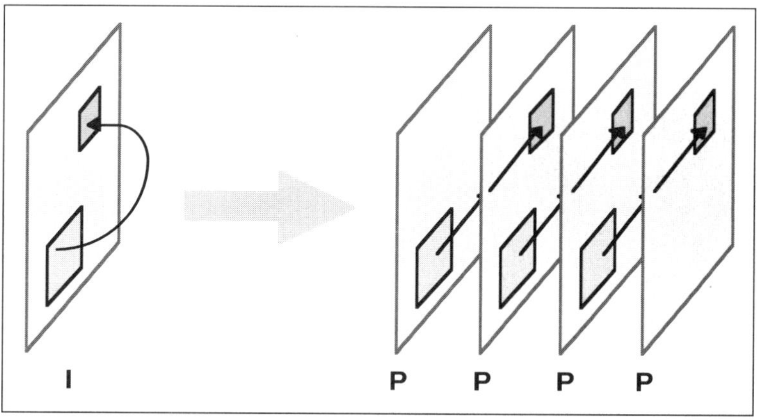

FIGURE 10.2.2 *I-frame decomposed into a sequence of P-frames*

In the next subsection we will present a solution to solve the time-delay. Thereupon, in a fractal video compression system all coarse frames will be treated as P-frames.

10.2.2 GHOST IMAGE

There is no reason to require the image display screen buffer and the decoding image buffer to be in the same dimension (though there may seem to be no reason to think that they should be different). However, in a fractal video compression system, it is much more convenient to allow the working image dimension to be larger than the display image dimension. For instance, we may assume that the working image has the same width as the display image but twice the height, and exactly the top half of the working image will be displayed. The invisible bottom half is called the *ghost image*.

Back to the I-frame problem, the equivalent 10 P-frames can be hidden in the ghost image portion of the 10 prior consecutive P-frames, as shown in Figure 10.2.3. Only in the last step is the hidden ghost image portion brought to the scene by some global transform.

In fact, the ghost image is a very powerful concept for video manipulation. It can be used not only in advanced content preparation as performed on the I-frame, but also in storage of temporary off-scene image and still image pieces. The key function of the ghost image is to shift some compressed video data ahead of time to achieve a constant bit-rate and constant frame-rate video compression, while consistently maintaining picture quality.

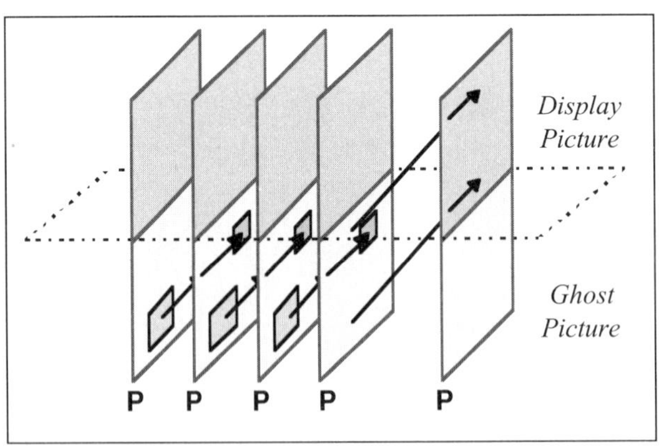

FIGURE 10.2.3 *Using ghost image to adjust timing*

10.2.3 ROBUSTNESS AND RESOLUTION INDEPENDENCE

In summary, a fractal video compression system compresses coarse frames only from their immediately preceding coarse frame using 1-to-1 and 2-to-1 local and global fractal codes. In comparison with all known technologies, this scheme stands out for its simplicity and unification in both the temporal and spatial directions.

Besides its proven high performance (assuming that there is adequate coverage of 2-to-1 contractive fractal codes), this scheme offers two extra features:

1. *Instant key frame.* A P-frame is coded from its previous frame, so a coded P-frame can be decoded only if the previous frame has already been decoded. To allow a user to view the video from the middle of a video clip, many I-frames are required (which are called *key frames*). However, in a fractal system this becomes unnecessary. Start at any compressed frame of the video clip—fractal contractivity will assure that without knowing the previous frames, the decoded video will have the same quality as if decoded from the beginning, except for a few frames at the beginning (which served as iteration frames). So no key frames need to be inserted for viewing the video clip from any selected frame in the middle.
2. *Resolution independence.* This unique property remains true to video for the same reason as that discussed in still imaging. Resolution independence provides an inherent flexibility to the video data so that the video data can be used simultaneously for various output devices.

10.3 VIDEO COMPRESSION STANDARDS

There are two emerging video industrial standards: H.263 for low bit-rate videoconferencing and MPEG-4 for generic and application-independent integrated video and audio—particularly high bit-rate entertainment video and motion pictures.

10.3.1 H.263 LOW BIT-RATE VIDEO COMMUNICATION

H.263 is an ITU (*International Telecommunication Union*) standard for low bit-rate video coding.

A digitized video clip is a sequence of picture frames. Each picture frame is coded as one luminance and two chrominance images: Y, C_B, and C_R (See Chapter 11 for color decomposition). There are five standardized picture formats: sub-QCIF, QCIF, CIF, 4CIF, and 16CIF, as shown in Table 10.3.1. All of them have aspect ratio 4:3, which implies a 12:11 pixel aspect ratio for all formats, except that the sub-QDIF picture format has a pixel aspect ratio 1:1. Pixel values are coded in the range of 1 through 255 and truncated to [16, 235] for luminance images and [16, 240] for chrominance images. The chrominance images are always in a quarter dimension of the luminance one. The sampling positions of luminance and chrominance samples are show in

TABLE 10.3.1 *Standard H.263 picture formats*

Picture Format	sub-QCIF	QCIF	CIF	4CIF	16CIF
Luminance Pixels per Line	128	176	352	704	1408
Luminance Lines per Frame	96	144	288	576	1152
Chrominance Pixels per Line	64	88	176	352	704
Chrominance Lines per Frame	48	72	144	288	576
Luminance Pixel Range	16 (Black)–235 (White)				
Chrominance Pixel Range	16–240				
Chrominance Zero Difference	128				
Luminance Pixel Range	128	96	64	48	48
Pixel Aspect Ratio	1 : 1	12 : 11			
Image Aspect Ratio	4 : 3				
Maximum Frame Rate	30 000/1001 ≈ 29.97 Hz				
GOB number per Frame	6	9	18		
Macroblock number per GOB	8	11	22	88	352

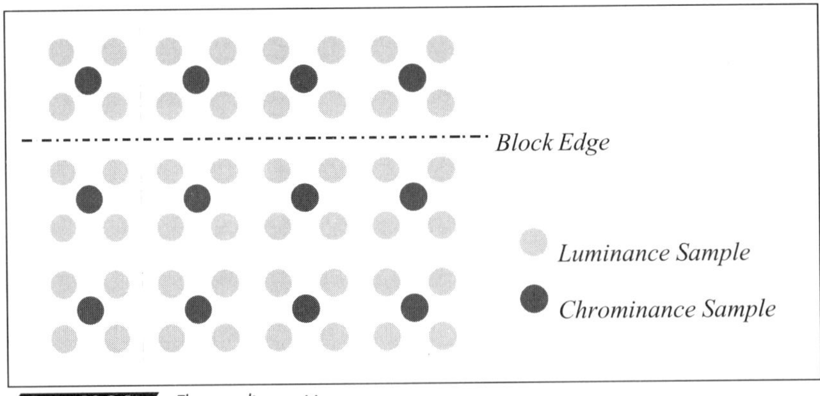

FIGURE 10.3.1 *The sampling positions*

Figure 10.3.1. The bilinear interpolation is used as the default up-sampling method.

Each picture is divided into groups of blocks. A *group of block* (GOB) comprises of k*16 lines, where $k = 1$ for sub-QCIF, QCIF, and CIF, $k = 2$ for 4CIF, and $k = 4$ for 16CIF.

GOBs are further divided into *macroblocks*. A macroblock relates to a 16×16 region of the luminance image Y and its spatial corresponding 8×8 regions from both chrominance images, C_B and C_R. A macroblock can be further split into six 8×8 *blocks*: four from Y, one from C_B, and one from C_R, as shown in Figure 10.3.2.

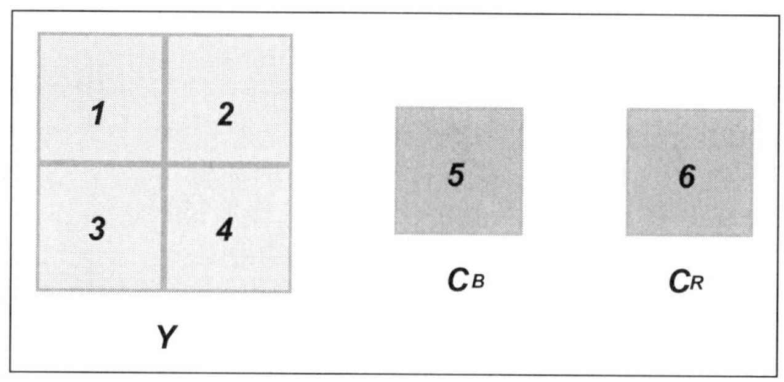

FIGURE 10.3.2 A macroblock

A picture frame is either an *INTRA* frame (i.e., I-frame, coded from itself) or an *INTER* frame (i.e., P-frame, coded with the prediction of motion compensation from the previous frame).

A set of compressed video data in H.263 format is arranged in a hierarchical structure with four layers: *picture frame*, *group of blocks*, *macroblock*, and *block*. Only the macroblock data structure will be presented in the next subsection since it illustrates the compression techniques that could be used in this format.

10.3.2 MACROBLOCKS

A macroblock in a P-frame always begins with a *coded indication bit* (COD): 0 if it is coded and 1 if it is a copy from the previous frame (thus, no further data for this macroblock). See Figure 10.3.3.

COD	MCBPC	CBPY	DQUANT	MVD	DCT data

FIGURE 10.3.3 Data string of a macroblock

Given a coded macroblock, a variable-length-code MCBPC determines the *macroblock types* (MT) and both chrominance block patterns (P5 & P6). Then, a variable-length-code CBPY defines all of the the four *Y*-block patterns (P1, P2, P3, P4). The code table of MCBPC is given in Table 10.3.2 for an I-frame and in Table 10.3.3 for a P-frame. And the code table of CBPY is given in Table 10.3.4.

TABLE 10.3.2 Variable-length-code table for MCBPC for I-frame

Index	MT	P5	P6	code	Index	MT	P5	P6	code
0	3	0	0	1	*4*	4	0	0	0001
1	3	0	1	001	*5*	4	0	1	000001
2	3	1	0	010	*6*	4	1	0	000010
3	3	1	1	011	*7*	4	1	1	000011

TABLE 10.3.3 *Variable-length-code table for* MCBPC *for P-frame*

Index	MT	P5	P6	code	Index	MT	P5	P6	code
0	0	0	0	1	4	2	1	0	0000100
1	0	0	1	0011	5	2	1	1	00000101
2	0	1	0	0010	6	3	0	0	00011
3	0	1	1	000101	4	3	0	1	00000100
4	1	0	0	011	5	3	1	0	00000011
5	1	0	1	0000111	6	3	1	1	0000011
6	1	1	0	000110	4	4	0	0	000100
7	1	1	1	000000101	5	4	0	1	000000100
8	2	0	0	010	6	4	1	0	000000011
9	2	0	1	0000101	7	4	1	1	000000010

TABLE 10.3.4 *Variable-length-code table for* CBPY

Index	P-INTRA	P-INTER	code	Index	P-INTRA	P-INTER	code
0	0000	1111	0011	8	1000	0111	00010
1	0001	1110	00101	9	1001	0110	000011
2	0010	1101	00100	10	1010	0101	0101
3	0011	1100	1001	11	1011	0100	1010
4	0100	1011	00011	12	1100	0011	0100
5	0101	1010	0111	13	1101	0010	1000
6	0110	1001	000010	14	1110	0001	0110
7	0111	1000	1011	15	1111	0000	11

There are five macroblock types:

1. (MT=0) *INTER*: A motion vector given to all blocks of the macroblock.
2. (MT=1) *INTER-Q*: A motion vector given to all blocks of the macro-block and the quantization parameter is adjusted.
3. (MT=2) *INTER-4V*: This is used only in the advanced prediction optional mode. Four motion vectors (one for each *Y*-block) are coded one after another.
4. (MT=3) *INTRA*: Coded using DCT.
5. (MT=4) *INTRA-Q*: Coded using DCT. In addition, the quantization parameter is adjusted.

There are two block patterns:

1. P=0: None non-INTRADC coefficient is presented.
2. P=1: Some non-INTRADC coefficient is presented.

In case the quantization parameter QUANT needs to be adjusted, a 2-bit adjustment DQUANT is added:

DQUANT =	00	01	10	11
QUANT +=	-1	-2	1	2

Then, the quantization parameter QUANT is always clipped to the ranges from 1 to 31, which correspond to a quantization factor QUANT*2.

In the INTER and INTER-Q cases, a common motion vector is shared by all blocks in the macroblock. The motion vector difference MVD is coded first as the horizontal component, and then the vertical component using the variable-length-code book, Table 10.3.5. The decoded quarter pixel resolution chrominance vector components are adjusted:

quarter pixel position	0	1/4	1/2	3/4	1
resulting position	0	1/2	1/2	1/2	1

In the INTER-4V case, four motion vectors are coded one after the other. A more detailed discussion is given in the next subsection.

TABLE 10.3.5 *Variable-length-code table for* MVD

MVD	*code*	*MVD*	*code*	*MVD*	*code*
0	1	±4	000001011s	±11.5	0000000**101**s
±0.5	01s	±4.5	000001010s	±12	0000000**100**s
±1	001s	±5	000001001s	±12.5	00000000**111**s
±1.5	0001s	±5.5	0000010001s	±13	00000000**110**s
±2	000011s	±6	0000010000s	...	00000000**xxx**s
±2.5	0000101s	±6.5	000000**1111**s	±15	00000000**010**s
±3	0000100s	±6.5	000000**1110**s	±15.5	000000000011s
±3.5	0000011s	...	000000**xxxx**s	+: s = 0	−: s = 1

The INTRA blocks are coded using the DCT transform, which is very similar to JPEG DCT (discussed in Chapter 6). In fact, the only difference is that instead of using a quantization matrix, a fixed quantization factor is given for DC terms and a unified quantization factor is assigned to all AC coefficients.

10.3.3 OPTIONAL MODES

In addition to the baseline H.263 coding standard, four negotiable coding optional modes are included for improved performance:

1. *Unrestricted Motion Vectors:* In this optional mode, motion vectors are allowed to point outside the picture, and the motion vectors' range is enlarged to $[-31.5, 31.5] \times [-31.5, 31.5]$.
2. *Syntax-Based Arithmetic Coding:* In this optional mode, *arithmetic coding* is used instead of *variable length coding.*
3. *Advanced Prediction:* This optional mode includes the *possibility of four motion vectors per macroblock* and the *overlapped block motion compensation.*

In the case of four motion vectors *MV1*, *MV2*, *MV3*, and *MV4* assigned to the four 8×8 *Y*-blocks of a macroblock, the candidate predictors *MVa*, *MVb*, and *MVc* of each of them are defined as indicated in Figure 10.3.4. The motion vectors for both chrominance blocks are derived from

$$MV5 = MV6 = \frac{MV1 + MV2 + MV3 + MV4}{8}, \tag{10.3.1}$$

with a half-pixel step truncation as shown below:

position in 16th	0	1	2	3	4	5	6	7	8	9	10	11	12	13	14	15
result in half	0	0	0	1	1	1	1	1	1	1	1	1	1	1	2	2

In the case of overlapped motion compensation, each pixel in an 8×8 *Y*-block is a weighted sum of three prediction values from three corresponding pixel values from the reference blocks generated by three motion vectors: the *current motion vector*, the *horizontal adjacent motion vector* and the *vertical adjacent motion vector*. The weighting values are given in Figure 10.3.5 with the common divisor of 8.

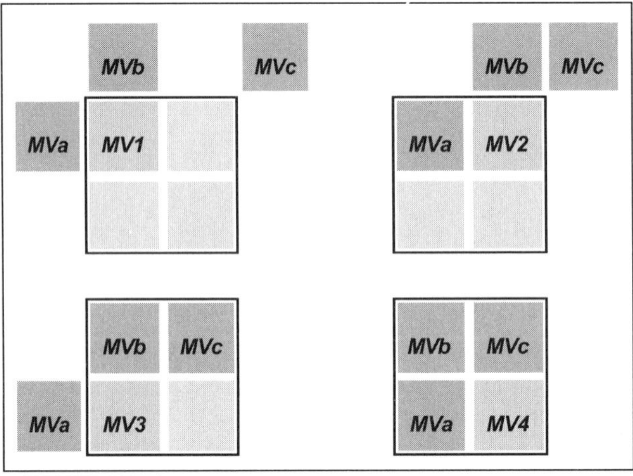

FIGURE 10.3.4 Redefinition of the candidate predictors for each *Y*-block of a macroblock

4	5	5	5	5	5	5	4
5	5	5	5	5	5	5	5
5	5	6	6	6	6	5	5
5	5	6	6	6	6	5	5
5	5	6	6	6	6	5	5
5	5	6	6	6	6	5	5
5	5	5	5	5	5	5	5
4	5	5	5	5	5	5	4

A

2	2	2	2	2	2	2	2
1	1	2	2	2	2	1	1
1	1	1	1	1	1	1	1
1	1	1	1	1	1	1	1
1	1	1	1	1	1	1	1
1	1	1	1	1	1	1	1
1	1	2	2	2	2	1	1
2	2	2	2	2	2	2	2

B

2	1	1	1	1	1	1	2
2	1	1	1	1	1	1	2
2	2	1	1	1	1	2	2
2	2	1	1	1	1	2	2
2	2	1	1	1	1	2	2
2	2	1	1	1	1	2	2
2	1	1	1	1	1	1	2
2	1	1	1	1	1	1	2

C

FIGURE 10.3.5 *Weighting values for overlapped motion compensation: (A) current; (B) horizontal; (C) vertical*

To the left 32 pixels of the 8×8 *Y*-block, the horizontal adjacent motion vector is the motion vector of the block at the left; and to the right 32 pixels of the block, the horizontal adjacent motion vector is the motion vector of the block at the right. To the top 32 pixels of the 8×8 *Y*-block, the vertical adjacent motion vector is the motion vector of the block above; and to the bottom 32 pixels of the block; the vertical adjacent motion vector is the motion vector of the block below. In the case when an adjacent motion vector is not available, the current motion vector is duplicated in its place (except for PB-frames).

4. *PB-frames:* A PB-frame consists of two pictures coded as one unit: one P-frame coded from the previous P-frame and one B-frame predicted from both the previous P-frame and the current P-frame. Temporally, the B-frame is located between the previous P-frame and the current P-frame.

In accommodating both the rapid technological advances and increasingly high market demands, ITU (*International Telecommunication Union*) is currently considering extending the H263 standard. There are two main extensions under consideration: H263+ (which is an incremental enhancement to H263) and H263L (which will accept some new coding algorithms).

10.3.4 MPEG-4

MPEG-4 is the emerging digital video standard for entertainment video and motion pictures. MPEG (*Moving Picture Experts Group*) is a group chartered by the ISO≠IEC (*International Organization for Standardization* and *International Electrotechnical Commission*). The goal of MPEG-4 is to *outperform the existing standards* (particularly to outperform H.263 at low bit-rates) and to *provide content-based coding* (for applications such as interactive TV, query-by-content data bases, shopping malls on the World-Wide-Web, and video games). In order to achieve this goal, two new aspects have been developed in MPEG-4: the *MPEG Syntax Description Language* (MSDL), and the *Synthetic–Natural Hybrid Coding* (SNHC). MPEG-4 also supports a few more SIF picture formats, as in

352×240 in 30Hz and 352×288 in 25Hz. The MPEG-4 work, started in Seoul in November 1993, has its completion scheduled for 1998.

The current MPEG-4 supports multilayer video objects. A video sequence can be decomposed into video objects. Each video object is represented and compressed separately as a complete video sequence of picture frames with a shape masked. Instances of video objects in a given time are called VOPs (*video object plane*). Thus, a video scene is composed of VOPs corresponding to a set of objects or layers carrying a semantic signification that a user can access or manipulate. The correlation of VOPs is then described by a sequence of alpha planes (that prioritize the video objects). Each of the VOPs is coded by using an algorithm that suits its content. The final video data are obtained by multiplexing the video data of each VOP. A verification model (VM) was established based upon the VOP structure. A VM may have an arbitrary shape; hence, a shape-coding algorithm based on a quadtree structure has been included, and a padding technique is used for the blocks on the VOP borders.

When a single VOP corresponding to the input frame is defined, the shape coding is not used, and the VM is very similar to H.263. In fact, currently the MPEG-4 baseline core technology is very much like H.263. However, many new technologies used to improve I-frame and P-frame have been submitted to the ad hoc working groups for recommendation.

At the present stage, core MPEG experiments focus mainly on four areas: *content-based coding, multifunctional coding, coding efficiency,* and *error resilience.*

Targeted to 2000, MPEG-7, a more general standard called *Multimedia Content Description Interface*, is in process. This standard is built on MPEG-4 by incorporating multimedia context and objectives.

In comparison to digital still imaging, direct digital video manipulation leaves much to be desired — due to the excessive data memory requirements that are beyond today's computer speeds. Currently, digital video is processed frame by frame as correlated still images. However, in the future, direct 3-D video data manipulation will possibly reshape our current thinking and form a new standard that our children will enjoy.

The budding spring entered into alliance with her. The dream of new life was teeming in the warmth of the slumbering air. The young green was wedding with the silver-grey of the olive-trees. Beneath the dark red arches of the ruined aqueducts flowered the white almond-trees. In the awakening Campagna waved the seas of grass and the triumphant flames of the poppies. Down the lawns of the villas flowed streams of purple anemones and sheets of violets. The glycine clambered up the umbrella-shaped pines, and the wind blowing over the city brought the scent of the roses of the Palatine.

— The New Dawn

Romain Rolland

Color Image 11

So far, grayscale images have been discussed. What about color images? A color vision model has been established in the science of colorimetry based on quantitative research of human light perception. In this model, any color is decomposed into three *primaries* (e.g., *red*, *green*, and *blue*). As a consequence, a color image can be split into three different color components, and each component can be treated as an independent grayscale image. Thus, compressing a color image is achieved by compressing three separate grayscale images. However, these color components are usually correlated, so further color redundancies should be extracted for even higher compression.

11.1 BASICS OF COLOR

Before studying color image representation, a brief background and some basic concepts in color science need to be introduced. The focus of this chapter is the color decomposition methods that suit image compression and representation.

11.1.1 COLORIMETRY

Light that is visible to our eyes is a form of electromagnetic energy consisting of a spectrum of frequencies having wavelengths ranging from about 400 nm (*nanometers*) for violet to about 700 nm for red, as illustrated in Figure 11.1.1. The color of an object is a function of the unabsorbed wavelengths of light reflected from it. Consequently, an object may reflect different colors under different viewing lights.

Much of the current understanding of color originates from Newton's treatise *Opticks*. Grassman (1854) developed a set of eight axioms that define the

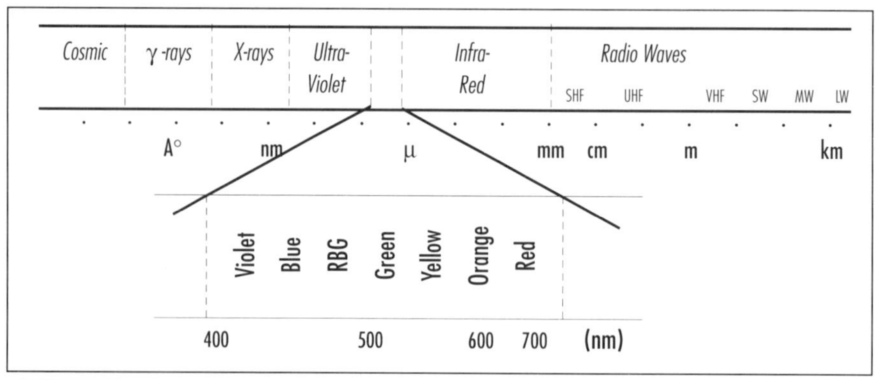

FIGURE 11.1.1 Visible spectrum of the electromagnetic wave

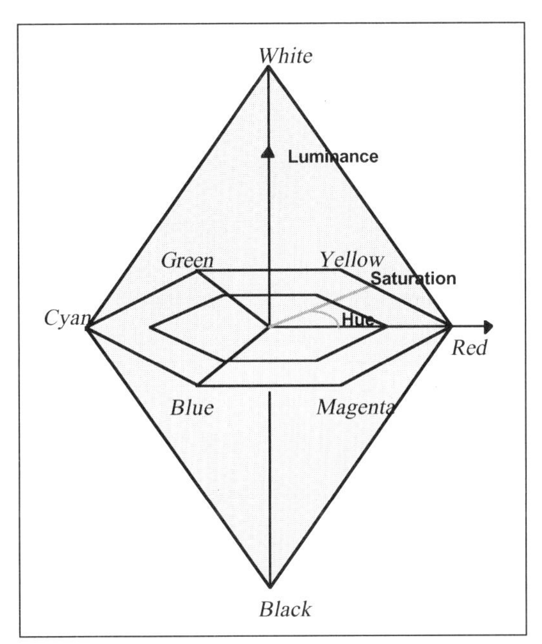

FIGURE 11.1.2 HLS representation of color space

trichromatic color matching system ([G]), which became the foundation of today's color science. Before describing these axioms, some terminology needs to be defined.

- *Luminance*, or *brightness*, is the amount of light received by the eye regardless of color, from very dim (*black*) to very bright (*white*). In some contexts, luminance and brightness have different meanings. The luminance is defined as the absolute light strength tested by equipment at an extremely local spot. But brightness often refers to the perceptual light strength, which depends on the surrounding lighting conditions. In our implementation of the compression system these two concepts are not distinguished. Only in Chapter 13 are their differences considered necessary for finding a visual image metric.
- *Hue* is the predominant spectral color in the light, as shown in Figure 11.1.2.
- *Saturation* indicates the spectral purity of the color in light. The saturation is increased by increasing the amount of the chromatic color.

Hue and *Saturation* form the *chrominance* part of a color.

The following is the set of Grassman's eight axioms.

1. *Primaries:* Any color can be matched by a mixture of some three colored lights X, Y, and Z.
2. *Luminance independence:* A color match is independent of luminance over a wide range.
3. *Luminance addition:* The luminance value of a mixture of colors is the sum of the individual luminance values.
4. *Color mixing:* Components of a color mixture cannot be resolved by the human eye.
5. *Color matching:* A color C can be stated in equation form as

$$C = x \cdot X + y \cdot Y + z \cdot Z. \quad (11.1.1)$$

   This means that the sum of x units of X, y units of Y, and z units of Z yields a match with the color C. The coefficients x, y, and z are called *tristimulus values*.

6. *Color addition law:* $C_1 = C_2$ and $C_3 = C_4 \Rightarrow C_1 + C_3 = C_2 + C_4$.
7. *Color subtraction law:* $C_1 = C_2$ and $C_1 + C_3 = C_2 + C_4 \Rightarrow C_3 = C_4$.
8. *Color transition law:* $C_1 = C_2$ and $C_2 = C_3 \Rightarrow C_1 = C_3$.

Many different coordinate systems have been employed for the specification of color. As a result of extensive tests with hundreds of observers, in 1931 the CIE (*Commission International d'Eclairage*) standardized three primaries: *red* (700 nm), *green* (546.1 nm), and *blue* (435.8 nm) based on the principle of color matching in colorimetry. Figure 11.1.3 illustrates a coordinate system using these three primaries.

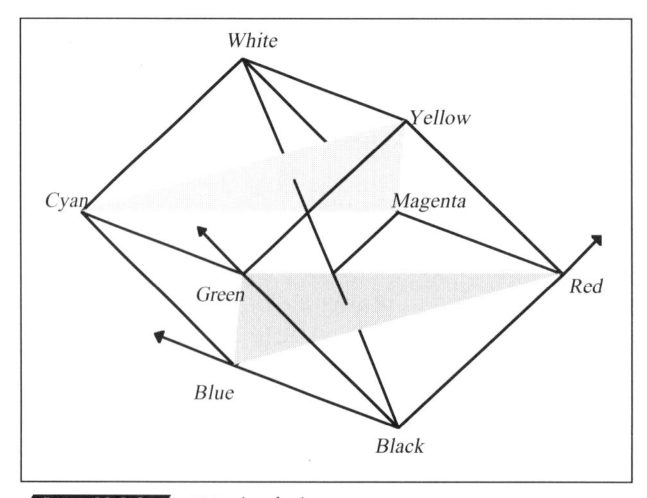

FIGURE 11.1.3 RGB cube of color space

11.1.2 DISCRETE INTENSITY VALUES

In digital imaging, color images are described in an array of tristimulus values (i.e., pixel vectors of three components) or equivalently, three arrays of values in three primaries (i.e., three image components of pixels). An array of values in a color component is called an *image channel*. In principle, a pixel channel value should be the intensity value of the image through a specified color filter:

$$C[i,j] = \iint I_{Filtered}(x,y)\Phi(x-i,y-j)\,dx\,dy\,, \tag{11.1.2}$$

where $I_{Filtered}(x, y)$ is the light intensity distribution filtered by a primary color filter, and $\Phi(x, y)$ is the human eye filter (also called the *relative luminous efficiency function*). As discussed in Chapter 2, $\Phi(x, y)$ is a bell-shaped surface. In digital imaging, discrete integer values are used in describing these values. Thus, the question of pixel value design is how these values can be defined so that the differences between any two adjacent values represent the same amount of brightness change with respect to the human eye.

According to Weber's law ([H3]), the sensitivity of the human eye to different luminance intensities can be characterized by the number

$$c = \frac{\Delta f}{f} = \frac{f' - f}{f}, \tag{11.1.3}$$

i.e., if the light intensity from f to $f+\Delta f$ is noticeably different, then the same noticeable difference should be observed from g to $g + g/f \cdot \Delta f$. As a result, logarithmic values are assigned in the digital case for balanced, visually equal steps.

Research shows that by choosing logarithmic, discrete pixel channel values such that any two adjacent values give just noticeably different light intensities, about fifty values will cover the entire visual spectrum. Consequently, 64 colors are often chosen by various output devices. For more accuracy, pixel channel values are designed to have integer values from 0 (*black*) to 255 (*white*).

11.1.3 COLOR DECOMPOSITION

In the commercial television industry a receiver phosphor standard was set by NTSC (*the National Television Systems Committee*). Its primary system RGB is related to the CIE primary system rgb by the following linear coordinate conversion:

$$\begin{pmatrix} R_{NTSC} \\ G_{NTSC} \\ B_{NTSC} \end{pmatrix} = \begin{pmatrix} 0.842 & 0.156 & 0.091 \\ -0.129 & 1.320 & -0.203 \\ 0.008 & -0.069 & 0.897 \end{pmatrix} \begin{pmatrix} r_{CIE} \\ g_{CIE} \\ b_{CIE} \end{pmatrix}. \tag{11.1.4}$$

In the development of the United States color television system, the NTSC formulated a color coordinate system for television signal transmission composed of three components, *YIQ*. In the PAL (*Phase Alternating Line*) and SECAM (*Sequential Couleur à Mémoire*) television systems used in many other countries, a similar coordinate system *YUV* was adopted. In both systems, the *Y* signal is the *luminance* of a color, which is the only value used for black-and-white television. The *I* and *Q* signals, or the *U* and *V* signals, jointly describe the *chrominance*, i.e., the *hue* and *saturation* attributes, of a color, which have been shown to be less sensitive than luminance to the human eye. In television signal transmission, only half-resolution chromatic data are sent. The *I* and *Q* signals are related to the *U* and *V* signals by a simple rotation of coordinates in color space:

$$\begin{aligned} I &= -U\sin 33^\circ + V\cos 33^\circ, \\ Q &= U\cos 33^\circ + V\sin 33^\circ. \end{aligned} \tag{11.1.5}$$

Their relationship to the primaries can be written as

$$\begin{pmatrix} Y \\ I \\ Q \end{pmatrix} = \begin{pmatrix} 0.299 & 0.587 & 0.114 \\ 0.596 & -0.274 & -0.322 \\ 0.211 & -0.523 & 0.312 \end{pmatrix} \begin{pmatrix} R \\ G \\ B \end{pmatrix},$$

$$\begin{pmatrix} Y \\ U \\ V \end{pmatrix} = \begin{pmatrix} 0.299 & 0.587 & 0.114 \\ -0.148 & -0.289 & 0.437 \\ 0.615 & -0.515 & -0.100 \end{pmatrix} \begin{pmatrix} R \\ G \\ B \end{pmatrix}. \tag{11.1.6}$$

Since a color image is always described as layers of several monochrome images, a color image compression system can be designed so that a color image is compressed as a set of separate layers of compressed grayscale images. However, there are some redundancies among these monochrome images. Different decompositions decorrelate the color redundancies differently. As a consequence, the performance of such a compression system depends upon the color image decomposition a great deal.

Given a grayscale image compression system, a color image can be compressed by coding each of its *RGB* images or by coding each of its *YUV* images. Experiments have shown that it is always more efficient to encode *YUV* images than to encode *RGB* images. In general, for most color images, for the same image quality, it takes about twice as much data to describe a color image in separated *RGB* images than is needed in separated *YUV* images.

YUV decomposition splits the luminance from the chrominance, extracting the luminance to the *Y* image and the chrominance to the *U* and *V* images. To a color image, the luminance is perceptually very important. As illustrated in Figure 11.1.4, for a fixed perceptual quality, the *U* and *V* images are highly redundant and can be compressed in high compression ratios. Color Plate 12 shows a sample image composed using two of the three decomposed components.

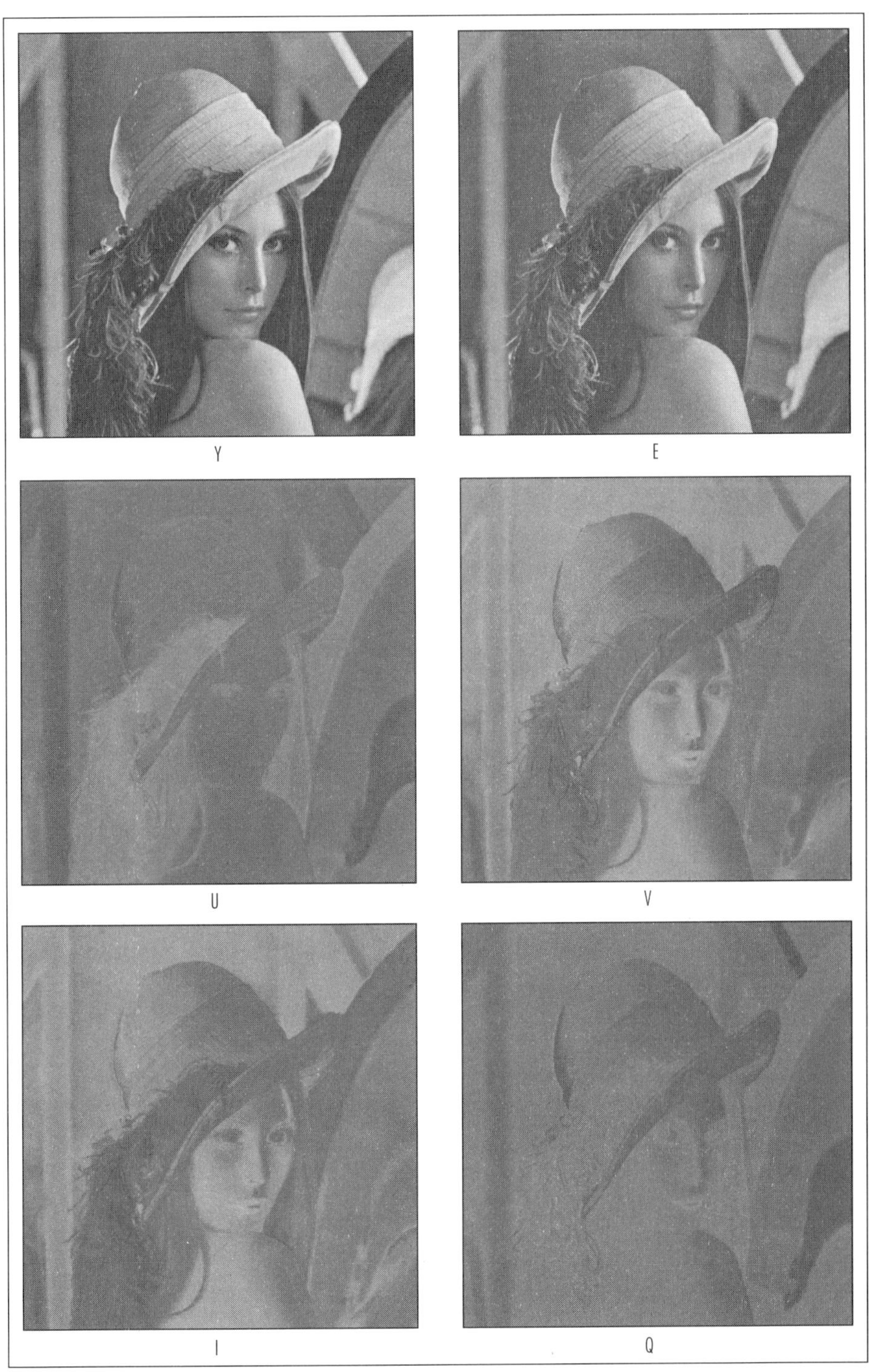

Figure 11.1.4 *Color decomposition used in television systems*

In our early approaches, the YUV decomposition was used to achieve high compression. In general, the compressed U and V signal data together take about half the size of the Y signal data. For simplicity of integer calculation, Y, U, and V signals are mapped into the discrete integer values 0, 1,..., 255, and the floating-point multiplication operations are replaced by integer multiplication with a shift:

$$\begin{pmatrix} Y \\ U \\ V \end{pmatrix} = \frac{1}{256}\left(\begin{pmatrix} 77 & 150 & 29 \\ -38 & -75 & 113 \\ 112 & -94 & -18 \end{pmatrix}\begin{pmatrix} R \\ G \\ B \end{pmatrix} + \begin{pmatrix} 128 \\ 128 * 257 \\ 128 * 257 \end{pmatrix}\right). \quad (11.1.7)$$

However, if the goal is purely compression, some alternative coordinate systems are recommended, for example, the *simplified color coordinate system LMN*:

$$\begin{pmatrix} L \\ M \\ N \end{pmatrix} = \begin{pmatrix} 0.25 & 0.50 & 0.25 \\ 0 & -0.50 & 0.50 \\ 0.50 & -0.50 & 0 \end{pmatrix}\begin{pmatrix} R \\ G \\ B \end{pmatrix}, \quad (11.1.8)$$

or its integer version

$$\begin{aligned} L &= (2G + R + B + 2)/4, \\ M &= (B - G + 257)/2, \\ N &= (R - G + 257)/2, \end{aligned} \quad (11.1.9)$$

and the *equalized color coordinate system EMN*

$$\begin{pmatrix} E \\ M \\ N \end{pmatrix} = \begin{pmatrix} 0.33 & 0.34 & 0.33 \\ 0 & -0.50 & 0.50 \\ 0.50 & -0.50 & 0 \end{pmatrix}\begin{pmatrix} R \\ G \\ B \end{pmatrix}, \quad (11.1.10)$$

or its integer version

$$\begin{aligned} E &= (G + R + B + 1)/3, \\ M &= (B - G + 257)/2, \\ N &= (R - G + 257)/2. \end{aligned} \quad (11.1.11)$$

A color image compression system that uses either LMN decomposition or EMN decomposition performs as well as the corresponding compression system that uses YUV decomposition, and numerically, LMN and EMN decompositions are even simpler than YUV decomposition. Hence, LMN decomposition will be used as the sample case. Figure 11.1.5 displays these color components of the sample image *Lena*. Indeed, in a compression system, if there is no special

Figure 11.1.5 *LMN color decomposition recommended for compression*

use for having the luminance image, then this image can be replaced by the green component G, i.e., the GMN decomposition.

In video compression standards, the CCIR (*International Radio Consultative Committee*) Recommendation 601 is used. In this recommendation, RGB images are converted into YC_BC_R images, as follows:

$$\begin{pmatrix} Y \\ C_B \\ C_R \end{pmatrix} = \begin{pmatrix} 0.257 & 0.504 & 0.098 \\ -0.148 & -0.291 & 0.439 \\ 0.439 & -0.368 & -0.071 \end{pmatrix} \begin{pmatrix} R \\ G \\ B \end{pmatrix} + \begin{pmatrix} 16 \\ 128 \\ 128 \end{pmatrix}, \tag{11.1.12}$$

$$\begin{pmatrix} R \\ G \\ B \end{pmatrix} = \begin{pmatrix} 1.164 & 0 & 1.596 \\ 1.164 & -0.392 & -0.813 \\ 1.164 & 2.017 & 0 \end{pmatrix} \begin{pmatrix} Y-16 \\ C_B-128 \\ C_R-128 \end{pmatrix}, \tag{11.1.13}$$

where Y takes values in the interval [16, 235], and C_B and C_R take values in [16, 240].

By convention, the decomposed color components are called *color screens*.

11.2 COLOR IMAGE COMPRESSION

Most image compression systems compress color images as several independently decomposed grayscale images. The most popular way to decompose a color image is to split this image into a primary luminance monochrome image and several secondary chrominance monochrome images, such as YUV decomposition and LMN decomposition.

The secondary images should not be treated identically to the primary one. In fact, the human eye is not as sensitive to variations of color as to subtleties of texture, yet it reacts more quickly to a subtle change in color than to a mistaken intensity.

11.2.1 COMPRESSION USING COLOR DECOMPOSITION.

Decomposing an RGB color image into one luminance image and two chrominance images is the method that has been used in most commercial applications, as well as the JPEG and MPEG imaging standards. Here, the LMN system is chosen for illustration. The same discussion could be easily extended to similar decomposition methods, such as YUV and YIQ.

There are two good reasons to decompose a color image in this manner. One reason is that for a black-and-white picture, this could use the luminance L screen only. This was an important function for feeding black-and-white TV sets and color TV sets the same broadcasting signal. The other reason for such color image decomposition is requirements for compression. *LMN* decomposition decorrelates the luminance data and chrominance data quite effectively.

Let us compare *L*, *M*, and *N* screen images with *R*, *G*, and *B* channel images. The screen images *L*, *R*, *G*, and *B* have similar complexities; that is, for a fixed quality, these four images generated from the same image usually compress to very similar file sizes, while screen *M* and screen *N* take much less space. For the same quality, each of them usually takes a file size of about a quarter of the data size of the luminance screen *L*. In conclusion, for a color image and for a given compression quality target, the compressed image data can be cut in half by compressing individually the screen images *LMN* instead of compressing individual channel images *RGB*.

While compressing the screen images *LMN* individually, a distortion rate needs to be assigned to each of them for the desired balanced quality. Should the numbers for *L*, *M*, and *N* screen images be the same or different?

Consider the inverse formulas

$$\begin{aligned} G &= (2L + M + N)/2 - 128, \\ B &= (-2L + 3M - N)/2 - 128, \\ R &= (-2L - M + 3N)/2 - 128. \end{aligned} \tag{11.2.1}$$

If the goal is to minimize the *RGB* error, the L^2-distortion of *RGB* pixels can be estimated as

$$\delta G^2 + \delta B^2 + \delta R^2 = 3\delta L^2 + \frac{11}{4}\delta M^2 + \frac{11}{4}\delta N^2 - \delta L \cdot \delta M - \delta L \cdot \delta N - \frac{5}{2}\delta M \cdot \delta N. \tag{11.2.2}$$

Therefore, if the *L* screen is searched with a distortion rate λ then the screens *M* and *N* should be searched with a distortion rate $\sqrt{12/11}\,\lambda = 1.0444\lambda$, which is too small to make any difference from using the same distortion rate λ for all screens *L*, *M*, and *N*.

If the *L*, *M*, and *N* screens are compressed sequentially—first *L*, then *M*, last *N*—then the correct procedure to minimize *RGB* error is given by further calculation from (11.2.2):

$$\begin{aligned} \delta G^2 + \delta B^2 + \delta R^2 &= 3\delta L^2 + \frac{11}{4}\delta M^2 + \frac{11}{4}\delta N^2 - \delta L \cdot \delta M - \delta L \cdot \delta N - \frac{5}{2}\delta M \cdot \delta N \\ &= \frac{8}{3}\delta L^2 + \frac{24}{11}\left(\delta M - \frac{1}{3}\delta L\right)^2 + \frac{11}{4}\left(\delta N - \frac{2}{11}\delta L - \frac{5}{11}\delta M\right)^2. \end{aligned} \tag{11.2.3}$$

To be precise, for a given distortion rate λ, the compression algorithm is described in the following steps:

1. Compress the screen L using the distortion rate λ. Let the decompressed image be denoted by L'.

2. Compress the screen image $P = M - 1/3(L - L')$ using the distortion rate $\sqrt{11/9}\lambda$. Let the decompressed screen image be denoted by P', and calculate $M' = P' + 1/3(L - L')$.

3. Compress the screen image $Q = N - 2/11(L - L') - 5/11(M - M')$ using the distortion rate $\sqrt{32/33}\lambda$. Let the decompressed screen image be denoted by Q'; then $N' = Q' + 2/11(L - L') + 5/11(M - M')$.

In practice, the distortion rates between the luminance and chrominance can be weighted quite differently according to specific applications—due to the fact that the human eye senses different colors differently. The eye is more sensitive to yellow and green than to blue and purple.

Also, the reaction to an image displayed on a screen monitor is totally different from the reaction to a printed image. An image on a dazzling, active screen surely needs far less color information than does a fine color print. In fact, in baseline JPEG and MPEG standards, the screens M and N are coded not only with a coarser quantization, but in quarter resolution as well.

In a fractal system, there is no reason to code the screens M and N in quarter resolution, for the fractal representation is always resolution independent. However, giving different distortion rates to the luminance and chrominance images accommodates human perceptual bias, for example, using 2λ for the screens M and N while the distortion rate λ is used for the screen L.

11.2.2 COMPRESSING CHROMINANCE FROM LUMINANCE

Looking at the screen images shown in Figures 11.1.3 and 11.1.4, the chrominance screens are still strongly related to their corresponding luminance images. They have similar geometric structures, though they are in different brightness and contrast. Locally, it makes sense to assume that the chrominance pieces can actually be coded from their corresponding luminance pieces with a brightness adjustment and a contrast adjustment. In other words, the chrominance images are fractally coded by taking the 1-to-1 corresponding location luminance screen pieces as their reference image pieces.

This algorithm performs very well in most natural scenic photography images. However, this algorithm can be hindered in some chrominance-dominant images. For any problem there is a cure. One always can swap the luminance image with a chrominance image or redefine the decomposition formulas.

11.2.3 FRACTAL VECTOR CODING

The correct way to compress a color image is to treat each color pixel as an inseparable entity.

In that algorithm, each color pixel of a color image is treated elegantly as a single vector of *RGB* or *LMN*. Based on the belief that fractal transformations identify similar objects within an image, there is no reason to think that the different color components of the same image piece should come from different objects.

Thus, the fractal vector coding system can be viewed as a collection of several fractal image compression systems, one for each color component. The same destination region of all color components must share the same reference region, as each color component maps to itself. Such an algorithm is also called a *fractal color image compression algorithm using commonly addressed reference regions.*

Exercise: Implement a fractal color image compression system using commonly addressed reference blocks.

Color Plate 13 shows a sample image compressed using various chrominance compression algorithms.

11.3 IMAGE DISPLAY TECHNIQUES

There are several techniques that are important to image display. Mainly, the question is how to display a high-quality image with a designated number of colors. Specifically, how well can a 24-bit true-color image be displayed on a 256-color or 16-color monitor, or printed on a 4-color or 1-color printer? In fact, by answering this question, the output image quality on inexpensive display hardware can be significantly improved.

This problem can be divided into two parts:

1. Find the right *representative colors*, which is equivalent to building a *color table.*
2. Map each pixel to a right representative—including spread error for better perceptual quality.

11.3.1 COLOR MAPPING TABLE

Let m be the designated color table size. Given a color image, what are the best m colors to describe this image?

A color is a vector of three tristimulus channel values in some color decomposition system. Consider all colors used by the image as the training set of color vectors—the construction of a color table becomes exactly the construction of a codebook in vector quantization. Thus, this problem can be solved using any of the VQ algorithms presented in Chapters 5 and 7.

For example, Equitz's PNN algorithm in Section 5.3.4 can be used to generate a color table. The bubbling algorithm presented In Section 7.2.1 gives a solution to the construction of a color mapping classification function once the color table is given. Color Plate 10 shows an image displayed using this algorithm.

For some images with a very unbalanced color histogram, such as that shown in Color Plate 11, a picture has a lot of blue but also many other colors [the above algorithm tends to give too few color representatives to the massive number of similar colors (*blue* in the sample case)]. The image color histogram needs to be considered and used as weights to design the color table. Therefore, the PNN algorithm needs to be improved to a weighted PNN algorithm.

In this algorithm, at any moment, every cluster is attached to a *weight*, which is the number of training colors mapped to this cluster. In the comparison of any two neighboring clusters, penalties that are proportional to their weights are added. As a consequence, heavy clusters have less chance to be combined, and more representatives are assigned to commonly used colors. By manipulating the penalty rate, the desired performance is achieved (see Color Plate 11).

11.3.2 IMAGE DITHERING

A false contour is the typical artifact in displaying a true color image in a color table with too few representatives. Dithering is the most commonly used technique to solve this problem.

The *error diffusion algorithm* was developed by Floyd and Steinberg [FS]. The result is quite satisfactory. As shown in Figure 11.3.1, the error (i.e., the difference between the exact pixel value and the approximated value actually displayed) is added to the values of the four adjacent pixels to the right of and below the active pixel: *7/16 of the error to the pixel to the right, 5/16 to the pixel immediately below, 3/16 to the pixel below and to the left, and 1/16 to the pixel below and to the right.*

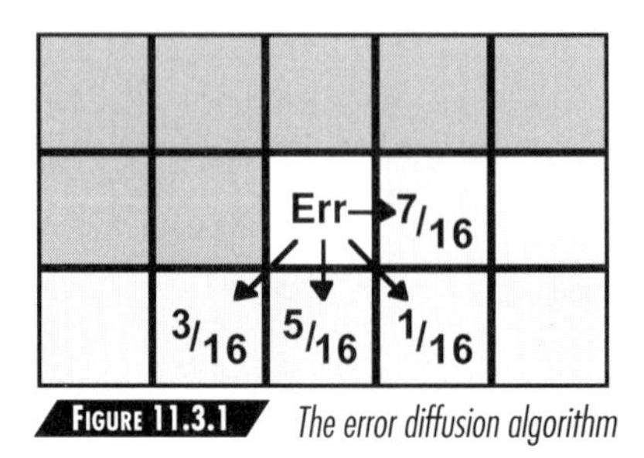

FIGURE 11.3.1 The error diffusion algorithm

FIGURE 11.3.2 Lena *dithered using the* Floyd-Steinberg error diffusion algorithm

Figure 11.3.2 shows the results of the Floyd-Steinberg algorithm applied to a grayscale image. In the color image case this algorithm can be applied to three individual *RGB* decomposed monochrome images.

He said only one word, then turned away, and left me behind. That word penetrated my heart, electrified my soul, and shook my body with a magnificence I had never experienced before.

— One Key One Lock

Anonymous

Entropy Coding 12

All image data compression algorithms, whether they be lossy or lossless, consist of two main components: *data modeling* and *code packing*.

In the first step (*data modeling*), the original data information is decomposed or transformed into new, equivalent forms that are suited for extracting data redundancy. Fractal model image pieces are described in affine transformations based upon the belief that these forms of fractal transformations give a simpler and more accurate way to describe an image.

The second step (*code packing*) actually stores the new data form with some predefined rules and conventions aimed at a much more compact data storage. In this case, transformations are stored in the indices or the quantized values of their coefficients and parameters (as in brightness means, contrast factors, spatial forms, reference addresses) in a packed format. Data in a packed format is called *code*.

Previous chapters have covered a great amount of information concerning image modeling. The code packing, for the use of an imaging model, is an equally important part of any compression system. This chapter will focus on algorithms and methods used in *code packing*. These algorithms and methods have been studied independently from image modeling. And most of them can be applied to all compression systems.

Because of the fact that images are often represented in 256×256 or 512×512 panels, in 4×4 or 8×8 square tilings, and in some different fractal code types and settings, most index or parameter arrays have a size not exceeding a few thousand entries in an image representation. Since adaptive techniques are not appropriate for small data strings, this chapter will study only the *fixed distribution entropy coding*. As an example, an exact adoption will be discussed later in Chapter 14, when a sample *future image format* is presented in more detail.

12.1 ENTROPY

A data set is a stream of *symbols*. A set of *codes* is a set of binary strings of 0's and 1's. With a certain criterion, a code is assigned to each symbol. The *length* of a code is the length of its string. Data packing takes these symbols and transforms them into codes. If the packing is effective, i.e., a good compression is achieved, the resulting codes will have a smaller total size than the stream of symbols. If we assume that the only known redundancy of a given stream of symbols is its probability distribution or frequency distribution, i.e., there is no adjacency relation between symbols when they are next to each other, such a model of symbols is called a *discrete memoryless source* (DMS) or a *zero-order Markov source*. In this case, the best packing method can be proven mathematically to be the *arithmetic coding algorithm*, based on Shannon's entropy theorem.

12.1.1 SYMBOLS AND CODES

More precisely, *data* is the message or information that is to be transmitted, stored, or expressed, which is formed as a string of symbols. *Symbols* are elements for describing the data, which could be letters, signs, numbers, marks, etc. For efficiency, or secrecy, or for other reasons, a *code* is assigned to each symbol, which in general is a string of 0's and 1's. So, instead of sending the data in symbols, the codes are transmitted. Hence, one can recover the original data if one knows the relation between the symbols and the codes, which is called the *codebook*. For example, let us consider a sequence of vowels:

$$\text{i, o, a, e, e, i, a, e,} \tag{12.1.1}$$

They could come from any sentence, such as *"It's gonna be the first game"* or *"In God's name he will ask her"* or *"Will Joan ever call me?"*

The symbol set is $S = \{\, a, e, i, o, u \,\}$, and the data is a string of eight symbols, $D = i\ o\ a\ e\ e\ i\ a\ e$. Here is a codebook, $C = \{\, 000, 001, 010, 011, 100 \,\}$, that converts the data string to

$$0\,1\,0\,0\,1\,1\,0\,0\,0\,0\,0\,1\,0\,0\,1\,0\,1\,0\,0\,0\,0\,0\,0\,1, \tag{12.1.2}$$

which uses 24 bits. Try another codebook: $C = \{\, 00, 01, 10, 110, 111 \,\}$. The data string is coded in only 17 bits:

$$1\,0\,1\,1\,0\,0\,0\,0\,1\,0\,1\,1\,0\,0\,0\,0\,1. \tag{12.1.3}$$

In comparing these two codebooks, it is obvious which one is more efficient for the data string.

What about $C = \{\, 01, 0, 10, 11, 110 \,\}$? A shorter code string of only 13 bits is obtained:

$$1\,0\,1\,1\,0\,1\,0\,0\,1\,0\,0\,1\,0. \tag{12.1.4}$$

Unfortunately, the data string cannot be recovered from this string. It could be *i o a e e i a e*, but it also could be *i o a e a e a e* or *i u i a e e i* or something else. Therefore, such a codebook is invalid.

When is a codebook valid? When does a codebook perform high compression? The first question has a simple solution. It is not difficult to prove that

> *A code book is valid if and only if no code is the first part of another code in the codebook.*

This is the *codebook condition*. In the previous example, the code 0 for *e* is the first part of the code 01 for *a*, and the code 11 for *o* is the first part of the code 110 for *u*. A simple and elegant solution to the second question was given by Huffman [H5]. This solution, called *Huffman Coding*, will be discussed in the next section.

12.1.2 THE SHANNON ENTROPY THEOREM

Let $S=\{ s_1, s_2, \cdots, s_n \}$ be a set of n symbols. Given $D=\{ d_1, d_2, \cdots, d_l \}$, a data set of l symbols in a sequence (the number l is also called the *data length* of D), the *probability distribution* of the symbol set S in the data D is the collection of positive numbers $P=\{ p_1, p_2, \cdots, p_n \}$, one for each symbol, defined by

$$p_i = \frac{\left|\left\{ d_k \in D \mid d_k = s_i \right\}\right|}{l}, \text{ for } i = 1, 2, \cdots, n. \tag{12.1.5}$$

Consequently, $\sum p_i = 1$. If the probability distribution is the only assumed redundancy information, the pair (S, P) is called a *zero-order Markov source*. The data sequence D is called a *zero-order Markov sequence*.

Using the above notations, the (*zero-order*) *entropy* of the data sequence D, introduced by Claude Shannon in the 1940s, who borrowed the word from thermodynamics, is defined to be

$$e = e(D) = -\sum_{i=1}^{n} p_i \cdot \log_2 p_i . \tag{12.1.6}$$

Let $C=\{ c_1, c_2, \cdots, c_n \}$ be a codebook. Following a simple calculation, the *average code length* using C to represent the data sequence D is

$$e_C(D) = \sum_{i=1}^{n} p_i \cdot |c_i|. \tag{12.1.7}$$

Here is the main theorem of the section:

SHANNON ENTROPY THEOREM:
Given a data sequence of symbols, for any codebook given to that symbol set, the average code length using the codebook to represent the data sequence is always greater than or equal to the zero-order entropy of the data sequence.

Proof: Let $D=\{ d_1, d_2, \cdots, d_l \}$ be a zero-order Markov data sequence of length l of symbols from the set $S=\{ s_1, s_2, \cdots, s_n \}$ with the probability distribution $P=\{ p_1, p_2, \cdots, p_n \}$. And let $C=\{ c_1, c_2, \cdots, c_n \}$ be a codebook of S. Then the theorem is equivalent to the formula

$$e_C(D) = \sum_{i=1}^{n} p_i \cdot |c_i| \;\geq\; e(D) = -\sum_{i=1}^{n} p_i \cdot \log_2 p_i . \tag{12.1.8}$$

The theorem will be proven inductively on the number of symbols n.

The case $n = 1$ is trivial.

For $n > 1$, let C_0 denote the set of codes whose first entry is 0, and C_1 the set of codes starting with 1. If one of the sets is empty, one can throw away the first entry of each code string and obtain a new set of codes C'. Since each code has a shorter length, replace the set of codes C with C'. Obviously, $e_{C'}(D) < e_C(D)$. Hence, if $e_{C'}(D) \geq e(D)$ has been proved, then $e_C(D) \geq e(D)$.

If both C_0 and C_1 are not empty, each contains fewer than n codes. Then let C_0^- and C_1^- denote the set of codes from C_0 and C_1 without the first entry, respectively. Splitting the symbols accordingly,

$$S_0 = \left\{ s_i \in S \;\middle|\; c_i \in C_0 \right\} \text{ and } S_1 = \left\{ s_i \in S \;\middle|\; c_i \in C_1 \right\}, \tag{12.1.9}$$

and two data sets D_0 and D_1 are obtained by restricting D to the subsets of symbols S_0 and S_1, respectively. Set

$$q_0 = \sum_{c_i \in C_0} p_i \text{ and } q_1 = \sum_{c_i \in C_1} p_i \,. \tag{12.1.10}$$

Clearly, $q_0 + q_1 = 1$, and the sets

$$P_0 = \left\{ \frac{p_i}{q_0} \in S \;\middle|\; c_i \in C_0 \right\} \text{ and } P_1 = \left\{ \frac{p_i}{q_1} \in S \;\middle|\; c_i \in C_1 \right\} \tag{12.1.11}$$

are the probability sets of D_0 and D_1 in symbols S_0 and S_1, respectively. Applying the theorem to (D_0, S_0, C_0^-, P_0) and (D_1, S_1, C_1^-, P_1), we have

$$e_{C_k^-}(D_k) = \sum_{c_i \in C_k} \frac{p_i}{q_k} \cdot \left(|c_i| - 1 \right) \geq e(D_k) = - \sum_{c_i \in C_k} \frac{p_i}{q_k} \cdot \log_2 \frac{p_i}{q_k}, \tag{12.1.12}$$

for $k = 0, 1$. Thus,

$$\sum_{c_i \in C_k} p_i \cdot |c_i| - \sum_{c_i \in C_k} p_i \geq - \sum_{c_i \in C_k} p_i \cdot \log_2 p_i + \sum_{c_i \in C_k} p_i \cdot \log_2 q_k \,, \tag{12.1.13}$$

for $k = 0, 1$. That is,

$$\sum_{c_i \in C_k} p_i \cdot |c_i| - q_k \geq - \sum_{c_i \in C_k} p_i \cdot \log_2 p_i + q_k \cdot \log_2 q_k \,, \tag{12.1.14}$$

for $k = 0, 1$. Adding the two formulas for $k = 0, 1$ together, we have

$$\begin{aligned} e_C(D) = \sum_{i=1}^{n} p_i \cdot |c_i| &\geq -\sum_{i=1}^{n} p_i \cdot \log_2 p_i + q_0 \cdot \log_2 q_0 + q_0 + q_1 \cdot \log_2 q_1 + q_1 \\ &= e(D) + \left(1 + q_0 \cdot \log_2 q_0 + (1 - q_0) \cdot \log_2 (1 - q_0)\right) \\ &\geq e(D), \end{aligned} \tag{12.1.15}$$

since the function $f(x) = x \cdot \log x + (1 - x) \cdot \log(1 - x)$ has a single critical point, which is a minimal point, at $x = 1/2$, according to Calculus 101. ♦

The *Huffman coding algorithm* finds a codebook that gives the smallest possible average code length. The *arithmetic coding algorithm*, by allowing codes with fractional length, will accomplish that the "average code length" is exactly the *entropy* of the data sequence D.

Similarly, one can define the *probability matrix* $P = \{ p_{ij} \}_{i,\, j=1,2,\cdots,n}$, where $P_i = \{ p_{i_1}, p_{i_2}, \cdots, p_{i_n} \}$ is the probability distribution after the symbol s_i, for all $i = 1, 2, \ldots, n$, i.e.,

$$p_{ij} = \frac{\left|\left\{ d_k \in D \mid d_{k-1} = s_i \ d_k = s_j \right\}\right|}{\left|\left\{ d_k \in D \mid d_k = s_i \right\}\right|}, \text{ for all } j = 1, 2, \ldots, n. \tag{12.1.16}$$

This implies that $\sum_j p_{ij} = 1$. The pair (S, P) is called a *first-order Markov source*. In this model, the data sequence D is called a *first-order Markov sequence*.

In the same manner, an *mth-order Markov source* is a model in which the probability of occurrence of a source symbol depends upon a finite number m of the preceding symbols. An mth-order Markov source has n^m states. Thus, the conditional probability set has n^{m+1} entries:

$$P = \left\{ p_{i_1 i_2 \cdots i_m, j} \right\}_{i_1, i_2, \cdots, i_m, j = 1, 2, \cdots n}, \tag{12.1.17}$$

where $p_{i_1 i_2 \cdots i_m, j}$ is the probability of the symbol s_j after m symbols $s_{i_1} s_{i_2} \cdots s_{i_m}$, for all $i_1, i_2, \cdots, i_m, j = 1, 2, \cdots, n$.

The higher-order Markov source can be restricted to a lower-order one. For example, the above first-order Markov source is restricted to a zero-order one by solving the systems of equations

$$\sum_{j=1}^{n} p_j = 1, \ \sum_{j=1}^{n} p_{ij} = 1, \text{ and } p_i = \sum_{j=1}^{n} p_j \cdot p_{ji}, \tag{12.1.18}$$

for all $i=1,2,\ldots,n$. It is not unusual to see higher-order Markov data. Yet in most real practices, a higher-order Markov data sequence is almost always converted to a lower-order one, e.g., a first-order Markov data string is often converted into a zero-order one by regrouping, differentiating, and using extra correlations inherited in specific cases. Nonetheless, the theory for the zero-order case is not too difficult to generalize to the higher-order cases.

Unless otherwise noted, all data sequences henceforth will be assumed to be zero-order Markov sequences.

12.2 HUFFMAN CODING

Given a set of symbols $S=\{ s_1, s_2, \cdots, s_n \}$ and given its probability distribution $P=\{ p_1, p_2, \cdots, p_n \}$, how can an efficient codebook $C=\{ c_1, c_2, \cdots, c_n \}$ be constructed? Similar attempts, now called the *Shannon–Fano coding algorithm*, were given independently by Shannon–Weaver [SW] and Fano [F2]. An elegant solution was discovered not too long after by Huffman [H5]. Here, the better one, the *Huffman coding algorithm*, will be presented.

12.2.1 THE HUFFMAN CODING ALGORITHM:

In a zero-order Markov source given by the symbol set $S=\{ s_1, s_2, \cdots, s_n \}$ and probability set $P=\{ p_1, p_2, \cdots, p_n \}$, the task is to find a good, efficient codebook $C=\{ c_1, c_2, \cdots, c_n \}$.

HUFFMAN CODING ALGORITHM:

The Huffman coding algorithm uses a *bottom-up tree merging method.* (You may be aware that the ideas of *bottom-up tree merging* and *top-down tree splitting* have been used in various contexts before in the book).

Step A-1: Find two symbols that have the smallest probabilities, e.g., $u, v \in S$. Let $\sigma=\sigma^n : \{1, 2, \cdots, n\} \to \{1, 2, \cdots, n\}$ denote a permutation such that $u=s_{\sigma(n-1)}$ and $v=s_{\sigma(n)}$. Then, define a new Markov source consisting of the set of symbols

$$S^{n-1} = \left\{ s_{\sigma(1)}, s_{\sigma(2)}, \cdots, s_{\sigma(n-2)}, t^{n-1} \right\} \tag{12.2.1}$$

and the set of probability distributions

$$P^{n-1} = \left\{ p_{\sigma(1)}, p_{\sigma(2)}, \cdots, p_{\sigma(n-2)}, p_{\sigma(n-1)} + p_{\sigma(n)} \right\}. \quad (12.2.2)$$

The n-symbol problem is reduced to an $(n-1)$-symbol problem of finding a codebook

$$C^{n-1} = \left\{ c_1^{n-1}, c_2^{n-1}, \cdots, c_{n-1}^{n-1} \right\} \quad (12.2.3)$$

for S^{n-1}.

In summary, the process of step 1 can be formulated as the following procedure

$$\left\{ S, P, C \right\} \Rightarrow \sigma^n \Rightarrow \left\{ S^{n-1}, P^{n-1}, C^{n-1} \right\} \quad (12.2.4)$$

Step A(2): Inductively, apply Step 1 to the $n-1$ symbol Markov sources $\{S^{n-1}, P^{n-1}, C^{n-1}\}$:

$$\left\{ S^{n-1}, P^{n-1}, C^{n-1} \right\} \Rightarrow \sigma^{n-1} \Rightarrow \left\{ S^{n-2}, P^{n-2}, C^{n-2} \right\}, \quad (12.2.5)$$

so that, the n–1 symbol problem is reduced to an n–2 symbol problem.

Step A(n): Repeat this process, after n steps, we obtain $\{S^1, P^1, C^1\}$, a data source of a symbol set S^1 that has only one symbol in it.

Step B(1): For a data source of a single symbol all we need to know is the source length, therefore there is no need for codes. In other words, the zero-bit empty code is assigned to the only symbol.

Step B(2): Now to a symbol set S^2 of only two elements, assign the one-bit code 0 to the first symbol and the one-bit code 1 to the other, i.e., $C^2 = \{ c_1^2 = 0, c_2^2 = 1 \}$.

Step B(3): Build the codes of $C^3 = \{ c_1^3, c_2^3, c_3^3 \}$ from the equations

$$c^3_{\sigma^3(1)} = c_1^2, \quad c^3_{\sigma^3(2)} = c_2^2 0, \quad \text{and } c^3_{\sigma^3(3)} = c_2^2 1. \quad (12.2.6)$$

Step B(n): Finally, construct the codes of $C = \{c_1, c_2, c_n\}$ from the equations

$$c_{\sigma^n(i)} = c_i^{n-1}, \text{ for } i = 1, 2, \cdots, n-2,$$
$$c_{\sigma^n(n-1)} = c_{n-1}^{n-1}0, \text{ and } c_{\sigma^n(n)} = c_{n-1}^{n-1}1 \,. \quad (12.2.7)$$

Let us illustrate this algorithm by using the same old example of the vowel string: **i, o, a, e, e, i, a, e.** The probability distribution is $P = \{ 0.25, 0.375, 0.25, 0.125, 0 \}$ for the symbol set $S = \{ a, e, i, o, u \}$. As shown in Figure 12.2.1, in the first step, vowels **o** and **u** are merged to a new symbol **x**, which has a probability 0.125.

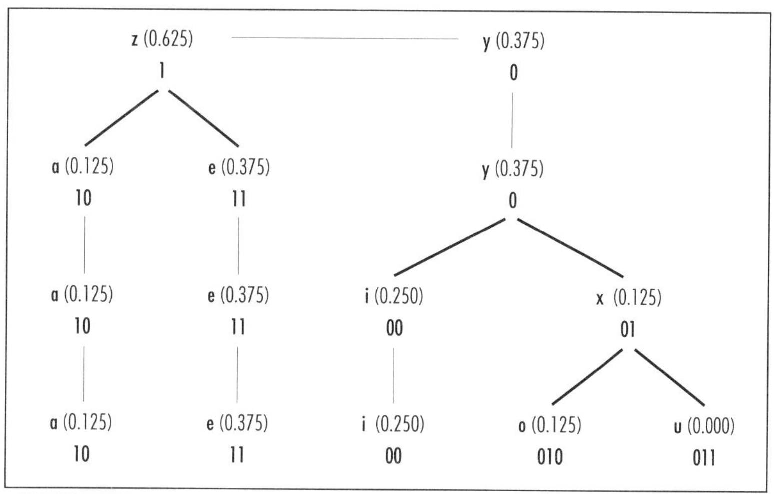

FIGURE 12.2.1 *Sample illustration of the* Huffman Coding Algorithm

In the next step, two with the smallest probabilities could be chosen as either **a** and **x**, or **i** and **x**. that is, symbols **i** and **x** are merged to **y**, which has a probability of 0.375. Thus, in the third step, the symbols **a** and **e** (or **a** and **y**) are merged to **z**.

Now assign **y** the code 0 and **z** the code 1, and then assign **a** the code 10 and **e** the code 11, and **i** the code 00 and **x** the code 01, and **o** the code 010 and **u** the code 011. This defines a codebook, C={ 10, 11, 00, 010, 011 }, which translates the original data string in 17 bits:

$$0\,0\,0\,1\,0\,1\,0\,1\,1\,1\,1\,0\,0\,1\,0\,1\,1. \quad (12.2.8)$$

If in the third step the symbols **a** and **y** are merged to **z**, the codebook will become C={ 10, 0, 110, 1110, 1111 }. It looks quite different from the previous codebook, but it gives a code string,

$$1\,1\,0\,1\,1\,1\,0\,1\,0\,0\,0\,1\,1\,0\,1\,0\,0, \qquad (12.2.9)$$

of the same length: 17 bits.

The entropy of the sample data sequence is 15.245. According to Shannon entropy, the sample data cannot be coded in fewer than 16 bits by using any zero-order entropy coding scheme.

The next subsection will show a C implementation.

12.2.2 C IMPLEMENTATION

Quite often, the art of describing an algorithm is very different from the art of writing a good program. The codebook is described in 0's and 1's in the algorithm, but there is no reason to store them in an array of bits. Actually, today's computers handle BYTE, WORD, or DWORD, at the same speed as when dealing with BIT. It is efficient to design the codebook structure to fit this situation. Especially in most applications, codes for symbols rarely exceed 16 bits, which can be held completely in a WORD. If 16 is too small, one can extend this number to 32 by using DWORD.

So a code is no longer a string of 0's and 1's but a structure of a WORD that stores the code and a BYTE that tells the length of the corresponding code in bits.

```
typedef struct Huffman_code
{
    WORD    code;
    BYTE    bits;
} HCODE;
```

For computational convenience, as a convention, the low bit will be read first, and high bits thereafter. For example, a code $\mathbf{x}$ = 011010010 will be stored as

```
HCODE   x;
x.code = 0x96;          // = 010010110 in binary
x.bits = 9;             // it has 9 valid bits
```

Without losing generality but limiting to integer calculation, the symbols will be assigned to the set $S = \{\ 1, 2, \cdots, n\ \}$ and the probabilities $P = \{\ p_1, p_2, \cdots, p_n\ \}$ are assumed to be integers divided by $2^{15} = 32{,}768$. Here is the routine of creating the Huffman codebook:

```
1   INT CreateHuffmanCodeBook(
2       WORD    N;              // number of symbols
3       WORD    *Prob;          // input probability array
4       HCODE   *Book;          // output Huffman code book
5   ){

6       WORD    i,k;
7       WORD    s1,s2;              // active symbols
8       WORD    p1,p2;              // active probabilities
9       BYTE    *merge1,*merge2; // merging symbols

10      GetMemory(merge1,(N<<1));
11      GetMemory(merge2,(N<<1));
```

```
     //  borrow code buffer for active flags
12       for( i=N; i>0; i-- ) Book[i].code = 1;

     //  loop of merging symbols
13       for( k=N-1; k>0; k-- )  // merge k+1 symbols to k
14       {
     //  choose s1 and s2 that have the smallest prob p1 and p2
15           p1 = p2 = 0x8888;
16           s1 = s2 = 0;
17           for( i=N; i>0; i-- ){
18               if( Book[i].code ){
19                   if( Prob[i]<=p2 ){
20                       if( Prob[i]<=p1 ){
21                           p2 = p1; s2 = s1;
22                           p1 = Prob[ s1=i ];
23                       }
24                       else p2 = Prob[ s2=i ];
25                   }
26               }
27           }
     //  merge s2 to s1
28           merge1[k] = s1; Prob[s1] += Prob[s2];
29           merge2[k] = s2; Book[s2].code = 0;
30       }


     //  set initial code for the last remaining symbol
31       Book[s1].code = Book[s1].bits = 0;

     //  loop of creating codes
32       for( k=1; k<N; k++ )  // get one more codes
33       {
34           s1 = merge1[k];
35           s2 = merge2[k];
36           Book[s2].code = Book[s1].code | (1<<Book[s1].bits);
37           Book[s2].bits = (++Book[s1].bits);
38       }

39       FreeMemory(merge2);
40       FreeMemory(merge1);
41       return(OK);
42   }
```

12.2.3 RUN-LENGTH ENCODING

After completing the codebook, the encoding process is straightforward. The following C code is self-explanatory.

```
1    INT HuffmanEncoder(
2        LONG     Data_length;      // input data length
3        WORD     *Data;            // input data buffer
4        HCODE    *Book;            // Huffman code book
5        LONG     *Code_length;     // output code length in bits
6        BYTE     *Code             // output code array (preset to 0)
7    ){

8        LONG     k,c;
9        BYTE     *codptr;          // output code cursor
10       BYTE     bitoff;           // bit offset cursor
```

```
11      codptr = Code;  bitoff = 0;
12      for( k=0; k<Data_length; k++ )
13      {
14          c = (Book[Data[k]].code<<bitoff);
15          codptr[0] |= (c&255);
16          codptr[1] = ((c>>8)&255);
17          codptr[2] = (c>>16);
18          bitoff += Book[Data[k]].bits;
19          while(bitoff >= 8){ bitoff -= 8; codptr ++; }
20      }

21      *Code_length = ((codptr - Code)<<3) + bitoff;
22      return(OK);
23  }
```

12.2.4 DECODING USING AN INVERSE CODEBOOK

The *run-length decoding algorithm* is a little bit trickier than the encoding algorithm. The standard method is to build a codebook tree and then to sort through it (cf. [N]). Here we introduce a faster technique: *decoding using an inverse codebook algorithm.*

Let `MAX_CL` denote the maximal code bit length. Thus, by reading `MAX_CL` bits, at least a symbol could be recovered. An inverse codebook can be made for all numbers from `0` to `(1<<MAX_CL)-1` by mapping each of them to the first symbol it will decode. The inverse codebook contains `(1<<MAX_CL)` entries as calculated in the routine below.

```
1   #define     MAX_CL      15
2   #define     M           (1<<MAXCL)-1

3   INT GetInverseCodeBook(
4       WORD    N;              // number of symbols
5       HCODE   *Book;          // Huffman code book
6       WORD    InvBook[M]      // inverse code book
7   ){
8       LONGk,b,c;

9       for( k = N; k>0; k-- ){
10          b = (1<<Book[k].bits);
11          for( c=Book[k].codes; c<=M; c+=b ) InvBook[c] = k;
12      }

13      return(OK);
14  }
```

Using the inverse book, the decoder can be designed as the exact reverse procedure of the previous encoder.

```
1   INT HuffmanDecoder(
2       HCODE   *Book;          // Huffman code book
3       WORD    *InvBook;       // inverse code book
4       LONG    *Code_length;   // current code position in bits
5       BYTE    *Code           // input code buffer
6       LONG    Data_length;    // decoded data length
7       WORD    *Data           // output decoded data buffer
8   ){
```

```
9        LONG    k,c;
10       BYTE    *codptr;          // reading code cursor
11       BYTE    bitoff;           // bit offset cursor

12       codptr = Code + (*Code_length>>3);
13       bitoff = (*Code_length&7);
14       for( k=0; k<Data_length; k++ )
15           c = codptr[0]|(codptr[1]<<8)|(codptr[2]<<16);
16           bitoff += Book[Data[k]=InvBook[(c>>bitoff)&M]].bits;
17           while(bitoff >= 8){ bitoff -= 8; codptr ++; }
18       }

19       Code_length = ((codptr - Code)<<3) + bitoff;
20       return(OK);
21   }
```

This decoder seems to take a lot memory for the inverse code book when the number `M` is huge. In an implementation, the number `M` (if it is huge) can be reduced by redesigning the symbol source set—by grouping the symbols with long codes together along with common prefixes and different extensions. Hence, the codebook is split into two: one codebook consists of the short-code symbols and long-code symbol prefixes, and the other codebook consists of the remaining long-code symbol extensions. And a data sequence will be split into two correspondingly equivalent sequences.

The detailed implementation is assigned as an exercise. In practice, for example, in the *Fax Group IV* algorithm (the binary run-length compression scheme adopted in standard facsimile transmission) the maximal code bit length could reach 20 bits, that implies an inverse table of 2MB. After using the *prefix-extension method*, five inverse code books can be constructed with a total memory less than 4KB.

12.3 ARITHMETIC CODING

Mathematically, the *arithmetic coding algorithm* is perfect for a zero-order Markov data sequence. The main disadvantage in comparison with Huffman coding is its decoding speed.

12.3.1 THE ARITHMETIC CODING ALGORITHM

To simplify discussion, the source set of symbols is assumed to be the integer set $S=\{1, 2, \cdots, n\}$. Let

$$f:\left\{1, 2, \cdots, n, n+1\right\}\rightarrow[0,1] \tag{12.3.1}$$

denote the *accumulate frequency function* of the probabilities

$$P=\left\{p_1, p_2, \cdots, p_n\right\}, \tag{12.3.2}$$

defined by

$$f(i) = \sum_{j<i} p_j \text{, for all } i = 1, 2, \cdots, n, n+1. \qquad (12.3.3)$$

ARITHMETIC CODING ALGORITHM:
Given $D = \{ d_1, d_2, \cdots, d_l \}$ a stream of symbols from $S = \{ 1, 2, \cdots, n \}$ of length l:

Step 0: Initialize the code c to zero, and start with the whole interval $[0, 1]$ as the next code incremental range. That is, if the interval length is denoted by r, set the initial values

$$c = 0, r = 1. \qquad (12.3.4)$$

Step 1: Pack the first datum d_1 by setting

$$c = f(d_1) \text{ and } r = p_{d_1} = f(d_1 + 1) - f(d_1). \qquad (12.3.5)$$

Step k: In general, pack the data d_k by setting

$$c = c + r \cdot f(d_k) \text{ and } r = r \cdot \left(p_{d_k} = f(d_k + 1) - f(d_k) \right), \qquad (12.3.6)$$

for $k = 2, 3, \cdots, l$.

Final: Store the code c in sufficient precision, e.g., in 2^p, such that

$$c \le \frac{[2^p c + 1]}{2^p} < \frac{[2^p c + 2]}{2^p} \le c + r. \qquad (12.3.7)$$

In fact, the code data will be the integer $[2^p c + 1]$ stored in exactly p bits.

For example, let us encode again the following stream of vowels:

$$\textbf{i, o, a, e, e, i, a, e.} \qquad (12.3.8)$$

The probabilities of **a**, **e**, **i**, **o**, and **u** are 25%, 37.5%, 25%, 12.5%, and 0%. Thus, the accumulate frequency will be

$$f = \{ 0, 0.25, 0.625, 0.875, 1, 1 \} = \{ 0, 4/16, 10/16, 14/16, 1, 1 \}. \qquad (12.3.9)$$

It is easy to use hexadecimal in the next discussion, since the common denominator is 8. Following the algorithm, the code and remaining range are listed step by step below:

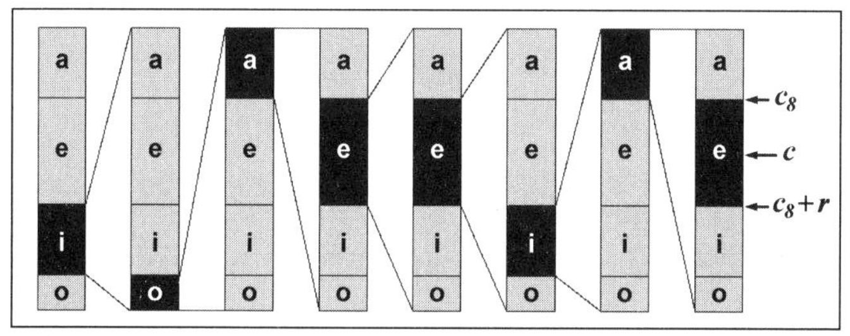

FIGURE 12.3.1 *Example of* Arithmetic Coding

	$c_0 = $ `0`,	$r = $ `1`,
i:	$c_1 = $ `0x0A/0x10`,	$r = $ `4/0x10`,
o:	$c_2 = $ `0xD8/0x100`,	$r = $ `8/0x100`,
a:	$c_3 = $ `0xD8/0x100`,	$r = $ `2/0x100`,
e:	$c_4 = $ `0xD88/0x1000`,	$r = $ `0x0C/0x1000`,
e:	$c_5 = $ `0xD8B/0x1000`,	$r = $ `0x48/0x10000`,
i:	$c_6 = $ `0xD8DD/0x10000`,	$r = $ `0x12/0x10000`,
a:	$c_7 = $ `0xD8DD/0x10000`,	$r = $ `0x48/0x100000`,
e:	$c_8 = $ `0xD8DE2/0x100000`,	$r = $ `0x1B/0x100000`.

Finally the 16-bit code is truncated:

$$c_8 \leq c = \texttt{0xD8DF/0x10000} < c_8 + r. \tag{12.3.10}$$

Notice that the code buffer may continue to have other types of codes. So the following bits could be any combination of 0's and 1's. Therefore, it is necessary to ensure that the followed bit string will not affect our result.

Since $c_8 + r = $ `0xD8DFA`, one stuffing bit 0 is required. The final code will have 17 bits, not 16:

$$c = \texttt{0xD8DE8/0x100000=0x1B1BD/0x80000}, \tag{12.3.11}$$

because the number 17 is the smallest number k to have a code c such that

$$c_8 \leq c < c + 1/(1 << k) \leq c_8 + r \tag{12.3.12}$$

and c is an integer over 2^k.

In this example, a string of 8 symbols is coded in 17 bits—a decimal number of 17-bit precision. What will it happen to a string of thousands of symbols? How can one deal with the seemingly incredible required precision? Also, in the example all probabilities are in fractions of a power of 2. How are the nondivisible cases managed? All of these are practical questions that will be answered in the next subsection.

12.3.2 C IMPLEMENTATION

Notice that in the example in the previous subsection, the first 8 bits `0xD8` are left unchanged since c_2. The active part of the code falls into a smaller and smaller interval, and as a consequence, the numbers in higher decimal positions of the code that reflect the early part of the data string hardly changes while more data are coded. Thus, the answer to the precision question is to temporarily store the early part of the code aside and focus on the active interval. The probabilities can be assumed to be fractions of some power of 2, since any number can be approximated in this form with arbitrarily small error. In practice, certain truncation criteria can be imposed in both the compression and decompression ends simultaneously. Although it will make codes system-specific, it has no impact on performance.

In the implementation presented here, the probabilities of the symbols will be given in a normalized accumulative manner, as a sequence of decreasing numbers from 16,384 ($=0\times4000$) to 0. The preparation routine of the *normalized accumulative probability table* is given below. By doing this, all division operations (which is the most time-consuming operation) are eliminated from the algorithm in both encoding and decoding.

```
1    INT AccumNormProbTable(
2        LONG *Prob;              // input probability table
3        LONG *AcumP;             // output prepared prob table
2        WORD MaxDatum;           // maximal data symbol
4        BYTE Norm                // = 0x4000 in our suggestion
5    ){

6        LONG    i;

     // Get accumulated probability
7        AcumP[MaxDatum] = 0;
8        for(i=MaxDatum-1; i>=0; i--)
9            AcumP[i] = AcumP[i+1]+Prob[i];

10       if(AcumP[0] > Norm){
11           AcumP[0] += MaxDatum;
12           for(i=MaxDatum-1; i>=0; i--)
13               Prob[i] = (Prob[i]*Norm+AcumP[0]-1)/AcumP[0];
14           for(i=MaxDatum-1; i>=0; i--)
15               AcumP[i] = AcumP[i+1]+Prob[i];
16       }

     // Normalization
14           for(i=MaxDatum-1; i>=0; i--)
15               AcumP[i] = (AcumP[i]*Norm)/AcumP[0];

16       return(OK);
17   }
```

Here is the C implementation of the arithmetic coding class:

```
1    class    ArithmeticEncoder {

2        public:

3        ArithmeticEncoder(
4            BYTE    *Code_begin // code buffer beginning
```

```
5           );

6       ~ArithmeticEncoder(
7           BYTE     **Code_end  // code buffer end
8           );

9       INT EncodeOneDatum(      // encode one datum at a time
10          WORD     datum       // packing datum
11          );

12      INT EncodeDataArray(     // encode a data array
13          WORD     *data;      // packing data array
14          LONG     data_length // data string length
15          );

16      WORD    *MaxDatum;       // maximal possible symbol
17      LONG    *Prob;           // probability tables

18      private:

19      LONG    *AcumP;          // acummulate prob in 0x4000th
20      BYTE    *CurCode;        // code buffer cursor
21      DWORD   Value;           // remain code value
22      DWORD   Range;           // active interval range
23  }
```

Initially, the code array cursor is set to the code beginning `Code_begin`, the code value `Value` is set to zero, and the current truncation accuracy is conventionally set to 2 to the power minus 12.

```
1   void    ArithmeticEncoder::ArithmeticEncoder(
2       BYTE *Code_begin
3   ){
4       AccumNormProbTable(Prob,AcumP,MaxDatum,0x4000);
5       CurCode = Code_begin;
6       Value  = 0;
7       Range = 0x01000000;
8       return;
9   }
```

In the end, dump out the remaining code array, stuff enough zeros so that future code array will have no effect while decoding the current code array, and return the code array pointer back to its parent routine for marking the code array ending.

```
1   void    ArithmeticEncoder::~ArithmeticEncoder(
2       BYTE **Code_end
3   ){
4       LONGa;

5       if( Range>0x01FFFF )
6           Value = (Value&0x0FFF0000) + 0x10000;
7       else if( Range>0x01FF )
8           Value = (Value&0x0FFFFF00) + 0x100;
9       else Value += 1;
```

```
10      if( Value&0x1000000 )
11      {
12          a = -1;
13              while(CurCode[a]==0xFF){ CurCode[a] = 0; a--; }
14              CurCode[a]++;
15          Value &= 0x00FFFFFF;
16      }

17      *(CurCode++) = (Value>>16);
18      if( Range<=0x01FFFF )
19      {
20          *(CurCode++) = ((Value>>8) & 0xFF);
21          if( Range<=0x01FF)
22              *(CurCode++) = (Value & 0xFF);
23      }

24      *Code_end = CurCode;
25      return;
26  }
```

To encode a datum `datum` is to map the current range to `[0, 16384]`. The inverse map at `AcumP[datum]` produces the code and the inverse of the interval `[AcumP[datum] , AcumP[datum+1])` is exactly the new range.

```
1   INT ArithmeticEncoder::EncodeOneDatum(
2       WORD datum
3   ){
4       LONG    a;

5       Range = ((Range*AcumP[datum])>>14) - (
6           a = ((Range*AcumP[datum+1])>>14) );
7       Value += a;

8       while( Range<0x00008000 )
9       {
10          if( Value&0x01000000 )
11          {
12          a = -1;
13              while(CurCode[a]==0xFF){ CurCode[a] = 0; a--; }
14                  CurCode[a]++;
15          }

16          *(CurCode++) = ((Value>>16)&0x00FF);
17          Range <<= 8;
18          Value = ((Value&0x0000FFFF)<<8);
19      }

20      return(OK);
21  }
```

To encode a data array is really a loop of encoding individual data entries.

```
1   INT ArithmeticEncoder::EncodeDataArray(
2       WORD    *data
3       LONG    data_length;
4   ){
```

```
5       LONG    k;

6       for(k = 0; k < data_length; k ++ )
7           EncodeOneDatum( data[k] );

8       return(OK);
9   }
```

As a reverse procedure, the decoder class is defined as follows:

```
1   class   ArithmeticDecoder {

2       public:

3       ArithmeticDecoder(
4           BYTE    *Code_begin // code buffer beginning
5           );

6       ~ArithmeticDecoder(
7           BYTE    **Code_end  // code buffer end
8           );

9       INT DecodeOneDatum(     // encode one datum at a time
10          WORD    *datum      // packing datum
11          );

12      INT DecodeDataArray(    // encode a data array
13          WORD    *data;      // packing data array
14          LONG    data_length // data string length
15          );

16      LONG    *AcumP;         // acummulate prob in 0x4000th

17      private:

18      BYTE    *CurCode;       // code buffer cursor
19      DWORD   Value;          // active code for reading
20      DWORD   Range;          // active interval range
21  }
```

Initially, set the code array cursor to the code beginning Code_begin, set Range interval to 2 to the power 12, and read the first 3 bytes of code value.

```
1   void ArithmeticDecoder::ArithmeticDecoder(
2       BYTE *Code_begin
3   ){
4       Value  = (Code_begin[0]<<16)
5              | (Code_begin[1]<<8) | Code_begin[2];
6       CurCode = Code_begin+3;
7       Range = 0x01000000;
8       return;
9   }
```

In the end, we point the cursor to the right position for the possible next data buffer.

```
1   void    ArithmeticDecoder::~ArithmeticDecoder(
2       BYTE **Code_end
3   ){
4       if( Range>0x01FFFF ) CurCode -= 2;
5       else if( Range>0x01FF ) CurCode --;
6       *Code_end = CurCode;
7       return;
8   }
```

To decode a datum is exactly to reverse the procedure by finding a datum that fits the specified interval.

```
1   INT ArithmeticDecoder::DecodeOneDatum(
2       WORD *datum
3   ){
4       LONG    a;
5       WORD    d=1;

6       while((a = ((Range&*AcumP[d])>>14)) >= Value)
7           d++;

8       Range = ((Range&*AcumP[--d])>>14)) - a;
9       Value -= a;
10      *datum = d;
11      while( Range < 0x00008000 )
12      {
13          Value = ((Value<<8)|*(CurCode++));
14          Range <<= 8;
15      }
16      return(OK);
17  }
```

The real speed bottleneck of the routine is in lines 6–7. From these two lines it is obvious that the most probable symbols should be put ahead of the less frequently used ones for fast speeds in this implementation.

To decode a data array is really a loop of decoding individual data entries.

```
1   INT ArithmeticDecoder::DecodeDataArray(
2       WORD    *data
3       LONG    data_length;
4   ){
5       LONGk;

6       for(k = 0; k < data_length; k ++ )
7           DecodeOneDatum( &data[k] );

8       return(OK);
9   }
```

12.3.3 INCREMENTAL ENTROPY CODER

Inspired by Pennebaker, Mitchell *et al.*'s IBM Q-coder [PMLA], here a new algorithm—one that removes all multiplication operations from the arithmetic coding scheme — is introduced. We call this method the *incremental entropy coding algorithm*.

By analyzing the arithmetic coding implemented in the previous section, note that in each step when a new datum, e.g., `datum`, is encoded, the code variable `Value` increases and the range interval `Range` is reset by executing

```
Value += Inc[datum+1];
Range = Inc[datum]-Inc[datum+1];
```

if denote `Inc[d]=Range*AcumP[d]/0x4000`, for all datum `d`, which is really equivalent to renormalizing the accumulated probability table by scaling the interval `[0,0x4000]` into `[0,Range]`. In the implementation, the interval `Range` has been kept in the interval `[0x8000,0x1000000]`. And multiplication operations are involved only in the renormalization step.

In the incremental entropy coding algorithm, a few more tables will be constructed ahead of time so that the renormalization step will be replaced by approximation. This time, the interval `Range` will be kept in the interval `[0x8000,0x10000)`.

Divide the interval `[0x8000,0x10000)` into n equal subintervals, for example $n = 16$, that is,

`[0x8000+i*0x0800,0x8000+(i+1)*0x0800)`, for $i = 0, 1, \ldots, 15$.

For each interval, an incremental table `*IncP[i]` is assigned using the same accumulated probability table setup routine `AccumNormProbTable()` as shown in the previous subsection by setting the normalization factor `Norm` to `0x8000+i*0x0800`. (Alternatively, a specifically designed routine could build all tables at the same time more efficiently.)

Now, in the encoder, if the current range interval `Range` is in the `i`th interval, the incremental table `*IncP[i]` will be used, i.e.,

```
Value += Inc[datum+1];
```

The new range interval will be set similarly, except for the `0` symbol, to which the overshoot range is given:

```
if(datum) Range = IncP[datum]-IncP[datum+1];
else Range -= IncP[1];
```

As a good choice and a default convention, the `0` symbol should be the most frequent symbol.

In each step, the idea probability table is given by

$$p_d = \frac{\texttt{IncP[d]-IncP[d+1]}}{\texttt{IncP[0]}}, \text{ for all } d; \tag{12.3.13}$$

and the one actually used is the following:

$$q_d = \frac{\texttt{IncP[d]-IncP[d+1]}}{\texttt{Range}}, \text{ for } d \neq 0;\ q_0 = \frac{\texttt{Range-IncP[1]}}{\texttt{Range}}. \tag{12.3.14}$$

The performance of this algorithm has the estimate

$$\begin{aligned}
\varepsilon &= \left(-\sum_d p_d \log q_d\right) - \left(-\sum_d p_d \log p_d\right) \\
&= -p_0 \log \frac{\texttt{Range-IncP[1]}}{\texttt{IncP[0]-IncP[1]}} + \log \frac{\texttt{Range}}{\texttt{IncP[0]}} \\
&= -p_0 \log\left(1 + \frac{\delta}{p_0}\right) + \log(1+\delta), \quad \text{where } \delta = \frac{\texttt{Range-IncP[0]}}{\texttt{IncP[0]}} < \frac{1}{2n} \\
&= -p_0 \left(\frac{\delta}{p_0} - \frac{1}{2}\left(\frac{\delta}{p_0}\right)^2 + \frac{1}{3}\left(\frac{\delta}{p_0}\right)^2 - \cdots\right) + \left(\delta - \frac{\delta^2}{2} + \frac{\delta^3}{3} - \cdots\right) \\
&\leq \frac{\delta^2}{2p_0} < \frac{1}{4n}, \quad \text{if } \delta \leq p_0.
\end{aligned} \tag{12.3.15}$$

In the case $n=16$, this algorithm increases no more than 0.016 bit per datum. For a mean buffer in an average of four bits per datum, the performance reduction is less than four percent.

The only constraint used in the performance deduction estimation is $\delta \leq p_0$, which can be secured by choosing $n \geq 2/p_0$. If one doesn't want to make too many tables, one always can increase p_0 by combining symbols.

The following is the complete C code.

```
1   class   IncrementalEncoder {

2       public:

3       IncrementalEncoder(
4           BYTE    *Code_begin         // code buffer beginning
5           );

6       ~IncrementalEncoder(
7           BYTE    **Code_end          // code buffer end
8           );

9       INT EncodeOneDatum(             // encode one datum at a time
10          WORD    datum               // packing datum
11          );

12      INT EncodeDataArray(            // encode a data array
13          WORD    *data;              // packing data array
14          LONG    data_length         // data string length
15          );

16      WORD    *MaxDatum;              // maximal possible symbol
17      LONG    *Prob;                  // probability tables

18      private:
```

```
19      LONG    *IncP[16];          // acummulate prob tables
20      BYTE    *CurCode;           // code buffer cursor
21      DWORD   Value;              // remain code value
22      DWORD   Range;              // active interval range
23      SHORT   Bits;               // count shift bits
24      SHORT   Next;               // next active table index
25  }
```

The initial setting is similar to that of arithmetic coding.

```
1   void    IncrementalEncoder:: IncrementalEncoder(
2       BYTE *Code_begin
3   ){
4       SHORT       i,j;

5       for(i=1;i<15;i++){
6           j = 0x8000+i*0x0800;
7           AccumNormProbTable(Prob,IncP[i],MaxDatum,j);
8       }
9       CurCode = Code_begin;
10      Value = 0;
11      Range = 0x008000;
12      Next = 0;
13      Bits = -1;
14      return;
15  }
```

The closing routine dumps out the remaining codes.

```
1   void    IncrementalEncoder::~IncrementalEncoder(
2       BYTE **Code_end
3   ){
4       LONG    a;

5       *Code_end = CurCode;
6       if(Bits<0) return;

7       Value += 0x7FFF;
8       while((Bits++)<8) Value <<=1;
9       }

10      if(Value&0x01000000 )
11      {
12          a = 0;
13          while(CurCode[a]==0xFF){ CurCode[a]=0; a--; }
14          CurCode[a]++;
15      }

16      *(*Code_end++) = ((Value>>16)&0x00FF);
17      return;
18  }
```

Next is the main encoding procedure.

```
1   INT IncrementalEncoder::EncodeOneDatum(
2       WORD datum
3   ){
4       LONG    a;
```

```
5       Value += IncP[Next][datum+1];
6       if(datum) Range=IncP[Next][datum]-IncP[Next][datum+1];
7       else Range -= IncP[Next][1];

8       while( Range<0x00008000 )
9       {
10          Range <<= 1;
11          Value <<= 1;
12          if( (++Bits) == 8) )
13          {
14          if(Value&0x01000000 )
15              {
16                  a = 0;
17                  while(CurCode[a]==0xFF){ CurCode[a]=0; a--; }
18                  CurCode[a]++;
19              }

20              *(CurCode++) = ((Value>>16)&0x00FF);
21              Value &= 0x0000FFFF;
22              Bits = 0;
23          }
24      }

25      Next = ((Range>>11)&0x000F);
26      return(OK);
27  }
```

Encoding a data array is equivalent to looping the above encoding routine over all individual data entries. The routine `EncodeDataArray()` is identical to the one given in the previous subsection.

As a reverse procedure, the decoder class can be given analogously. We skip the decoder and leave it to the reader as an exercise.

12.4 FRACTAL BILLIARDS

A fresh view of arithmetic coding was ingeniously formulated by Barnsley. The idea was mentioned in his co-authored book *Fractal Image Compression* [BH], and further study has been done at Iterated Systems [BDX]. Barnsley presented the arithmetic coding algorithm as a chaotic dynamic system: a *fractal billiards game.*

12.4.1. FRACTAL BILLIARDS BOARD

Back to the example of a vowel string: **i, o, a, e, e, i, a, e,**

Let us take the unit square $[0, 1]\times[0, 1]$ with the diagonal reflection line $x=y$ drawn first. The diagonal line is called the *diagonal switch line.* Again let

$$f:\left\{1, 2, \cdots, n, n+1\right\}\rightarrow[0,1] \tag{12.4.1}$$

denote the *accumulate frequency function* of the probabilities:

$$P = \{ p_1, p_2, \cdots, p_n \}, \tag{12.4.2}$$

which is defined by

$$f(i) = \sum_{j<i} p_j \text{, for all } i = 1, 2, \cdots, n, n+1. \tag{12.4.3}$$

For each symbol $1 \leq i \leq n$ on the unit square, the stripe $[f(i), f(i+1)) \times [0, 1]$ of width p_i is called *the ith symbol zone*, and the straight line segment between the point $(f(i), 0)$ and $(f(i), 1)$ is called the *ith probability curve*. Given any point c on the baseline $y=0$, i.e., from the point $(c, 0)$ let's hit the ball straight ahead vertically. Now the ball bounces following the next three rules:

1. *If a vertical ball hits a probability curve, it will go horizontally.*
2. *If a horizontal ball hits the diagonal switch line, it will go vertically.*
3. *If a ball hits the unit square wall, it will bounce back in the opposite direction.*

If you start to hit the ball at the arithmetic code point

$$c = \texttt{0xD8DE8/0x100000=0.8471450805664}, \tag{12.4.4}$$

the magic happens: The orbit of the ball will hit the probability curves right in the order **i**, **o**, **a**, **e**, **e**, **i**, **a**, **e**, ... (See Figure 12.4.1.) What is this telling us?

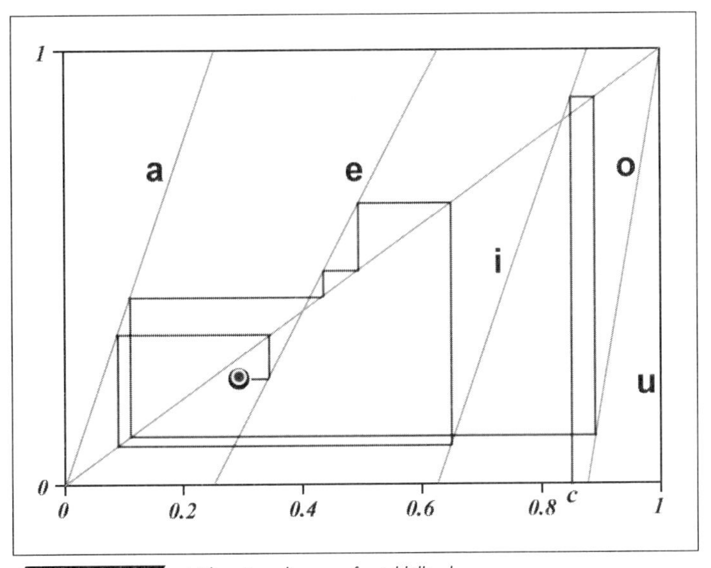

FIGURE 12.4.1 *Arithmetic coding as a fractal billiards*

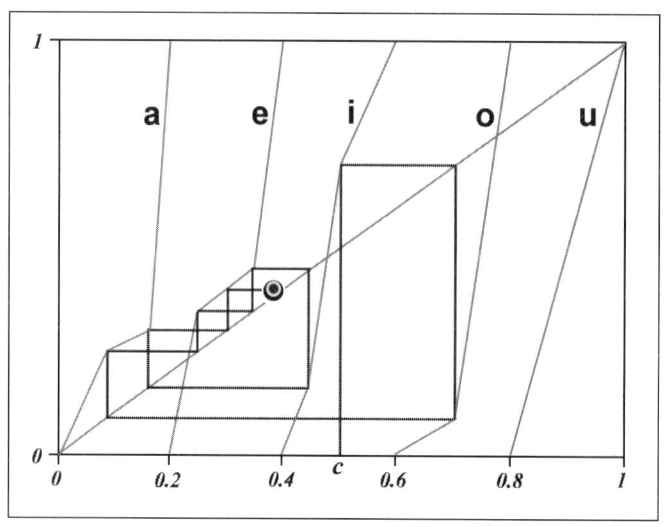

FIGURE 12.4.2 New design of the probability curves

> *The arithmetic coding is equivalent to finding the starting point on the baseline such that the orbit of the ball hit from that point will meet the probability curves in the exact order of the decoded data sequence of symbols.*

Therefore it is understandable that the longer we want the ball to stay on the track, the higher the precision of the starting point that is required.

Let us modify the probability curves on the billiards board. If we replace the straight line segments by any *monotonically increasing curves* between the points $(f(i), 0)$ and $(f(i), 1)$, for $i = 1, 2, \cdots, n$, the whole game is still valid. Figure 12.4.2 pushes this idea to another extreme.

By bending the probability curves, we are able to let the starting point to be wherever we like, e.g., $c = {}^1/_2$, and the orbit of the ball still hits the probability curves in the desired sequential order. What does this mean? This opens another dimension for lossless data compression. It simply says that you can either spend more bits in the precision of the baseline starting point, or you can have more bits to describe sophisticated probability curves.

12.4.2. FIRST-ORDER MARKOV SOURCE

Not only this fractal billiards board handles the zero-order Markov source. It works naturally for higher-order Markov data sequences as well. *In the first-order Markov source, the probability curves are fractal curves.*

Given a first-order Markov source of the symbol set $S = \{ 1, 2, \cdots, n \}$ and probability set $P = \{ p_{ij} \}_{i, j=1,2,\cdots,n}$, the zero-order probabilities are given by the equations

$$\sum_{j=1}^{n} p_j = 1,\ \sum_{j=1}^{n} p_{ij} = 1,\ p_i = \sum_{j=1}^{n} p_j \cdot p_{ji},\ \text{for } i = 1, 2, \cdots, n, \tag{12.4.5}$$

which is equivalent to the linear equations

$$\begin{pmatrix} p_{11} - 1 & p_{21} & \cdots & p_{n-1,1} & p_{n1} \\ p_{12} & p_{22} - 1 & \cdots & p_{n-1,2} & p_{n2} \\ \vdots & \vdots & \ddots & \vdots & \vdots \\ p_{1,n-1} & p_{2,n-1} & \cdots & p_{n-1,n-1} - 1 & p_{n,n-1} \\ 1 & 1 & \cdots & 1 & 1 \end{pmatrix} \begin{pmatrix} p_1 \\ p_2 \\ \vdots \\ p_{n-1} \\ p_n \end{pmatrix} = \begin{pmatrix} 0 \\ 0 \\ \vdots \\ 0 \\ 1 \end{pmatrix}. \tag{12.4.6}$$

After solving the zero-order probabilities, we name the accumulate frequency functions as before, except that we add the conditional ones as well:

$$f_i = \sum_{j<i} p_j,\ f_{ki} = \sum_{j<i} \cdot p_{kj} \tag{12.4.7}$$

for $i = 1, 2, \ldots, n+1$ and $k = 1, 2, \ldots, n$. Hence, the ith probability curve connects the points $(f_i, 0)$ and $(f_{i+1}, 1)$. Because the probability of hitting i is the stripe width p_i, and the probability of hitting j after hitting i is f_{ij}, the jth probability subzone inside of the ith probability zone will have a width $p_i \cdot p_{ij}$. Therefore, the ith probability curve must go through the following two points:

$$(f_i + p_i \cdot f_{ij},\ f_j) \text{ and } (f_i + p_i \cdot f_{i,j+1},\ f_{j+1}). \tag{12.4.8}$$

And between these two points, as shown in Figure 12.4.3, lies the probability curve of the state after hitting first i then hitting j, which should be the same as simply after hitting j, for the first-order Markov source. This implies that the affine mapping w_{ij} that maps the rectangle $[f_i, f_{i+1}] \times [0, 1]$ to the rectangle

$$[f_i + p_i \cdot f_{ij},\ f_i + p_i \cdot f_{i,j+1}] \times [\, f_j,\ f_{j+1}] \tag{12.4.9}$$

should map the jth probability curve to the above segment of the ith probability curve, as shown in Figure 12.4.3.

This can be written as the formula

$$w_{ij}\begin{pmatrix} x \\ y \end{pmatrix} = \begin{pmatrix} \dfrac{p_i p_{ij}}{p_j} & 0 \\ 0 & p_j \end{pmatrix} \begin{pmatrix} x \\ y \end{pmatrix} + \begin{pmatrix} f_i + p_i f_{ij} - \dfrac{p_i p_{ij}}{p_j} f_j \\ f_j \end{pmatrix}, \tag{12.4.10}$$

for $i, j = 1, 2, \ldots, n$. In conclusion, *the fractal billiards board of the first-order Markov source is obtained from its degenerate zero-order billiards by repeatedly applying the iterated function system* $\{ w_{ij} \}_{i,\,j=1,2,\cdots,n}$.

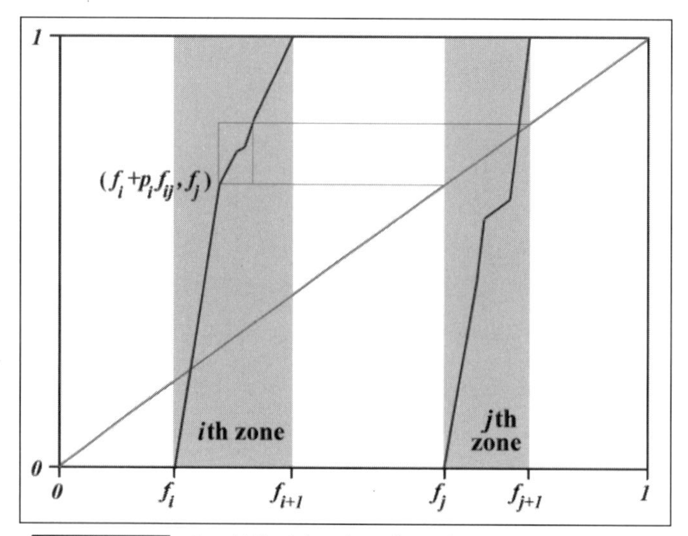

FIGURE 12.4.3 *Fractal billiards for a first-order Markov source*

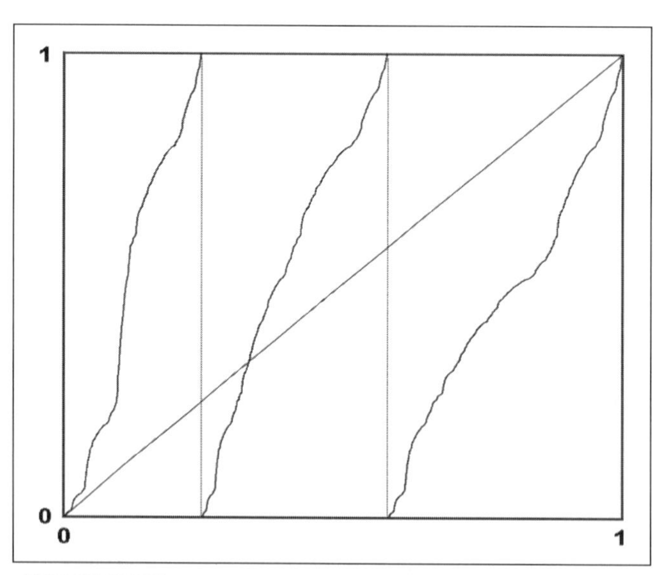

FIGURE 12.4.4 *Fractal billiards of a first-order Markov source*

Figure 12.4.4 shows an example of three symbols with the probability matrix

$$P = \begin{pmatrix} 0.4 & 0.1 & 0.5 \\ 0.2 & 0.3 & 0.5 \\ 0.2 & 0.5 & 0.3 \end{pmatrix}. \tag{12.4.11}$$

TABLE 12.4.1 *Fractal affine transforms for a first-order Markov source*

i	j	sx	tx	sy	ty
1	1	0.40000	0.00000	0.25000	0.00000
1	2	0.07500	0.08125	0.33333	0.25000
1	3	0.30000	−0.05000	0.41667	0.58333
2	1	0.26667	0.25000	0.25000	0.00000
2	2	0.30000	0.24167	0.33333	0.25000
2	3	0.40000	0.18333	0.41667	0.58333
3	1	0.33333	0.58333	0.25000	0.00000
3	2	0.62500	0.51042	0.33333	0.25000
3	3	0.30000	0.70000	0.41667	0.58333

Solving the system of equations, the zero-order probability set is calculated: $P = \left(\frac{1}{4}, \frac{1}{3}, \frac{5}{12}\right) = (0.25, 0.33\dot{3}, 0.41\dot{6})$. Thus, an IFS of nine fractal affine transforms is presented in Table 12.4.1, where

$$w_{ij}(x, y) = (s_x \cdot x + t_x, s_y \cdot y + t_y). \tag{12.4.12}$$

They build the fractal billiards for the first-order Markov source.

From this example, you can understand better now why a fractal modeling has been viewed as a first-order prediction, while other conventional modeling methods are considered to be zero-order.

In fact, a design of a fractal billiards board is a design of a compression system. A customized compression system for a specific data type can be realized by designing a customized billiards board to change the odds in favor of the data type.

12.4.3. THE BURROWS–WHEELER TRANSFORM

Given a string of data, could a fractal billiards board that is specially designed for this data string be characterized in a description that is simpler than coding the data string directly? The answer is surprisingly beautiful. The Burrows-Wheeler transform [BW] of any data string gives exactly a description of its associated fractal billiards board.

Given a string of symbols of length n:

$$S_0: a_0, a_1, \cdots, a_{n-1}, \tag{12.4.13}$$

we treat the string as if the last symbol wraps back around to the first. Thus, there are exactly n strings that have the exact same wrapped strings:

$$S_i: a_i, a_{i+1}, \cdots, a_{n-1}, a_0, a_1 \cdots, a_{i-1}, \text{ for } i = 0, 1, \cdots, n-1. \tag{12.4.14}$$

Now we assign an order to the symbol set and sort these n strings in the corresponding alphabetic order:

$$S_{i_0}, S_{i_1}, \ldots, S_{i_{n-1}} \text{ for } i=0, 1, \cdots, n-1, \tag{12.4.15}$$

which mathematically is equivalent to an n-permutation:

$$\sigma = \begin{pmatrix} 0 & 1 & \ldots & n-1 \\ i_0 & i_1 & \ldots & i_{n-1} \end{pmatrix}. \tag{12.4.15}$$

The *Burrows–Wheeler transform* (BWT) of the string S_0 is a string of n symbols—the last n symbols of the sorted n strings, one from each string:

$$BWT(S_0): a_{\sigma(0)-1}, a_{\sigma(1)-1}, \cdots, a_{\sigma(n-1)-1}, \tag{12.4.17}$$

where the index notation has a modulo-n convention, i.e., $a_{-1}=a_{n-1}$ and $a_n=a_0$.

Why is BWT interesting? The trivial answer is that *it is reversible*; the exciting answer is that *it gives the exact description of the fractal billiards board*, and the useful answer is that *it extracts pattern redundancy and therefore is powerful in lossless data compression.*

BURROWS-WHEELER THEOREM:

The cyclic class of a string S_0 is completely characterized by its Burrows–Wheeler transformed BWT(S_0).

As a consequence, *the string S_0 is completely determined by its BWT, BWT(S_0), and its alphabetic position in the cyclic class $\sigma^{-1}(0)$.*

We leave the direct proof to you as an exercise. Or you can refer to Nelson's paper [N2]. The following construction of a fractal billiards board gives implicitly a constructive proof of the theorem.

To simplify the discussion, the early sample string S_0=**ioaeeiae** is used again for illustration. As shown in Figure 12.4.5, the BWT of this string is $BWT(S_0)$=**oiaeaeei**.

Now we use $BWT(S_0)$ to design the fractal billiards board for the string S_0. Notice that the strings S_0 and $BWT(S_0)$ are formed by exactly the same symbols. Order these symbols alphabetically, and we have the string

$$Order(S_0): a_{\sigma(0)}, a_{\sigma(1)}, \ldots, a_{\sigma(n-1)}. \tag{12.4.18}$$

In our example, $Order(S_0)$=**aaeeeiio**. They are marked on both edges of the billiards board as shown in Figure 12.4.6. The bouncing points of the fractal billiards board are defined as the paired coordinates of $BWT(S_0)$ and $Order(S_0)$:

$$(a_{\sigma(0)-1}, a_{\sigma(0)}), (a_{\sigma(1)-1}, a_{\sigma(1)}), \cdots, (a_{\sigma(n-1)-1}, a_{\sigma(n-1)}). \tag{12.4.19}$$

In our example they are (**o,a**), (**i,a**), (**a,e**), (**e,e**), (**a,e**), (**e,i**), (**e,i**), and (**i,o**).

	Order							BWT
S_2:	a	e	e	i	a	e	i	o
S_6:	a	e	i	o	a	e	e	i
S_3:	e	e	i	a	e	i	o	a
S_4:	e	i	a	e	i	o	a	e
S_7:	e	i	o	a	e	e	i	a
S_5:	i	a	e	i	o	a	e	e
S_0:	i	o	a	e	e	i	a	e
S_1:	o	a	e	e	i	a	e	i

FIGURE 12.4.5 Burrows–Wheeler Transform

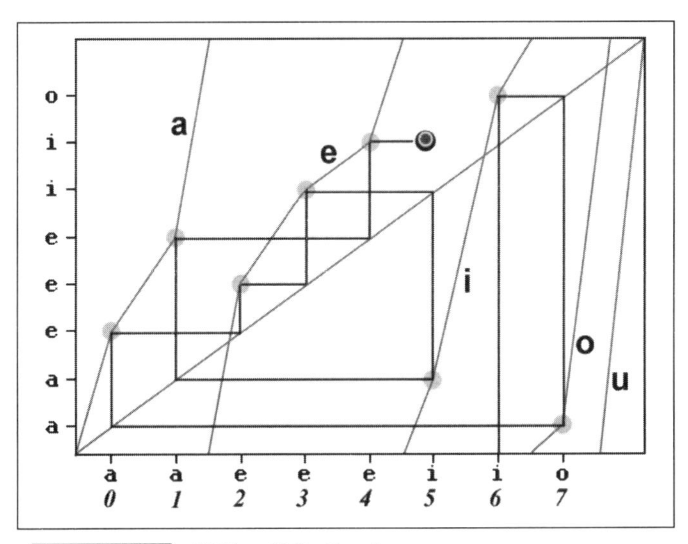

FIGURE 12.4.6 BWT fractal billiard board

They are plotted one after another in sequential order in the first available location (i.e., vertically from the bottom up and horizontally from the left) at which there are no previously plotted points in both horizontal and vertical directions. In the example, the points (**o**,**a**), (**i**,**a**), (**a**,**e**), (**e**,**e**), (**a**,**e**), (**e**,**i**), (**e**,**i**), and (**i**,**o**) are plotted in Figure 12.4.6.

To make it look the same as Figures 12.4.1 and 12.4.2, the probability curves can be added as anything reasonably connecting those points. On this fractal billiards board, the string S_0 is completely determined by the board and the initial shooting point $\sigma^{-1}(0)$. In the example, $\sigma^{-1}(0) = 6$.

Now we have seen the coding of a data string in both arithmetic coding and BWT to describe a fractal billiards board and an initial shooting point. Can BWT be used in lossless data compression as well? The answer is clear from the illustration in both Figure 12.4.7 for images and Figure 12.4.8 for text. In Figure 12.4.8 the letter `m` is used five times as the first letter of words in the original text string. As a consequence, we have five consecutive spaces □□□□□ in its BWT string. In fact, the mandatory comparison between Nelson's version BWT and PKWare's commercial product PKZip has shown quite encouraging results, even from the commercial perspective of this new method.

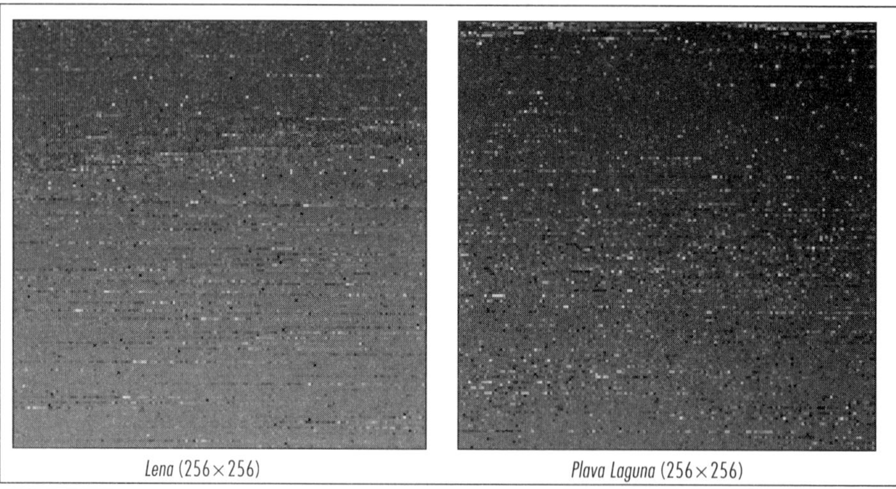

FIGURE 12.4.7 *BWT transformed images (provided by Ruifeng Xie)*

Original String

```
He□said□only□one□word,□then□turned□away,□and□left□m
e□behind.□That□word,□penatrated□my□heart,□electrifi
ed□my□soul,□and□shook□my□body□with□a□magnificence□I
□had□never□experienced□before.□
```

BWT String (in alphabetic order: `□ , . H I T a b c d ···`)

```
.e.h,,ddey,rIydatdkddyd,edy,nteyltddyed□□□□hms□□ehr
□w□□□nnleeeaneeinrrnocmHnrhlnctiblb□hcipvpnn□iieeat
□T□tesfafrnrhwoue□n□□□□□eeeiaaore gobo□□hwwfs□xetoo
oetua□□□farai□ec□otea□□□eammmld
```

FIGURE 12.4.8 *BWT transformed text*

12.5 NORMAL DISTRIBUTIONS

Given a data string of symbols, if the source model is given, whether if it is efficient or not, the data string can be encoded and then decoded. But what is the right source model that should be used? How should we parameterize the source models so that the decoder can identify which is which? Specific types of zero-order Markov sources will be studied in this section.

For zero-order Markov sources, assuming that the symbol set is known by both the encoder and the decoder, the probability table could be sent along with the coded data. However, when the data string is not too long or the symbol set too large, the overhead of the probability table may become significant. What can we do? One choice is to leave the probability table fixed for both encoder and decoder. Nonetheless, it may be impossible to find one to fit all. So in practice, we have discovered that almost always the probability table can be narrowed down to a few specific choices. After these choices are indexed, a table index is sent instead of an actual table.

This section will present some classes of probability tables that are often indexed together. Some of those probability distributions have not only been demonstrated to be useful, but they have been proven intrinsically meaningful in many theoretical probability studies. They are probabilities generated from ideal random processing models.

In practice, when we randomly subtract image pixel values from their neighboring pixels, the obtained difference values are very close to zero most of the time. Beneath the randomness there is a well-behaved pattern as shown in the histogram (Figure 12.5.1) that is generated by 1,000,000 differences of randomly chosen *a* adjacent pixel pairs and *b* arbitrary pixel pairs from the testing image *Lena*. For this reason we will work on the data strings that are jumping around zero in this section.

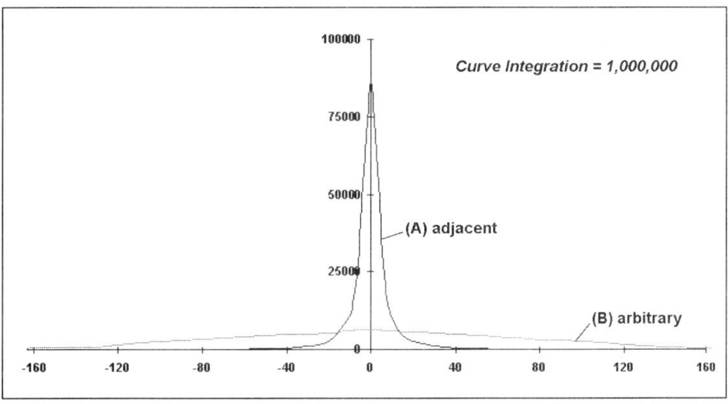

FIGURE 12.5.1 *The histogram of random difference values of: (A) adjacent pixel pairs; (B) arbitrary pixel pairs from* Lena

12.5.1 VARIANCE OF A PROBABILITY

Given any symbol set, without loss of generality, it can be assumed to have the form $\{-N, -N+1, \cdots, -1, 0, 1, \cdots, N-1, N\}$, for some large enough N. A *probability set* of this set is a finite-length sequence $\{p(i)\}_{i=-N, -N+1, \cdots, -1, 0, 1, \cdots, N-1, N}$ of nonnegative numbers with $\sum_{i=-N}^{N} p(i) = 1$. Extending this to the continuous situation, a *probability function* can be defined as a nonnegative function, or more generally, a generalized distribution function $\pi: [-N, N] \longrightarrow [0, \infty)$ with $\int_{[-N,N]} \pi(x)dx = 1$. Then, the discrete case becomes a special case of the continuous one. By truncating the events in $[i-1/2, i+1/2)$ to i, we have $p(i) = \int_{[i-\frac{1}{2}, i+\frac{1}{2})} \pi(x)\, dx$. The interval $[i-1/2, i+1/2)$ is called an *amplitude interval*. Given a string of data $\{x(k)\}_{k=1,2,\cdots,L}$ of length L bouncing around the value 0, for instance, assume that they are from $\{-N, \cdots, -1, 0, 1, \cdots, N\}$ of a probability $\{p(i)\}_{-N \le i \le N}$. The *variance* of the data string, characterizing how badly they are away from the origin 0, is defined to be their average distance from the origin:

$$\sigma = \sqrt{\frac{1}{L} \sum_{k=1}^{L} x(k)^2}\,. \tag{12.5.1}$$

As a consequence, the *variance* of a probability can be deduced from the formula

$$\sigma^2 = \sum_{i=-N}^{N} i^2 \cdot p(i) = \int_{[-N,N]} x^2 \cdot \pi(x)\, dx\,. \tag{12.5.2}$$

It is not difficult to understand that the low variance gives a low entropy data string—which yields a high compression ratio. The example shown in Figure 12.5.1 has the variance $\sigma = 11.55$.

12.5.2 NORMAL DISTRIBUTIONS

Probability distributions similar to Figure 12.5.1 have been seen all over the place in various areas of science. They have been seen when a stone is dropped into water, they have been heard when a drum set the air vibrating, and they have been felt when one sits next to a fireplace. These distributions, which have been presented as the solutions to heat or wave equations in thermo-, hydro-, or aerodynamics, will be used to model our digital data distributions. Basically, there are three well-known normal distributions: the *Gaussian*, *Laplacian*, and *gamma distributions*. These will be discussed. They have the following formulas:

$$G(x) = \alpha e^{-\beta x^2}, \quad L(x) = \alpha e^{-\beta|x|}, \quad \Gamma(x) = \frac{\alpha}{\sqrt{|x|}} e^{-\beta|x|}, \tag{12.5.3}$$

respectively. From the probability condition and the variance formula, we can calculate the parameters α and β in terms of the variance σ as follows:

$$G(x) = \frac{e^{-(x^2/2\sigma^2)}}{\sqrt{2\pi}\,\sigma}, \quad L(x) = \frac{e^{-\sqrt{2}|x|/\sigma}}{\sqrt{2}\,\sigma}, \quad \Gamma(x) = \frac{\sqrt[4]{3}\, e^{-\sqrt{3}\cdot|x|/2\sigma}}{\sqrt{8\pi\sigma|x|}}, \tag{12.5.4}$$

respectively, as shown in Figure 12.5.2. In principle, given any data string, its variance is calculated for identifying probability models from the above formulas. Then the arithmetic coding scheme is used to choose one of the three models and to code the actual data string in the chosen model. The variance (which will be used as the probability table index) and model type are stored together with the packed data string.

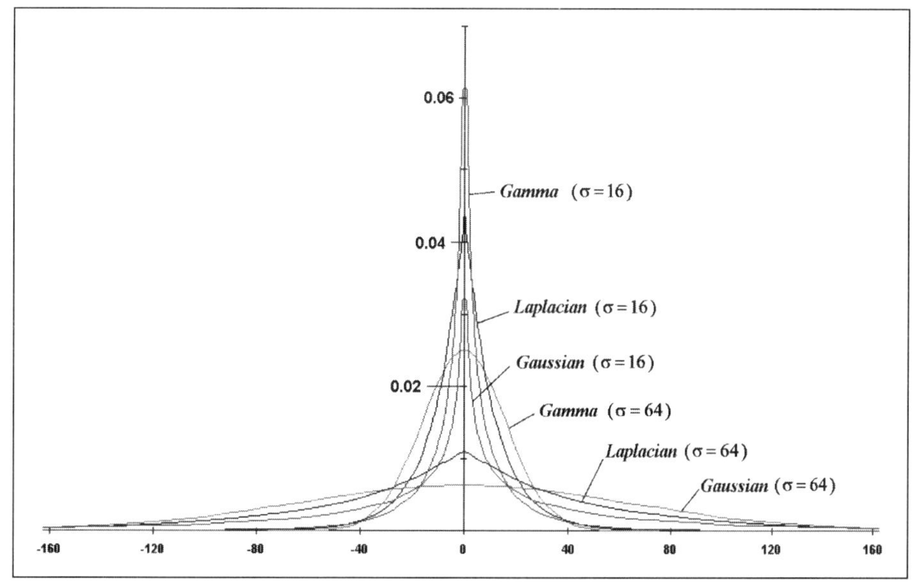

FIGURE 12.5.2 The Gaussian, Laplacian, and gamma distribution curves. The display variances: $\sigma = 16$, $\sigma = 64$

In real applications, all calculations can be approximated by using only integer operations. The results are remarkably similar to the high-precision decimal ones. The tables shown in Chapter 6 were generated by the following integer implementation.

12.5.3 C IMPLEMENTATION

Given a data string `data[i],i=0,1,...,Length-1` of length `Length` of integers bouncing around zero, its approximated integer variance square can be calculated by the following subroutine:

```
1    INT  CalculateVarianceSquare(
2        SHORT    *Data;        // data string
3        LONG     Length;       // data string length
4        LONG     *Sigma2       // variance square * 16
5    ){
6        LONG     i,s;
7        LONG     cnt[MAX_VALUE]; // histogram counter

     //  Counting histogram:
8        memset(cnt,0,MAX_VALUE*sizeof(LONG));
9        for(i=0;i<Length;i++)
10       {
11            if(Data[i]>=0) cnt[Data[i]]++;
12            else cnt[-Data[i]]++;
13       }

     //  Calculate the variance square
14       s = 0;
15       for(i=1;i<MAX_VALUE;i++) s += i*i*cnt[i];

     //  Now we truncate the variable sigma2 up to 1/16
16       *Sigma2 = (s<<4)/Length;

17       return(OK)
18   }
```

Instead of storing a long integer, all we eventually need is a single byte:

```
1    INT LabelVarianceSquareInByte(
2        LONG     Sigma2;       // variance square * 16
3        BYTE     *VrSqB        // variance square byte
4    ){
5        LONG     i,k;

     //  Find minimal bit length k of the number Sigma2
6        i = Sigma2;
7        k = 1;
8        while((i>>=1)) k++;
9        if(k>15) return(FAIL);
                 // if k>15, these models will not work anyway!
```

```
    //  Take only the top 5 bits, by actually store 4 of them,
    //  since the very first one is 1 always
10      if(k>5) i = (Sigma2>>(k-5))&15;
11      else i = (Sigma2<<(5-k))&15;

    //  get the variation square byte formed by two 4-bits
12      *VrSqB = (k<<4)|i;

13      return(OK);
14  }
```

The actual variance value used for indexing the distribution table is given by the following restoring routine:

```
1   INT GetVarianceSquareFromByte(
2       BYTE    VrSqB;      // variance square byte
3       LONG    *Sigma2 // variance square * 16
4   ){
5       LONGk;

6       k = (VrSqB>>4);
7       if(k<=5) *Sigma2 = ((VrSqB|16)>>(5-k));
8       else *Sigma2 = (((VrSqB<<1)|33)<<(k-6));

9       return(OK);
10  }
```

The Gaussian, Laplacian, and gamma frequency tables used for arithmetic coding are generated from that given variance square number:

```
1   INT GetNormalDistribution(
2       LONG    Sigma2; // variance square * 16
3       LONG    *freq   // frequency table
4   ){
5       LONG    k;

6       freq[0] = 2000;
7       for(k=1;k<MAX_VALUE;k++)
8       {
9           freq[k] = freq[-k] =

#if GAUSSIAN
10          = 2000*((double) exp((double) -k*k*8/Sigma2));

#elif LAPLACIAN
11          = 2000*((double) exp(-sqrt((double) 32/Sigma2)*k));

#elif GAMMA
12          = 2000*((double) exp(-sqrt((double) 24/Sigma2)*k)
13                  / sqrt((double) k));
#endif

14          if(!freq[k]) freq[k] = freq[-k] = 1;
15      }
16      return(OK);
17  }
```

Additional knowledge of a signal's data structure can always be translated into the entropy coding design. The theories and algorithms presented in this chapter can always be used with flexibility. For example, multi-dimensional data can be handled by multi-frequency tables; classified image types can use classified distribution tables.

The *adaptive entropy coding* schemes are not covered in this chapter for the following reasons: In still image compression, parameter data sizes are hardly long enough for an adaptive scheme to perform well. And in video compression, decompression speed is critical. Combining the compression gain and the decompression speed deduction, as we have seen so far, it is really not worthwhile to use adaptive schemes. Technically, adaptive coding is equivalent to dynamically changing fixed-coding codebooks.

With all art expression, when something is seen, it is a vivid experience, sudden, compelling, and inevitable. The visualization is complete, a seemingly instant review of all the mental and imaginative resources called forth by some miracle of the mind-computer that we do not comprehend. For me this resource is not of things consciously seen or transcriptions of music recollections; It is, perhaps, a summation of total experience and instinct. Nothing modifies or replaces it.

—Examples

Ansel Adams

Visual Image Metric

13

In the image visual quality assessment, subjective testing is the only *reliable* method that has been used as the ultimate measurement of image quality acceptance (for there is no satisfactory mathematical model serving this task). In almost every compression system, where a mathematical model must be used, the L^2-metric has been chosen as the de facto model for its simplicity. This metric, except for its mathematical elegance as a metric, has nothing to do with the human visual system. A few metrics, e.g., the *Hausdorff metric*, have been applied primarily to simple binary images in fractal imaging. The main difference between such an advanced metric and the L^2-metric is that the advanced metric not only looks at the local image color intensity distortions but also looks at the spatial image perturbations. By doing so, the metric is enhanced by measuring images as 2-dimensional geometric objects in a 3-dimensional space instead of measuring them only in a parallel grid set of 1-dimensional intersections. However, human visual science has never been considered as a factor in all the aforementioned image metric models.

Human vision is a complex subject that has been studied in many disciplines in science. From psychological ophthalmology to optical physics, our interests span the full aspect of understanding the human visual mechanism. Despite the efforts that have been expended, our knowledge of this subject is quite primitive and shallow—although it has been widely applied to almost every aspect of our daily lives (as in visual art, fashion design, commercial advertisement, military camouflage, magic illusion, and so on). Until today, most studies were done in a subjective statistical manner by using MOS (*mean opinion score*), which is equivalent to using the human brain as a tool for *quantification*. Further progress in this area requires a sound model for the human visual system for methodological study. There is always a time in any scientific discipline that waits for a *qualitative leap* while accumulating *quantitative studies*.

In this chapter a general metric model is not intended, but an improved metric model is presented for objective digital image quality assessment by incorporating some important properties from human vision science. This should close the gap between using universal mathematical models and using pure subjective testing. This chapter is an attempt to push the research onto a higher plateau—further improvements are expected in the near future.

13.1 CHARACTERISTICS OF HUMAN VISION

In order to model the human visual mechanism, the first step is to understand it. There are two major characteristics of human vision: *memorial selectivity* and *environment adaptability*.

Memorial selectivity includes all prejudices, biases, inclinations, and preferences that are caused by our accumulated historical knowledge and experiences. It is the result of the brain's previous interpretation.

Environment adaptability consists in the perceptual adjustment in space, time, and color. Among these three aspects, *color* can be considered as a sensual dimension perpendicular to space and time that includes color hue and intensity brightness. Environment adaptability is a part of biophysical adaptation native in the human visual system.

All human subjective visual behaviors are basically caused by these two characteristics.

13.1.1 MEMORIAL SELECTIVITY

Memorial selectivity follows the following basic rule:

> **THE FAMILIARITY RULE:**
> A subject, an object, or a fact is determined, identified, or recognized when some certain quantitative threshold of the matching portion between the received information and our memory has been exceeded. Within the received information, familiar data are processed preferentially.

Based on this rule, or hypothesis, two types of human experiences can be explained.

Here is the first experience. Take a paragraph in English of about 250 letters, blank out 50% of the letters as shown in Figure 13.1.1*a*, and let a group of people fill in the missing letters. It seems that it should be a hopeless task. Then start randomly and gradually to add one of the missing letters (every 10 seconds) back to the paragraph, and see when a group of people can fill in the remaining blank letters successfully. With 40% blank (Figure 13.1.1*b*), there are people that can actually tell exactly what the paragraph says. If you are one of them you must be familiar with the works of a famous English playwright. Then, some people made good guesses with 20% blank (Figure 13.1.1*c*). With 10% blank there are no difficulties in filling in the blank letters for most native English speakers. Within 5% blank (Figure 13.1.1*d*), some people who knew nothing about English could fill in the gaps by acknowledging the pattern. Indeed, the power of knowledge is really the ability of guessing.

Similar exercises can be carried out with images. Here is a trick we learned from David Mumford [M7] at a conference. Take a look at the picture shown in Figure 13.1.2*a*, and ask people to draw the face contour, which does not exist in the image. The result will be Figure 13.1.2*b*. It is interesting to see how our brain sees things that are not there. It is also amazing how our brains interpret

A	◊o◊◊◊◊◊o◊◊◊◊◊ to ◊e◊◊t◊at◊is◊t◊e◊◊u◊s◊◊◊◊:‖Wh◊◊h◊r◊◊ti ◊ ◊o◊l◊◊◊◊n◊t◊◊◊◊◊◊d t◊ s◊◊f◊r◊◊◊e ◊◊i◊◊◊◊◊◊◊◊◊◊◊◊◊ws ◊ ◊◊ou◊◊◊geo◊◊◊◊◊◊◊◊u◊◊,◊◊r◊◊◊o ◊a◊e◊a◊◊◊◊◊g◊ins◊ ◊ ◊◊a◊◊◊ ◊◊ro◊◊◊e◊◊◊◊◊n◊ ◊y op◊◊s◊◊◊◊e◊d◊◊h◊m?◊◊◊ ◊◊e◊◊◊o◊◊l◊◊p; ◊◊o ◊o◊◊; ◊n◊◊b◊ a s◊◊◊◊◊◊◊ s◊y ◊e ◊◊◊
B	To◊◊◊◊◊◊o◊◊◊◊◊ to ◊e◊ t◊at◊is◊t◊e◊qu◊s◊◊◊◊:‖Whe◊h◊r◊'ti ◊ ◊o◊l◊◊ ◊n◊t◊◊◊◊◊◊d t◊ su◊fer◊◊◊e ◊◊i◊◊◊ ◊n◊◊◊◊◊◊◊ws ◊ ◊◊ou◊◊◊geo◊◊ f◊◊◊u◊◊,◊◊r◊◊◊o ◊a◊e◊a◊◊◊ ag◊ins◊ ◊ ◊◊a ◊◊ ◊ro◊◊le◊◊◊◊n◊ ◊y op◊◊s◊◊g e◊d◊◊h◊m?◊◊◊ ◊◊e◊ ◊o◊◊l◊◊p; ◊No ◊o◊◊; ◊nd by a s◊◊◊◊ ◊◊ s◊y ◊e ◊◊◊
C	To◊◊e,◊or◊not to ◊e: t◊at is◊the qu◊sti◊n:‖Whe◊h◊r 'ti ◊ nobl◊r in th◊ mi◊d t◊ su◊fer‖◊he slin◊s an◊◊arr◊ws o f◊ou◊rageou◊ for◊une,‖Or◊◊o ◊ake ar◊s ag◊inst ◊ sea ◊f ◊rou◊le◊,‖And ◊y op◊osi◊g e◊d th◊m? ◊◊ die◊ to sl◊◊p; ‖No mo◊e; ◊nd by a s◊e◊p t◊ say we e◊d
D	To be,◊or not to ◊e: that is the question:‖Whether 'ti s nobler in th◊ mind to suffer‖◊he slings an◊ arrows o f◊outrageous fortune,‖Or ◊o take arms against ◊ sea of trouble ,‖And by opposing e◊d them? To die: to sle◊p; ‖No more; ◊nd by a s◊eep t◊ say we end

FIGURE 13.1.1 *Fill in the missing letters and spaces represented by ◊*

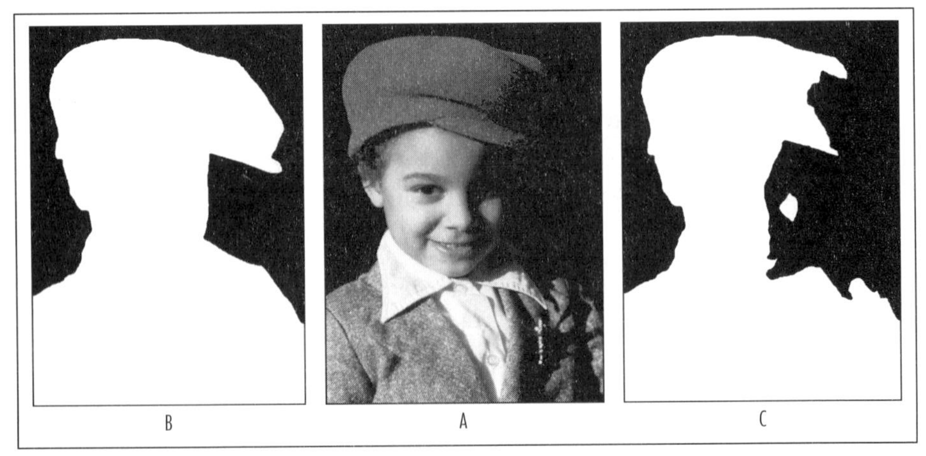

FIGURE 13.1.2 *How did our brain assume things that are not there: (B) Boundary in our mind; (A) A boundaryless face; (C) Boundary by computer*

missing pieces based on past visual experiences. And Figure 13.1.2*c* shows the actual image boundary processed by a computer.

All of the above examples demonstrate that when visual information matches our memory over a certain threshold, we rarely hesitate interpreting the whole picture even before the picture is complete.

Studies have shown that people have a strong tendency to overemphasize the things they know about. They were born with the general belief that the things they know are the things more important than those things they don't know. The result of this is that people unconsciously enhance the "contrast" when the "contrast" exceeds some threshold.

For example, when you meet a group of foreign friends who talk about you in their language, when your name is mentioned you immediately recognize it among the other words you do not understand. If you put a child's favorite toy among new toys that he has never played with before, this familiar toy will stand out to the child. Now, if you use a toy that looks like his but is not his, he will generally claim it is his. Similar phenomena occur when we view images. Our previous knowledge tends sometimes to force us to see structures beyond the image itself. As pictured in Figure 13.1.3, adapted from Marr's work *Vision* [M4], originally created by Stevens [S5], people see rays and circles that are not explicit in the images.

It is almost impossible to give a familiarity model, because not only is it very individually dependent, but also it never stops evolving. Culture, nature and lifestyles all give each of us different memories and therefore different perceptions. A lumberjack's descendant may see hundreds of greens and a sailor's son may sense intuitively that blue is the richest color in the whole color spectrum to his eyes. Therefore, at the moment, memorial selectivity components are overlooked in the presented visual metric modeling.

We say that this is "almost" but not "absolutely" impossible, because of the impossibility in modeling such diversity. However, we are living in a small world, and there are many things that are commonly shared: the same sky, similar star charts, etc. The common part of memorial selectivity can still be studied and modeled. Actually, some results in this area have been used to create human

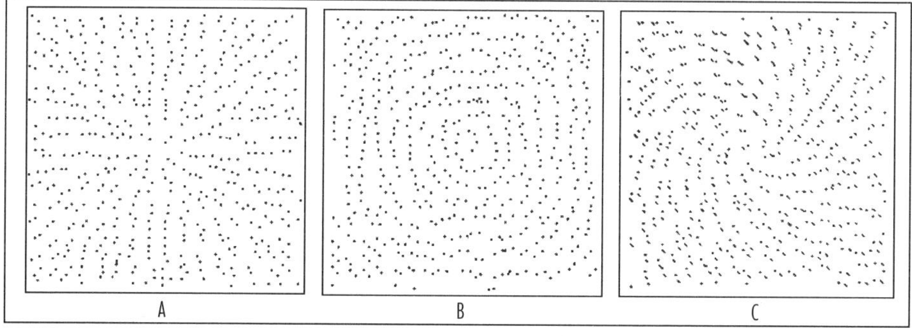

FIGURE 13.1.3 *Structures that are seen not by our eyes but by our brain: (A) Rays; (B) Circles; (C) Whirls*

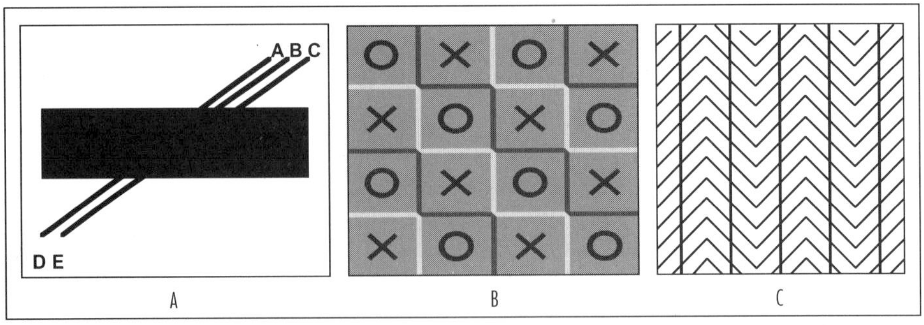

FIGURE 13.1.4 *Some geometric illusions: (A) Which of A, B, C connect D or E? (B) Which blocks are pop-out, X's or O's? (C) Are those lines really straight and parallel?*

illusion in magic shows and were also used in WW-II British Navy camouflage. As a consequence, this familiarity property in our visual system (generally considered an advantage) could be turned into an inborn deficiency.

Figure 13.1.4 presents a few more simple illusions. All of them were created against our visual perception, which is trained to understand 3-dimensional space using a pair of 2-dimensional signal receivers. The answers to *a* and *c* can be measured. As for *b*, you may try to view it by turning the page upside down and sideways and then comparing the results. Is it astonishing? Our subconscious actually assumes that the light is always above and shadows are always beneath.

In summary, memorial selectivity is a part of brains' prejudices caused by the limitation in their knowledge and experience.

13.1.2 ENVIRONMENT ADAPTABILITY

Environment adaptability is a part of the human survival instinct. It is one of the properties encoded in our genes when we were created. Everything lives in relation to its environment. When the environment changes, our body adapts itself accordingly and automatically within its limits. Some of those adjustments happen in a fraction of a second, and some of them take many generations to evolve. In general, the adaptation time fits the rate of environmental change. When we sit or move, our heartbeat rate changes without our noticing it. But it may take many generations for a group of people to migrate from flat land to a high mountainous region.

Have you ever noticed an oppressively bad smell when you walk into a room, but then a couple minutes later it is no longer noticed? How long would it take your eyes to adjust in a darkened movie theater? When you wash a red car for an hour, you then suddenly realize that when you look away how pale your immediate surroundings are? If you understand these visual reactions, you will never try to tune the hue of your TV set again, since the hue difference would not be noticed after few seconds of your eyes' adjustment. For the same

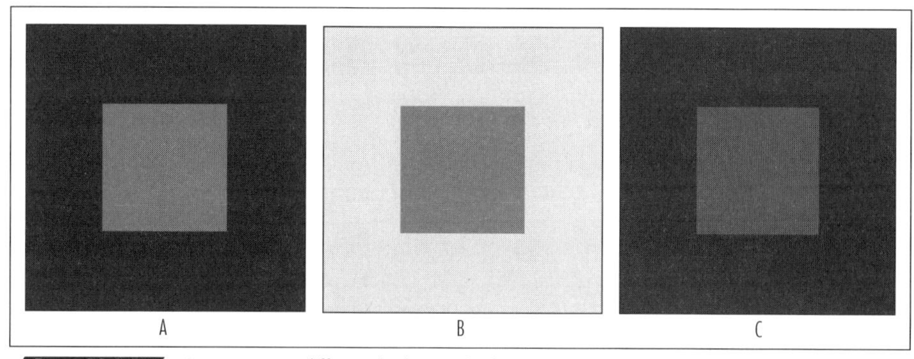

FIGURE 13.1.5 *Gray square in different background colors: (A & B) Numerically the same; (C) Perceptually the same*

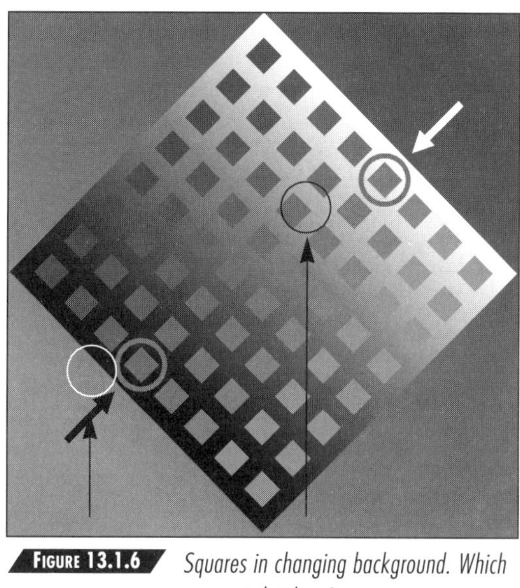

FIGURE 13.1.6 *Squares in changing background. Which square is brighter?*

reason, it is not difficult to tell what is *red*, *blue*, or *green*, but it is seemingly difficult to define what is *white*.

Based on the above discussion, the next several figures will demonstrate more examples of these phenomena. In Figure 13.1.5*a* and *b*, the same gray square is put into both a black background and a white background. As our eyes adjust, the picture *a*, the one in the black background, becomes perceptually much brighter. One actually thinks that the center block of picture *b* has the same brightness as the center block of picture *c*. Figure 13.1.6 shows the same

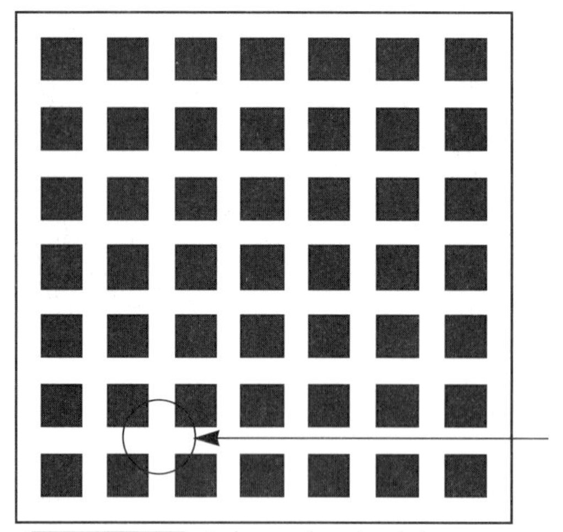

FIGURE 13.1.7 *Dark spots; bouncing dark spots*

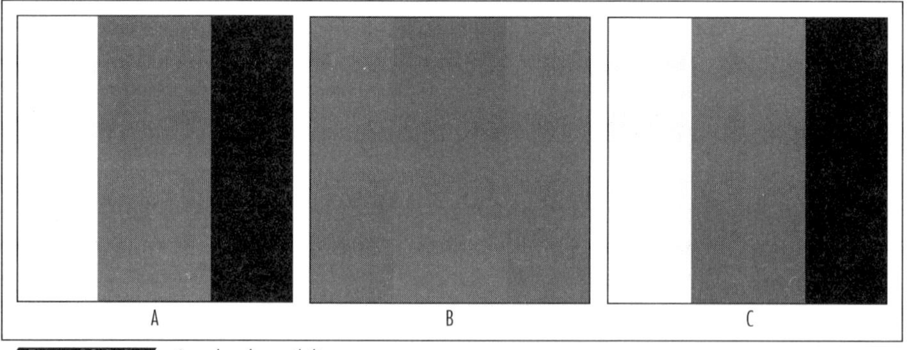

FIGURE 13.1.8 *Gray level sensibility test*

phenomenon. Can you believe that the four corner squares have actually two different colors? Can you tell which two share the same color?

Figure 13.1.7 illustrates another phenomenon. Can you see the dark jumping spots in the intersections of the horizontal and vertical white bands? The dark spots are observed because the neighboring average intensity value in a white band crossing the intersection is brighter than it is between two black squares.

Figures 13.1.8 and 13.1.9 show some relative brightness sensibility tests. In Figure 13.1.8, pictures *a* and *b* have the same middle gray zone. However, we are much more sensible to *b* than to *a*. Actually, one cannot tell the difference between image *a* and image *c*, which is the vertical flipping of *a*. Figure 13.1.9 illustrates how lower-level noise is covered by higher levels. Image *b* is obtained

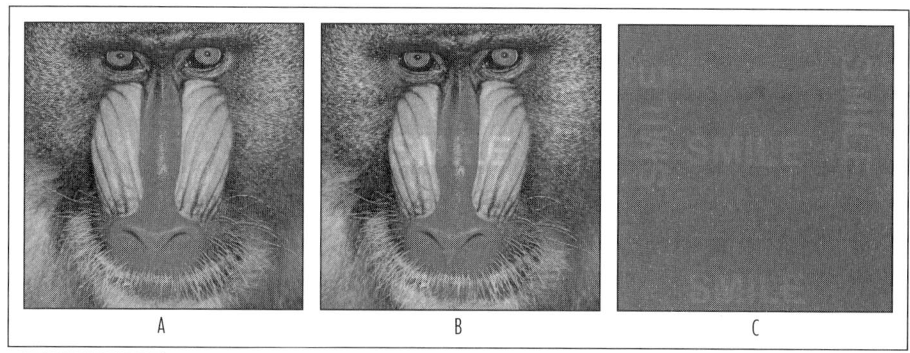

FIGURE 13.1.9 *Noise significance test*

from image *a* by adding image *c*. Among the four added "smile" words, the only "smile" one can see is the "smile" on the face, where the image is relatively smooth.

In conclusion, environment adaptability is a part of our physical and biological survival nature for fitting the environment.

13.2 A VISUAL METRIC MODEL

Digital image distortion has been measured widely by the mean square signal-to-noise ratio, which mathematically is equivalent to the intensity direction Hausdorff metric. While it is mathematically simple, the measurement is known to be unsatisfactory. Before giving an image metric model, let us first show how wrong this mean square error metric can be.

In fact, as shown by Figures 13.2.1 and 13.2.2, under this measuring system, a slight motion with no apparent change of visual appearance could cause a huge distortion error, while severely losing detail, as shown by Figure 13.2.3, may increase only a small amount of distortion. The most amazing comparison is the fact that the totally blurred image Figure 13.2.4 has even less error than the one shown in Figure 13.2.2.

13.2.1 VISUAL CRITERIA

Although the visual perception of the human eye is far more complex, the following characteristics are among the most important factors that influence the human visual system.

1. *The background brightness around the focal point dominates the human eye pupil size, which determines different color intensity sensitivity curves.*

FIGURE 13.2.1 *Original image: A restaurant in downtown Atlanta*

FIGURE 13.2.2 *A 3-pixel left motion generates error: L2 = 6.9663, PSNR = 31.27*

This has been illustrated in Figure 13.1.9. Here is another good example: When facing the sun, a flying bird in the sky is nothing but a black shadow. However, when looking at a bird sitting on a lake, one can see its colorful feathers in detail.

2. *The surrounding contrast and texture variation around the focal point determine different color intensity sensitivity curves.* This has been displayed in Figure 13.1.9. What this really says is a trivial fact: a bird in the sky is more visible than a bird in a tree.

FIGURE 13.2.3 *A detail losing picture with error: L2 = 3.3586, PSNR = 37.61*

FIGURE 13.2.4 *A blurred image with error: L2=6.3063, PSNR=32.14*

3. *The shape of a pattern is far more critical than the accuracy of color brightness and exact physical location.* The human brain has bias on its past experiences, which leads to an extreme sensitivity to patterns, in particular, the geometric patterns. This has been discussed in the memorial selectivity subsection. Figure 13.2.5 gives a more direct proof.

Figure 13.2.5 shows that it is far more important to maintain the geometrical entity than to preserve more pixels in their right values. In *b* 100 pixels have been switched from black to white and vice versa, and in *c* 100 additional pixels

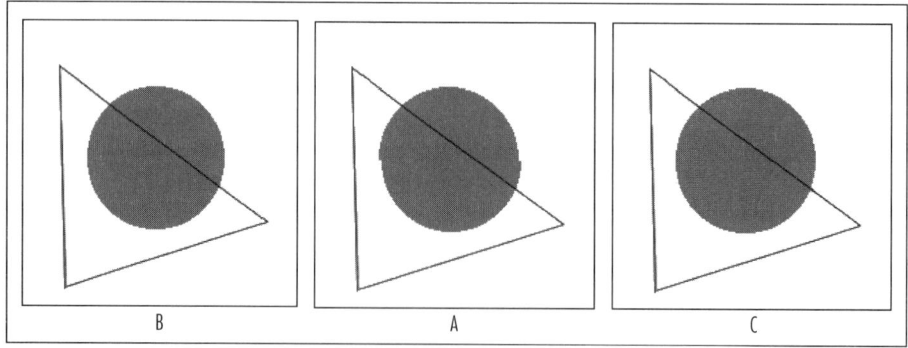

FIGURE 13.2.5 Geometry pattern vs. wrong pixel numbers: (B) 100 wrong pixels; (A) Original; (C) 200 wrong pixels

have their color switched. Image *c* looks indistinguishable from *a*, while *b*'s flaw seems quite obvious.

Next we will start to give an image metric model by studying these attributes one at a time. For simplicity, in this chapter, an image *P* will be considered as a function on a rectangular support as defined in Chapter 2; that is,

$$P: \Delta = [0, W] \times [0, H] \longrightarrow [0,1]. \qquad (13.2.1)$$

13.2.2 FOCUS AREA

Let $(x_0, y_0) \in \Delta$ be a given *focal point*, which can be any point in the image support. The *focus area* of the focal point is the area that affects the pupils when the eyes focus at the focal point. The focus area is weighted proportionally to the influence of the corresponding location. The location weight function is called the *focus area function*. The focus area is a subjective concept. It depends not only on the media focused on, but also the viewing distance, viewing environment, and viewer's eyesight. Mathematically, the focus area function can be modeled as a function

$$\Phi: (-\infty, +\infty) \times (-\infty, +\infty) \longrightarrow [0,1] \qquad (13.2.2)$$

with the following properties:

1. *At the focus point the function* $\Phi(0,0)$ *reaches its maximal value.*
2. *The function* Φ *has compact support.*
3. *The function* Φ *decreases monotonically in any ray from the focus center* (0,0).
4. *The function* Φ *is symmetric with respect to both axes.*

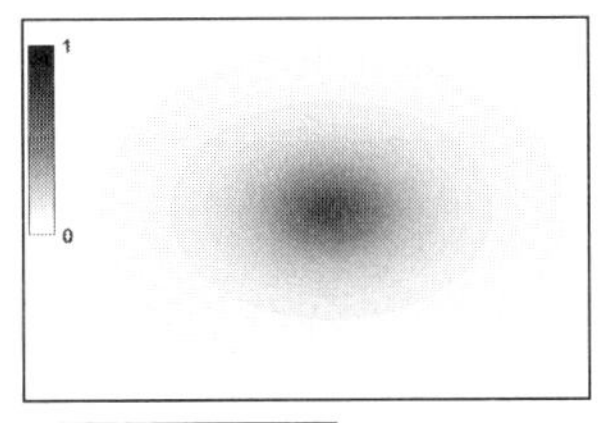

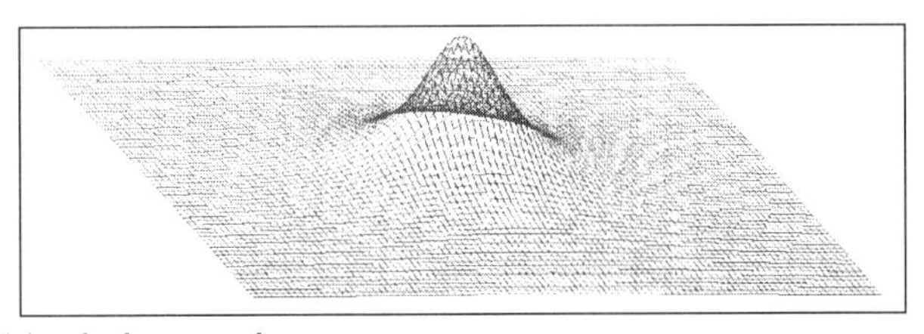

FIGURE 13.2.6 Gaussian, a modeling for focus area function

The *Gaussian* is a good example (see Figure 13.2.6) of a focus area function.

In general, for any focal point x_0,y_0 the focus area function $\Phi_{(x_0,y_0)}$ is defined by

$$\Phi_{(x_0,y_0)}(x, y) = \Phi(x - x_0, y - y_0) \text{ for all } x \text{ and } y. \tag{13.2.3}$$

13.2.3 BACKGROUND BRIGHTNESS

Let x_0,y_0 be the focal point and Φ the corresponding focus area function. The *background brightness* of the image P at the point x_0,y_0 is defined to be

$$B(x_0, y_0) = \frac{\iint_\Delta P(x,y) \cdot \Phi_{(x_0,y_0)}(x,y)\, dxdy}{\iint_\Delta \Phi_{(x_0,y_0)}(x,y)\, dxdy}. \tag{13.2.4}$$

For each background brightness b an *intensity rescaling function*

$$\rho_b = \rho((b, \bullet): [0,1] \longrightarrow [0,1] \tag{13.2.5}$$

can be selected to represent the true perceptiveness of the color intensity under this background circumstance. The construction of ρ is based on the following facts:

1. *It is more sensitive to the color close to its background.*
2. *It is more sensitive in the dark end than in the bright end.*

As a consequence, a natural choice of ρ is a multiplication of these two facts:

$$\rho(b, x) = \delta(b) \cdot \Psi(x - b), \tag{13.2.6}$$

where

$$\delta : [0,1] \longrightarrow (0,1], \tag{13.2.7}$$

a monotonically decreasing function, is the focus area adjustment, and

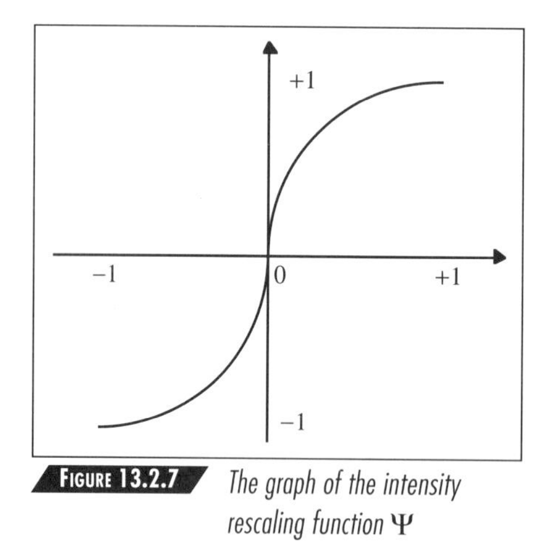

FIGURE 13.2.7 The graph of the intensity rescaling function Ψ

$$\Psi : [-1, 1] \longrightarrow [-1, 1], \tag{13.2.8}$$

antisymmetric, i.e., $\Psi(-x) = -\Psi(x)$, with monotonically decreasing derivative in the interval [0, 1], as pictured in Figure 13.2.7, is a universal intensity rescaling function.

13.2.4 BACKGROUND VARIATION

The background contrast and texture variation provides a shelter for the errors. In a high-contrast rough region, errors are more acceptable than they would be in a flat, smooth area. Such a variation is called the *background variation* of the focal point (x_0,y_0) and defined by the formula

$$V(x_0, y_0) = \left[\frac{\iint_\Delta (\rho_{B(x_0,y_0)}(P(x,y)))^2 \cdot \Phi_{(x_0,y_0)}(x,y)\, dx dy}{\iint_\Delta \Phi_{(x_0,y_0)}(x,y)\, dx dy} \right]^{1/2}. \tag{13.2.9}$$

The background variation will determine the *difference functions* that will be used. Let d be the family of difference functions with the background variations as its parameters. Based on the fact that for each given variation v the corresponding difference function d_v has the most sensitive interval around v, the difference functions can be well illustrated, as shown in Figure 13.2.8.

Let Q be another image. The *vertical distortion of the image Q from the image P at the focal point* (x_0, y_0) can be defined as

$$D^P_{(x_0,y_0)}(Q) = \left[\frac{\iint_\Delta (d_{V(x_0,y_0)}(|\rho_{B(x_0,y_0)}(Q(x,y)) - \rho_{B(x_0,y_0)}(P(x,y))|))^2 \cdot \Phi_{(x_0,y_0)}(x,y)\, dx dy}{\iint_\Delta \Phi_{(x_0,y_0)}(x,y)\, dx dy} \right]^{1/2}. \tag{13.2.10}$$

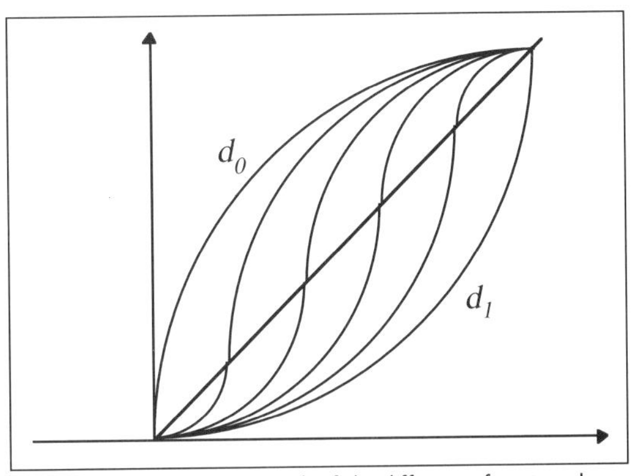

FIGURE 13.2.8 *The graph of the difference functions* d

13.2.5 ELASTIC MOTION

To analyze how geometric structures are preserved, the concept of elastic motion will be introduced. Mathematically, an *elastic motion* of the image support is simply a *homeomorphism* (i.e., *continuous bijective function*) of it. Let

$$\varepsilon : \Delta = [0, W] \times [0, H] \longrightarrow \Delta = [0, W] \times [0, H] \tag{13.2.11}$$

be an elastic motion. Its *distortion at the point* (x_0,y_0) is defined by the formula

$$E_{(x_0,y_0)}(\varepsilon) = \left[\frac{\iint_\Delta \left\| \varepsilon(x,y) - (x,y) \right\|^2 \cdot \Phi_{(x_0,y_0)}(x,y)\, dx\, dy}{\iint_\Delta \Phi_{(x_0,y_0)}(x,y)\, dx\, dy} \right]^{1/2}. \tag{13.2.12}$$

Let α be a positive number (called the *spatial distortion penalty ratio*). It will be used as a penalty constant factor to weight the horizontal spatial distortion against the vertical intensity error.

The *distortion of the image Q from the image P at the point* (x_0,y_0) is defined by considering all possible elastic motions as the following:

$$\Delta^P_{(x_0,y_0)}(Q) = \min_{\substack{\text{all possible} \\ \text{elastic motions } \varepsilon}} \left\{ \left[\left(D^P_{(x_0,y_0)}(Q \circ \varepsilon) \right)^2 + \left(\alpha\, E_{(x_0,y_0)}(\varepsilon) \right)^2 \right]^{1/2} \right\}. \tag{13.2.13}$$

13.2.6 DEFINITION OF METRIC

Factoring the global location weight function, the metric will be defined. In fact, the *distortion of the image Q from the image P* is defined to be

$$\Delta^P(Q) = \left[\frac{\iint_\Delta \left(\Delta^P_{(x,y)}(Q)\right)^2 \Gamma(x,y)\, dx\, dy}{\iint_\Delta \Gamma(x,y)\, dx\, dy} \right]^{1/2}. \tag{13.2.14}$$

The corresponding *signal-to-noise ratio of the image Q from the image P* is defined to be

$$SNR^P(Q) = \frac{\log(\Delta^0(P)) - \log(\Delta^P(Q))}{\log(\Delta^0(P))}. \tag{13.2.15}$$

The error between the image P and the image Q is defined to be

$$\Delta(P,Q) = \left[\frac{\iint_\Delta \left(\min(\Delta^P_{(x,y)}(Q), \Delta^Q_{(x,y)}(P))\right)^2 \Gamma(x,y)\, dx\, dy}{\iint_\Delta \Gamma(x,y)\, dx\, dy} \right]^{1/2}. \tag{13.2.16}$$

13.3 AN IMPLEMENTATION FOR DIGITAL IMAGES

As an example, let us look at the digital image case.

Given an image P, the color intensity interval is remapped from the interval $[0, 1]$ to the discrete integers $\{0, 1, \cdots, 255\}$; the image support $[0, W] \times [0, H]$ is sampled to its integer grid $\{0, 1, \cdots, w-1\} \times \{0, 1, \cdots, h-1\}$; and the image f is described in its filtered sample values $\{P[i, j]\}_{0 \le i < w, 0 \le j < h}$.

The focus area function Φ, as previously described, depends on the image resolution and the viewing angle. Considering a normal viewing distance of about twice the image height at a resolution of, for example, 800×600 and adding the computational simplicity, the default choice of the function Φ is given as the following:

$$\Phi : \{\cdots, -1, 0, 1, \cdots\} \times \{\cdots, -1, 0, 1, \cdots\} \longrightarrow [0, 1], \tag{13.3.1}$$

$$\Phi(x,y) = \begin{cases} 1, & \text{if } x = y = 0; \\ 1/2, & \text{if } 0 < |x| + |y| \le 2; \\ 1/4, & \text{if } |x| + |y| > 2, |x| \le 3, |y| \le 3; \\ 0, & \text{otherwise}. \end{cases} \tag{13.3.2}$$

Values of Φ are given in Figure 13.3.1.

0	0	0	0	0	0	0	0	0	0	0	0	0
0	0	0	0	0	0	0	0	0	0	0	0	0
0	0	0	0	0	0	0	0	0	0	0	0	0
0	0	0	¼	¼	¼	¼	¼	¼	¼	0	0	0
0	0	0	¼	¼	¼	½	¼	¼	¼	0	0	0
0	0	0	¼	¼	½	½	½	¼	¼	0	0	0
0	0	0	¼	½	½	1	½	½	¼	0	0	0
0	0	0	¼	¼	½	½	½	¼	¼	0	0	0
0	0	0	¼	¼	¼	½	¼	¼	¼	0	0	0
0	0	0	¼	¼	¼	¼	¼	¼	¼	0	0	0
0	0	0	0	0	0	0	0	0	0	0	0	0
0	0	0	0	0	0	0	0	0	0	0	0	0
0	0	0	0	0	0	0	0	0	0	0	0	0

FIGURE 13.3.1 *The discrete focus area function* Φ

We choose the identity map to be the default focus area function. The intensity rescaling function

$$\Psi : \{-255,\cdots,-1,0,1,\cdots,255\} \longrightarrow \{-255,\cdots,-1,0,1,\cdots,255\} \quad (13.3.3)$$

is a linear approximation of the graph in Figure 13.2.7 and is defined as follows:

$$\Psi(x) = \begin{cases} 4x, & \text{if } |x| < 17; \\ 3x \pm 16, & \text{if } 17 \le |x| < 37; \\ 2x \pm 52, & \text{if } 37 \le |x| < 58; \\ x \pm 109, & \text{if } 58 \le |x| < 88; \\ [\frac{x}{2}] \pm 153, & \text{if } 88 \le |x| < 134; \\ [\frac{x \pm 1}{3}] \pm 175, & \text{if } 134 \le |x| < 200; \\ [\frac{x}{4}] \pm 192, & \text{if } |x| \ge 200. \end{cases} \quad (13.3.4)$$

Specifically, the function Ψ has a look-up table:

```
SHORT  Psi[256] = {
  0,  4,  8, 12, 16, 20, 24, 28, 32, 36, 40, 44, 48, 52, 56, 60,
 64, 67, 70, 73, 76, 79, 82, 85, 88, 91, 94, 97,100,103,106,109,
112,115,118,121,124,126,128,130,132,134,136,138,140,142,144,146,
148,150,152,154,156,158,160,162,164,166,167,168,169,170,171,172,
173,174,175,176,177,178,179,180,181,182,183,184,185,186,187,188,
189,190,191,192,193,194,195,196,197,197,198,198,199,199,200,200,
```

```
201,202,202,202,203,203,204,204,205,205,206,206,207,207,208,208,
209,209,210,210,211,211,212,212,213,213,214,214,215,215,216,216,
217,217,218,218,219,219,220,220,220,221,221,221,222,222,222,223,
223,223,224,224,224,225,225,225,226,226,226,227,227,227,228,228,
228,229,229,229,230,230,230,231,231,231,232,232,232,233,233,233,
234,234,234,235,235,235,236,236,236,237,237,237,238,238,238,239,
239,239,240,240,240,241,241,241,242,242,242,242,243,243,243,243,
244,244,244,244,245,245,245,245,246,246,246,246,247,247,247,247,
248,248,248,248,249,249,249,249,250,250,250,250,251,251,251,251,
252,252,252,252,253,253,253,253,254,254,254,254,255,255,255,255 };
```

The difference functions d_v will be defined by using the function Ψ, considering the similarity of the graphs shown in Figures 13.2.7 and 13.2.8.

For any $v \in \{0, 1,..., 255\}$ the difference function d_v is defined as follows:

$$d_v(x) = \begin{cases} \frac{v}{255}\Psi\left(255 \cdot (\frac{x}{v} - 1)\right), & for\ x \leq v; \\ v + \frac{255 - v}{255}\Psi\left(255 \cdot (\frac{x - v}{255 - v})\right), & for\ x \geq v. \end{cases} \tag{13.3.5}$$

The elastic motions can be restricted to the local affine maps. For faster computational speeds, these elastic motions will be limited to local translations. The spatial distortion penalty ratio α will be set to 4, which is believed to be a reasonable number.

Assuming the global location weight function Γ as the identity, our metric is well determined.

In the remainder of the section, a detailed C language implementation will be offered for the digital images. The main function will be named as

```
INT     GetImageDistortion(
            BYTE    Image1[HEIGHT][WIDTH],
            BYTE    Image0[HEIGHT][WIDTH],
            float   *distortion_error )
```

where `Image0` is the original image, and `Image1` is the distorted one. They both are assumed to be monochrome images of the same dimension given by the constants `WIDTH` and `HEIGHT`.

```
1   #define ALPHA   4

2   LONG    *Square,PreSquare[2048];

3   SHORT   *Psi,PrePsi[512] = {
4           -255,-255,-255,-255,-255,-254,-254,-254,
5           ... ... ... ...
6           -32,-28,-24,-20,-16,-12,-8,-4,
7           0,4,8,12,16,20,24,28, 32,36,40,44,48,52,56,60,
8           ... ... ... ...
9           254,254,254,254,255,255,255,255};

10  SHORT   DfMap[256][256];
```

```
11  SHORT   **Phi,*PrePhi[7];
12  SHORT   PrePrePhi[7][7] = {1,1,1,1,1,1,1,
13          1,1,1,2,1,1,1, 1,1,2,2,2,1,1,1,2,2,4,2,2,1,
14          1,1,2,2,2,1,1, 1,1,1,2,1,1,1,1,1,1,1,1,1,1 };

15  INT     GetImageDistortion(
16              BYTE    Image1[HEIGHT][WIDTH],
17              BYTE    Image0[HEIGHT][WIDTH],
18              float   *distortion_error
19  ){

20      SHORT   i,j,x,y,rx,ry;
21      LONG    m,n,Dxy;

    //  Make square table:
22      Square = PreSquare + 1024;
23      for(x=0;x<1024;x++) Square[x] = Square[-x] = x*x;

    //  Make Phi table:
24      Phi = PrePhi + 3;
25      for(x=0;x<7;x++) PrePhi[x] = &PrePrePhi[x][3];

    //  Make Psi table:
26      Psi = PrePsi + 256;

    //  Make difference function DfMap table:
27      for(y=0;y<256;y++)
28      {
29          PrePhi[x] = &PrePrePhi[x][3];
30          for(x=0;x<y;x++)
31          DfMap[y][x] = (Psi[(x*255+(y>>1))/y-255]*y+128)/255;
32          for(x=y;x<256;x++)
33          DfMap[y][x] = (Psi[(x-y)*255+((255-y)>>1))/(255-y)]
34              *(255-y)+ 128)/255 + y;
35      }

36      *distortion_error = 0;

37      for(y=0;y<HEIGHT;y++)
38      for(x=0;x<WIDTH;x++)
39      {

    //  Get background brightness:
40          m = n = 0;
41          for(j=-3;j<=3;j++)
42          for(i=-3;i<=3;i++)
43          if((i+x)>=0 && (i+x)<WIDTH
                && (j+y)>=0 && (j+y)<HEIGHT){
44              m += Phi[j][i]*Image0[j+y][i+x];
45              n += Phi[j][i];
46          }
47          bb = (m + (n>>1))/n;

    //  Get background variation:
48          m = 0;
49          for(j=-3;j<=3;j++)
50          for(i=-3;i<=3;i++)
51          if((i+x)>=0 && (i+x)<WIDTH
```

```
52              && (j+y)>=0 && (j+y)<HEIGHT){
53              m += Phi[j+3][i+3]
54                  * Square[Psi[Image0[j+y][i+x]-bb]];
55          }
56          bv = (m + (n>>1))/n;

    //  Loop all local translations within a radius 8:
57          Dxy = 0x100000;
58          for(ry=-7;ry<=7;ry++)
59          for(rx=-7;rx<=7;rx++)
60          {

        //  1.   Check if it's valid one on the image support:
61              for(j=-3;j<=3;j++)
62              for(i=-3;i<=3;i++)
63              if((i+x)>=0 && (i+x)<WIDTH
64                  && (j+y)>=0 && (j+y)<HEIGHT){
65                  if((i+x+rx)<0 || (i+x+rx)>=WIDTH
66                      || (j+y+ry)<0 || (j+y+ry)>=HEIGHT)
67                      goto NEXTEMOTION;
68              }

        //  2.  Get spatial distortion:
69              Err = ALPHA * (Square[rx]+Square[ry]);

        //  3.  Get vertical distortion:
70              m = 0;
71              for(j=-3;j<=3;j++)
72              for(i=-3;i<=3;i++)
73              if((i+x)>=0 && (i+x)<WIDTH
74                  && (j+y)>=0 && (j+y)<HEIGHT){
75                  m += Phi[j][i]*Square[DfMap[bv][
76                      Psi[Image1[j+y+ry][i+x+rx]-bb] -
77                      Psi[Image0[j+y][i+x]-bb] ]];
78              }
79              if( (z = m/n + Err)< Dxy ) Dxy = z;
80          }
83          *distortion_error += Dxy;
82      }

83      *distortion /= WIDTH*HEIGHT;
84      return(0);
85  }
```

The modeling presented is not complete. How does this metric system apply to real-world images? This is left as a project for the reader to complete. Our goal is to initiate new research directions in designing computer process that distinguish images as the human eyes do. This model will be refined in many ways—some additional visual attributes will be incorporated, some components will be switched for better ones—after many parameters, coefficients, and penalty weights are derived from more conclusive subjective statistical and experimental research.

In fact, ideas discussed above reflect only a small portion of the research in visual image metric area that relate to digital image processing. Many other interesting works have been done in both academic and commercial sectors.

For example, the *image discrimination model* ([AB]) is a visual model based on the phenomenon shown in Figure 13.1.9 (image noise significance test). This model believes that any image piece is associated to a noise masking function which is capable of hiding up to a certain degree of noise. As one of the many results, Watson developed *DCTune*, an enhanced DCT image compression system ([W2]) that dynamically selects quantization tables block-by-block based on the anticipated sensitivity of the human eye to the target portion of the image. Reports show that this system yields perceptually better compressed JPEG images.

There are many other commercial imaging application features linked to image visual assessment that are being researched: image attribute metadata search and retrieval in a digital library [RRK], object and pattern recognition and forewarning in security control or medical diagnosis imaging [EAW], and even in medical treatment of visually impairment [PFP].

So far, most works in the area are driven by intuitive observations and statistics analyses of psychovisual experiments. More convincing and systematic theories are yet to be developed.

I wanted the moments of my life to follow and order themselves like those of a life remembered. You might as well try and catch time by the tail.

—Nausea

Jean-Paul Sartre

Future Image Format

14

As technology matures and the market becomes more demanding, some compressed image formats will be widely accepted and inevitably standardized. Every year, new image standards are created. Some are recommended by formal standardization organizations: e.g., Fax Group IV by CCITT (The International Telegraph and Telephone Consultative Committee), JPIG, JPEG, and MPEG by ISO (International Organization for Standardization) and IEC (International Electronical Commission), and most are actually built as de facto standards by individual companies: e.g., BMP by Microsoft, GIF by CompuServe, TIFF by a consortium headed by Aldus, PostScript by Adobe Systems, PhotoCD by Kodak, FIF by Iterated Systems and more recently FlashPix by Kodak in collaboration with Hewlett-Packard, Live Picture, and Microsoft [FPX].

Serving basic industrial and commercial needs, most imaging format standards are not a concern of research. However, the impact of standardization on future research is enormous. A good image format standard could enhance communication, eliminate duplicative work, and accelerate new research and development. Nonetheless, a format standard that is efficient will also cut general interest in further technological evolution, which implies less money and less resource funding to format-related research areas.

The image format issue is presented here not because of its industrial requirements for applications, but because it is a carrier of technology. By becoming involved in the complete process of format design, research scientists can ensure that the format standardization process will have more positive impact on the future of technological progress.

This chapter will present general concerns in creating an image format and will discuss all the image attributes, data components, and file structures that will be embraced by future technology and production innovations based on current knowledge. Two specific examples will be shown. The intention is to provide image format building pieces. It is up to the reader, based on his or her own need, to put these components together to create a complete image format.

The two concrete sample formats are given for two different purposes. First is an *unpacked fractal transformation format*, which can serve as a common playground or an intermediate step for research into fractal image compression and representation. Second is a *progressive image slicing format* that shows a practical solution for particularly hot applications: *images on the information superhighway.*

14.1 FORMAT DESIGN

Since the fractal architecture presented implies all well-known compression methods, e.g., DCT, VQ, subband, and wavelet as well, the format that will be introduced will intrinsically support all of them. Notice that a given image could be expressed in many different forms in this image format, according to different compression techniques and different progressive criteria. For illustrative purposes, a 6" × 4" 200 dpi monochrome image will often be used.

14.1.1 DESIGN GOAL

The general goal of a digital image format can be abridged to one sentence: *A format will enable the industry to deliver desktop computer solutions that make it easy, enjoyable, and commonplace for people to use digital color photographic images in their offices and homes.* More precisely, people want to use digital images in compositions, to include them in their communications, and to make them part of their recorded memories.

Composition implies easy image accessing, editing, and manipulation. *Communication* implies progressive transmission, quick display speeds, and output device flexibility. *Memory recording* implies data compression and high image quality. Putting all of these elements together, we have the following list.

- *General Purpose*—Most photographic images are stored either in 8-bit monochrome or 24-bit RGB colors. However, a general format should be applicable to practically all continuous-tone images, including black-and-white binary facsimile images and 16-bit X-ray medical images. In fact, it should also not be restricted by image dimension, color spaces, aspect ratios, intensity value precision, etc.
- *Progressiveness*—This is mainly for Internet transmissions. In that environment it is convenient to put image data into a structure of slices. In this slice structure, if the image data file is cut anywhere in the middle, the front cut portion of the image data forms a legitimate image file itself. The data slices contained in this new image file give a lower-quality image description of the same image.
- *Compression*—This is the key component of transmission and storage. In an ideal situation, a progressive image file can be cut in any data file size, so that the obtained image quality is at or near the state-of-the-art compressed image quality with regard to the compression ratio.
- *Idempotency*—This is a critical feature for any coded file format to be considered as a format for original image storage. Idempotency means that a pair of encoder and decoder exists so that the decoder decodes images in this format to images in some pixel format, and the encoder can convert these decoded images back identically to their original forms. This property is essential to keep an image from deteriorating after repeated editing.
- *Easy Access*—Often an image will go through some sort of geometric transformation, such as cutting and pasting, dragging and dropping, flipping and rotating. Sometimes a huge image may need a tiling geometry for the convenience of memory managing and subimage organizing. A good image format should allow for these procedures with inherent ease.

- *Extendible to Video*—Video always seems to be the next step beyond stills. A video image format must be considered as a natural extension of the still image format.
- *Resolution and Device Independence*—Fractal image representation gives an image description without any output resolution, which is exactly what the industry demands. From a quick thumbnail screen display to a large poster printing, various resolution representations of the same image are often necessary. Currently, a multiresolution layer image format, Kodak's PhotoCD format, is designed for this purpose. A resolution-independent image format automatically reduces the multiple duplicate and redundant image data layers.
- *Fast Display*—Unlike compression procedures, decompression speeds always need to be fast. A good image format should anticipate the decompression procedure and arrange for easy data reading.
- *New Technology*—An industrial standard should not close the door to possible emerging technologies, including new hybrid techniques.

In practice, some items in the above list of goals may contradict one another. When this happens, priorities will be followed. For example, for the sample image format, one can say that web pages and photo archives are primary, and relatively, medical images and satellite pictures are secondary. Similarly, continuous-tone still images are primary, while video clips are secondary.

14.1.2 FILE ORGANIZATION

An image file is always formed sequentially by *header*, *body*, and *tailer*. The header consists of two parts: the *standard header*, and the *optional header*. As the names indicate, the standard header carries the image information that is always presented, and the optional header contains additional information about the image. More optional information is allowed to be put at the end of the data file as an optional tailer. For a progressive format, the tailer information will be received last. Consequently, it must be irrelevant to the image reconstruction, and it is more likely to be some background descriptive information.

In a progressive image format, the image data body is divided into slices as drawn in Figure 14.1.1. The thinner the slices are, the more progressive steps are needed. In general, a data slice contains only one specific image code data type or parameter for a designated screen. When progressiveness is not required, everything is constructed together in a single slice.

A video file format is given by grouping slices into frames, as shown in Figure 14.1.2. Without loss of generality and by adding empty slices, all frames are assumed to have the same number of slices.

The image slices are not limited to data or codes representing color intensity value information. In fact, the slicing structure gives much more than the

Table 14.1.1 *Image file organization;* p *—number of panels,* n *—number of slices*

Standard Header		
Optional Header		
	1st Slice Data	1st Slice Tag
		1st Slice of 1st Panel Data
		1st Slice of 2nd Panel Data
		... ...
		1st Slice of *p*th Panel Data
	2nd Slice Data	
	... ...	
	*n*th Slice Data	
Optional Tailer		

Table 14.1.2 *Video file organization;* k *—number of slices per frame,* p *—number of total slices must be a multiplier of* k

Standard Header		
Optional Header		
	1st Frame	1st Slice Data
		2nd Slice Data
		... ...
		*k*th Slice Data
	2nd Frame	(*k*+1)th Slice Data
		(*k*+2)th Slice Data
		... ...
		2*k*th Slice Data
	3rd Frame	
	... ...	
	*n*th Frame	
Optional Tailer		

progressiveness property. Taking advantage of the structure, a composed image can be defined by adding an *opacity operation slice*. As additional slices, *image editing information* can be embedded as well. An *image masking slice* can indicate a marked image region including transparency masking. Some *watermark slices* can be inserted to provide image property control. The concept of slicing simply adds flexibility to image file organization.

The slices that are not essential to the image data are called the *peripheral slices*. The peripheral slices are used mainly to provide some extra system-specific information data for helping a particular application system.

14.2 HEADER AND TAILER

The image file header and tailer contain *nonimage data*. In reality, the definitions of what is image data and what is not fall into a gray area. For example, sometimes a Huffman table is considered as nonimage data, because it has been put in the header. However, without it the image cannot be reconstructed. An appropriate classification of image header and tailer data is into three different categories: *instruction property sets*, *recommendation property sets*, and *information property sets*.

The instruction property sets provide instructive information that is required by the decompression system, e.g., image dimension, packing method, and file version number. The recommendation property sets suggest the default output choices for better quality, e.g., postfilter information, printing γ-correction, and colormap table. The information property sets are purely informative, including the records of when and where the picture was taken, who took the picture, what aperture and speed were used, who owns the copyright, etc.

Before getting into the details of the header data, two geometric concepts need to be introduced: *image resolution* and *image tiling*.

14.2.1 RESOLUTION AND PANEL

Does every digital image representation have an *inherited resolution*? For all representations in pixel form the answer is obviously yes. And it is also true for any fractal representation whose affine transformation coefficients and parameters are integers in some spatial grid. Therefore, in practice we always can assume that there is an intrinsic resolution.

However, this resolution has no reason to be connected to its output dimension (which can be in a totally different measurement system). The image *output dimension* is device dependent: On-screen it can be measured in pixels, and in print it needs to be in its physical size in inches (") or centimeters (cm).

The *physical size unit* can be described in two bytes: The first byte is either 'm' for *meter*, '"' for *inch*, or "p" for *pixel*; and the second byte, from –127 to 128, is the power of 10 of the physical size unit in meters, inches, or pixels, respectively.

For convenience of image processing, an image is often *folded out* (i.e., expanded by duplicating the image boundary pixels) to or embedded in a larger image of some fixed multiple dimension. Thus, the image representation given in a file is really the representation of the larger image. Conventionally, the larger image is processed, and the real one is cropped right before displaying it. Thus, the *final output dimension* of the image will be cropped, the *cropped origin* will be the coordinates of the top left corner of the cropped image, and the *cropped width* and the *cropped height* will define the cropped image dimension. The large foldout image dimension is called the image *working dimension*.

For convenience of image handling and memory management, an image, especially a really large one, is necessarily cut into small pieces, called *panels*. In general, an image is cut into panels in equal dimensions, excepting possibly the panels located in the last row or the last column — such an image panel cutting is called an *image tiling*. There are cases where it is desirable to cut an image into image panels of differing dimensions. For instance, one may want to avoid dividing an image across its center.

14.2.2 Standard header

The instruction property sets must be given in the standard header except when defaulted choices are expected. The standard header, also called *universal header*, has a fixed size in most file formats.

The following data fields are usually found in the standard header:

- *File Identifier* — A one to four byte constant that distinguishes our file format from other file types. It is also called the file *signature*.
- *Version Number* — A file version number is often useful to declare to which version of a data format specification a data stream adheres. Most data formats evolve over time to include new features, and in some cases to drop old features.
- *Output Dimensions* — As discussed in the previous subsection, the output dimensions are *physical unit index*, *final output dimension*, *cropped origin*, and *working dimension*. By adjusting the physical size unit, all these fields can be assumed to be WORD or DWORD integers.
- *Inherited Resolution* —It consists of the number of pixels per row and the number of pixels per column. Equivalently, it can be replaced by the pair of the horizontal dot-per-unit number and the vertical dot-per-unit number.
- *Panel Partition* — The working image can be either simply cut in an image tiling or covered by a collection of rectangular boxes. In the first case, assumed to be the default case, we need the common *panel width* and *panel height* in either the physical size unit or inherited resolution grid. In the more general case, we set the above entries to zero and store the collection of boxes in the optional header.
- *Working Color Setting* — Working color decomposition is defined by its *screen types*, which are characterized by a constant or a character symbol: *Y* for monochrome image; *YUV, LMN, EMN, RGB, HSV*, etc. for 3-channel

color image; *YUVA*, *RGBA*, etc. for additional opacity screen; *YUVW*, *CMYK*, etc. for 4-channel color image; and so on.

- *Tag Number*—The number of tagged fields in the optional header.
- *Entropy Pack Version*—The version of the set of *entropy packers* used, including the *bit order* in a byte.
- *Header Length*—The total header data size, which is equal to the relative location of the beginning of the data from the beginning of the file.
- *File Length*—The total file size, including the header.
- *Lossless Slice Number*—The maximum number of slices.
- *Lossless Length*—The original lossless file size. Hence, this number is always greater than or equal to the file length. Equality holds if and only if the file is a lossless representation.

From the above list one may see that some fields are actually redundant. For instance, if a packing method is given, the current file length could be unnecessary. However, it is convenient to have both for different application uses. It is up to the image file creator to maintain data information consistency.

Table 14.2.1 shows a very simple example of a header structure to a 6" × 4" 200 dpi monochrome image by following our data file list. This example is given purely for illustration and is not intended to describe any specific image format in the current market.

All the items that are not universally required are left to the optional header.

TABLE 14.2.1 *Sample file header*

Bytes	Data (in hex)	Description
0 - 2	66 69 66	Image file id. = "fif"
3	01	File version no. = 1
4 - 5	22 FD	Size unit in 1/1000 inch
6 - 9	70 17 A0 0F	Image physical size 6" × 4"
10 - 11	05 05	Pixel size 5 × 5 in 1/1000 inch
12 - 15	B0 04 20 03	Image size 1200 × 800 in pixel
16 - 19	90 01 90 01	Image is tiled into 6 400 × 400 panels
20 - 23	5A 00 00 00	Only *Y* screen image (monochrome)
24 - 25	01 00	Using *Version 1.00* packer
26 - 27	00 00	No optional tagged fields
28 - 29	64 00	Maximal up to 100 slices
30 - 33	00 80 25 00	Captured lossless file size = 240KB
34 - 35	28 00	File header size = 40 bytes
36 - 39	00 00 05 00	Current file size = 80KB

14.2.3 OPTIONAL HEADER OR TAILER

Optional headers and tailers are usually designed in a tagged structure as shown in the following:

- *Tag Index*—It tells what this field is exactly.
- *Tagged Field Length*—Tagged field data length.
- *Tagged Field Data*—Tagged field data.

In many applications the tagged information data could be more reasonably put at the end of the file. In this case, these data become *optional tailers*—although an optional header *must* also be given for each tailer as follows:

- *Tag Index*—It tells what this field is exactly, and if it is a tailer.
- *Tagged Field Offset*—Tagged field data offset from the file beginning.
- *Tagged Field Length*—Tagged field data length.

In general, a WORD (2 bytes) for the tag index, a DWORD (4 bytes) for the tag field offset, and a DWORD (4 bytes) for the tag field length are given. However, for all of these fields it is recommended that they be stored in variable byte form—this will be described in the next section. The tagged field data has a size of *tagged_field_length* bytes.

Nondefault instructive property sets are always stored in the optional header. We group them into several classes of property sets.

The spatial geometry property sets:

- *Image Orientation*—The default setting is from left to right, then from the top down. In some applications, for certain practical reasons, a different coordinate system may be used. In that case, images could be stored from the bottom up, or from right to left. When this happens, it can be specified in the optional header.
- *General Panel Structure*—As discussed in the standard header, one may use a more image-dependent panel cutting by using a collection of covering rectangular boxes.
- *Shaped Image*—Images do not necessarily have to be rectangular. A nonrectangular image can be described by its boundary curve or a masking screen. In either case, the shape definition data can be stored in the optional header.

The color intensity property sets:

- *Image Intensity Depth*—Conventionally, the default intensity bit number is assumed to be 8 for each screen. For computer graphic images, as well medical images, this number is often different.

- *Customized Color Formulas*—For better compression or for some other practical reason, a set of nonstandard color converting formulas is used in the image description, which is specified in the optional header.

  The coding property sets:

- *Tables*—Some special Huffman tables and/or quantization tables could be efficiently included in the image data file, in particular for some irregular images.
- *Reference Images*—This image could be represented with the help of some universal seed image or FTT.

The video property sets: If the file is a video file, additional information must be given:

- *Frame Structure*—This includes the *frame number*, and the *slice number per frame*. Optionally, the collection of frame beginning offsets can be stored as indices for quick access.
- *Frame rate*—The number of frames per second should be displayed.

14.2.4 RECOMMENDATION PROPERTY SETS

The recommendation property sets are usually stored in the optional header for better decoded image quality. For example:

- *More Version Numbers*—An *encoder version number* is useful for tracking in which compression system and in what period of time an image file was originally made. Thus, an appropriate decoder can be chosen to view it. A *decoder version number* is used to tell whether any old decoders can still read and process this image.
- *Decoded Image Filter*—Part of the image decompression is image rendering. Some rendering filter information, such as the depth of block edge smoothing, could be helpful to obtain a high-quality image. The filter field could be just a parameter of some universal filter and could be an individual filter specially attached to some particular images.
- *Display Color Table*—For fewer than 24-bit output devices, as in a color ink-jet printer, or a 256-color super-VGA monitor, a color image invariably needs to be converted into a colormapped image by color clustering or/and color dithering. The color table can be indexed after color components of a working image, e.g., YUV, or color components of a decoded image, e.g., RGB. Including a color table in the image data file not only speeds up decompression but also may give a special color table that is specially tuned to the image.
- *Output Color Setting*—This is the field telling what output device the image is made for and therefore should be used. This does not preclude the use of different devices, but it gives a hint for possible adjustment. For instance, an

image could be given for printing, and one may still want to view it on a monitor. A contrast adjustment will improve the image's appearance. In some cases, this field may have only some general recommendations, such as output *color channel composition* and color *γ-correction*.

- *Parent and Child Images*—In real applications an image is often not isolated. It may belong to a bigger image as a local close-up picture. And some interesting small pieces of it may actually be stored as separate individual pieces. Knowing this information, this image could be improved by consulting other related images.
- *Password*—An advanced password is always melted into the image data to ensure the difficulty of deciphering original images. The password field will contain a combination string that is generated by a *locker* after reviewing the given password and the unlocked image data, so that the high-quality image can be obtained only when the password is entered. Without the password, a deformed image or low-quality image may still be displayed. The *locker combination string* is used for a locker to prevent unauthorized use. This string may also be used to hide the actual password to protect key loss and to enable a universal key.
- *Seal and Watermark*—Some indispensable message in the image data may be embedded in the image file by the creator of the image. Only an authorized software detector can erase or display this message.

14.2.5 INFORMATION PROPERTY SETS

Regarding a beautiful image, there are three types of questions that are commonly shared: the contents, the intellectual rights, and the file source. These questions imply the three major types of information property sets: the *content property set*, the *intellectual property set*, and the *file source property set*.

Most of the information property set data will be stored in the text format—the *Unicode format* has been encouraged.

The *contents property sets* emphasize the contents of the image, such as

- *Image Identifier*—A unique number is assigned to this image, in general based on the image contents, that is used for image cataloguing and management.
- *Date, Place, and Event*—These are the actual date, place, and circumstances where the image was originally captured.
- *People and Things*—A list of the descriptions of all major objects in the image, including the relative locations of the object centers in the image. This will help database searching and will also help to create application buttons where further information concerning those objects may be available.
- *Titles and Notes*—These fields include the image subject title and some additional text description notes. The roll or group title could be added as well. If the image belongs to a roll of film or a group of themes, the roll or group title could be added, respectively.

The *intellectual property sets* are mainly the following:

- *Copyright*—This is the complete copyright statement.
- *Legal Brokers*—The persons or organizations that hold the legal right to grant permission or restrict use of the original image or the digital image file.
- *Authorship*—The image creator, e.g., a photographer.

The *file source property sets* tell exactly how the image was created.

- *Origin*—Was this image captured directly using a digital camera? Or is it a secondhand image obtained by either a film or print scanner, or a video grabber? Or is it computer generated?
- *Software Specification*—In the procedure of digitization some specific software must be used. A specification list will include the software product name, software manufacturer, product release version, etc.
- *Software Setting*—This describes how the software was used to produce the image. For example, which sharpening filter was used or what the parameters were.
- *Hardware Specification*—Lists the hardware that was used, including manufacturer, model, serial number, etc. Some specs of the hardware configuration should be included as well.
- *Hardware Setting*—This is different for a camera or for a scanner. For a digital camera, it includes aperture size, shutter speed, and the surrounding light. For a scanner, the first question is whether it scans from a reflective surface, transparency, or negative. Then one may specify what kind of print or film was used, how the print was printed, and how the film was developed.

This list of headers and tailers is long enough. The intention is not to enumerate all possible header or tailer fields but to present an outline of what can be and what has been put in the header or tailer of an image format.

14.3 IMAGE DATA SLICE

The body of the image format is formed in slices. Each image *slice* is formed by a *slice tag*, which is a sequence of identification characters that defines the slice. The *slice data* consists of consecutive data blocks for each panel. Each *panel data block* is formed by the *panel-slice data length*, which is the remaining size of this panel data block, and the *panel-slice data* (cf. Figure 14.1.2).

14.3.1 SLICE TAGS

Both *slice tag* and *panel-slice data length* can be stored in *variable byte form*.

Variable byte form stores data information in actual 7-bit per byte by leaving the highest bit of each byte as the ending signal: 0 for continuing and 1 for ending. For example, any number can be stored in the following form:

```
                0, 1, ... , 127:   80 h, 81 h, ... , FF h;
128, ... , 255, 256, ... , 16,383:   00 80 h, ... , 7F 80 h, 00 81 h, ... , 7F FF h;
    16,384, ... , 2,097,152, ...:   00 00 80 h, ... , 7F 7F FF h, ....
```

In fact, in this form, two bytes per field is expected most of the time without limiting the extreme cases that could be automatically adjusted for fewer bytes or for more bytes.

What kind of slice tags are necessary for all possible image data presented in this book? Excluding peripheral slices, the following is a list of categories that the slice tags need to specify:

- *Slice Grid*—The smallest natural resolution that carries the full information of the slice; e.g., it can be the block dimension in the simple fractal quadtree decomposition codec.
- *Active Screen*—There is no reason to mix different image screens together. In most implementations it will be one of the luminance or chrominance screens as discussed in Chapter 11.
- *Additive Weight*—This determines the relation between the newly created image and the image created from the previous screens. With weight 0%, the old image is fully replaced by the new one; with weight 100%, the current slice image is simply added as the additional terms on the top of the previous one.
- *Reference Type*—The reference regions may come from a seed image, a previous image, a current image, a final decoded image, or in some classified wavelet forms or texture patterns. Thus, the fractal codes obtained from a slice data may need to enact only one iteration, many iterations only when that slice is decoded, or many iterations every time an additional slice is added.

Then, within the same category, more fractal code classification information can be specified conventionally in the tag field. For example, in the simple fractal scheme presented in Chapter 3, one can specify the spatial contraction factor, the allowed set of spatial forms, the contrast adjustment value, the quantization table used, etc.

14.3.2 A FRACTAL TRANSFORMATION SLICE

An unpacked fractal affine transformation slice could be defined as a collection of fractal transformations. The slice header will determine the following properties:

- *Screen Information*—Is it a *Y* screen? or a *U* screen? or something else?
- *Destination Region*—Is it a square? What is the width? Etc.
- *Reference Type*—Is it from itself or a seed image? Is it in a coordinate system originated relative to its destination region?
- *Reference Type*—Is it from itself or a seed image?
- *Spatial Symmetry s-form*—Is it default? indexed? or variable?
- *Spatial Tilting t-form*—Is it default? indexed? or variable?
- *Spatial Deformation s-form*—Is it default? indexed? or variable?
- *Intensity Scaling γ-factor*—Is it default? indexed? or variable?
- *Intensity Offset β-value*—Is it indexed? quantized? or variable? Or is it actually a mean value?

Then, as an example, the fractal code for each transformation can be given in 24 bytes as shown in Table 14.3.1.

TABLE 14.3.1 *An unpacked fractal code*

Type	Bytes	Fields	Description
WORDS	0 - 3	`dx, dy`	Destination region origin
WORDS	4 - 7	`rx, ry`	Reference region origin
SHORT	8 - 9	`c`	γ-value in index or variable
SHORT	10 - 11	`b`	β-value, or the mean of region
SHORTS	12 - 19	`s[4]`	Spatial deformation s-form
SHORTS	20 - 23	`t[2]`	Spatial tilting t-form

This format is inefficient, but it serves as a good intermediate format for research as well as for any interactive fractal imaging systems. Further code packing could be done totally based on these.

14.3.3 A SIMPLE SBA PROGRESSIVE SLICE

Now let us see a very simple progressive image format: the *significant-bit additive progressive format* (SBA form).

Given any integer, it can be uniquely decomposed into powers of 3 in the following form:

$$a_0 + 3a_1 + 3^2 a_2 + 3^3 a_3 + 3^4 a_4 + \cdots,$$

for some $a_0, a_1, a_2, a_3, a_4, \cdots \in \{-1, 0, 1\}$.

Given an image screen, its *self-difference screen* is defined to be a screen of the same dimension obtained by subtracting 128 from the first pixel, by subtracting from any other pixels of the first row their preceding pixels, and by subtracting from the remaining pixels in the image the pixels exactly above them, as illustrated in Figure 14.3.1. For any pixel, its subtracting pixel is also called its *reference pixel.*

Given an 8-bit monochrome image, its intensity value covers a range from 0 to 255. Image pixels are always read in scan-line order.

The first slice consists of the power indices of the self-difference screen using Table 14.3.2. The recovered value is the decoded value of the first slice screen. There are nine possible indices to be packed.

The *residual screen* is the difference of the recovered screen from the original screen. The second slice packs the next power level coefficients of the residual screen using Table 14.3.3.

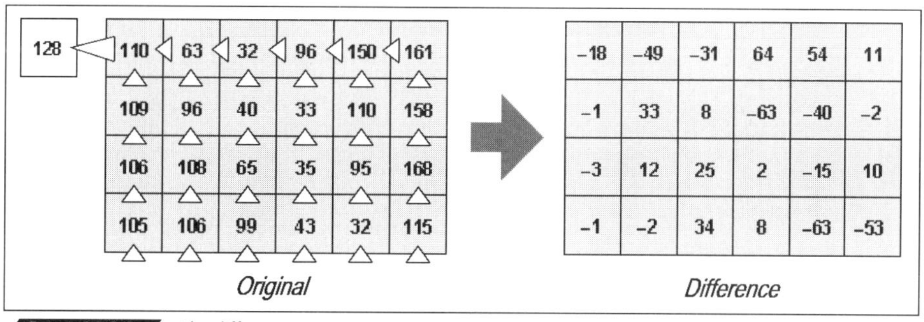

FIGURE 14.3.1 *The difference screen*

TABLE 14.3.2 *First slice index table*

value range	[±255, ±122]	[±121, ±41]	[±40, ±14]	[±13, ±5]	[−4,4]
power index	±6	±5	±4	±3	0
next power level	4	3	2	1	1
recovered value	±243	±81	±27	±9	0

TABLE 14.3.3 *Second slice index table*

if current level	4	3	2	1	0	done
power index value range	±1 for [±121, ±41]	±1 for [±40, ±14]	±1 for [±13, ±5]	±1 for [±4, ±2]	±1 for ±1	skip
power index value range	0 otherwise	0 otherwise	0 otherwise	0 otherwise	0 for 0	skip
new power level	3	2	1	0	done	done
recovered value	±81	±27	±9	±3	±1	0

Keep creating new slices by updating the new residual screen — always pack the next power level coefficients of the residual screen using Table 14.3.3, until every pixel is down. Consequently, at most six slices will be created.

Table 14.3.4 illustrates an example of the actual process. And Figure 14.3.2 shows the visual result applied to the image 512×512 *Lena*. The recovered image quality of the first slice is quite high. It is visually transparent except for a few spots in high-frequency areas. The recovered image of the first two slices is visually lossless — so it is pointless to put more in the figure.

TABLE 14.3.4 *An example of SBA slicing*

8-bit	120	128	135	139	142	137	136	118
Difference	-8	9	7	2	5	-7	-1	-19
1st slice (level)	-3(2)	3(2)	3(2)	0(2)	3(2)	-3(2)	0(2)	-4(3)
recovered	119	128	137	137	146	137	137	110
residual	1	0	-2	2	-4	0	-1	8
2nd slice (level)	0(1)	0(1)	-1(1)	1(1)	-1(1)	0(1)	0(1)	1(2)
recovered	119	128	134	140	143	137	137	119
residual	1	0	1	-1	-1	0	-1	-1
3rd slice	1(0)	0(0)	1(0)	-1(0)	-1(0)	0(0)	-1(0)	0(1)
recovered	120	128	135	139	142	137	136	119
residual	0	0	0	0	0	0	0	-1
4th slice	-	-	-	-	-	-	-	-1(0)
recovered	120	128	135	139	142	137	136	118

A

B

FIGURE 14.3.2 Lena *in SBA format: (A) After the 1st slice, 56KB; (B) After the 2nd slice, 108KB*

The original 512×512 *Lena* has a size of 262,144 bytes, the first three slices have almost equal sizes: 56,218, 51,798 and 51,928 bytes. The remaining slices have a total size of 4455 bytes. This routine not only gives the progressiveness but also offers a simple lossless compression. In this case the lossless compression ratio is $262{,}144{:}164{,}399 \approx 1.6{:}1$. The data of slices are packed using simple arithmetic coding. Some slight improvement may be expected when a more advanced 2-dimensional entropy codec is involved.

The next challenge is clearly how to slice the image data into *thinner* (i.e., smaller in data size) data slices.

14.4 A SUBBAND LOSSY-TO-LOSSLESS CONTINUUM

In Chapter 9 we presented the wavelet subband coding algorithm. Notice that the hierarchical structure is natural for progressiveness. A good coding implementation developed by Said and Pearlman ([SP]) will be presented in this section.

14.4.1 REVERSIBLE S TRANSFORM

A wavelet subband coding algorithm consists of two major steps: *quadrature filter dyadic decomposition* and *coefficient data packing*. Any quadrature filter defined in Chapter 9 can be used here to achieve the progressiveness of decompression. However, for a lossless representation, the transform must be numerically *reversible*.

For example, the 2-dimensional S transform is generated by the 1-dimensional transform defined as follows:

$$l[x] = p[2x] + p[2x+1], \text{ and } h[x] = p[2x] - p[2x+1], \qquad (14.4.1)$$

which converts a string of length $2w$ into two strings of length w. One is called a *low pass*, and the other is called a *high path*. Notice that both $l[x]$ and $h[x]$ are even or odd at the same time. So if the last bit of $l[x]$ is thrown away, the original values can still be recovered. Actually, the determinant of the transform is 2, which implies one bit overhead. Therefore, the last bit of $l[x]$ is ignored by saying that the *step* of the value $l[x]$ is 2. At the same time, the step of $h[x]$ remains 1.

In general 2-dimensional cases, given an image $\{P[x, y]\}$ of width $2w$ and height $2h$, the transform will transform the image into four screens of dimension $w \times h$:

$$\begin{aligned}
LL[x,y] &= P[2x,2y] + P[2x+1,2y] + P[2x,2y+1] + P[2x+1,2y+1], \\
HL[x,y] &= P[2x,2y] + P[2x+1,2y] - P[2x,2y+1] - P[2x+1,2y+1], \\
LH[x,y] &= P[2x,2y] - P[2x+1,2y] + P[2x,2y+1] - P[2x+1,2y+1], \\
HH[x,y] &= P[2x,2y] - P[2x+1,2y] - P[2x,2y+1] + P[2x+1,2y+1]. \qquad (14.4.2)
\end{aligned}$$

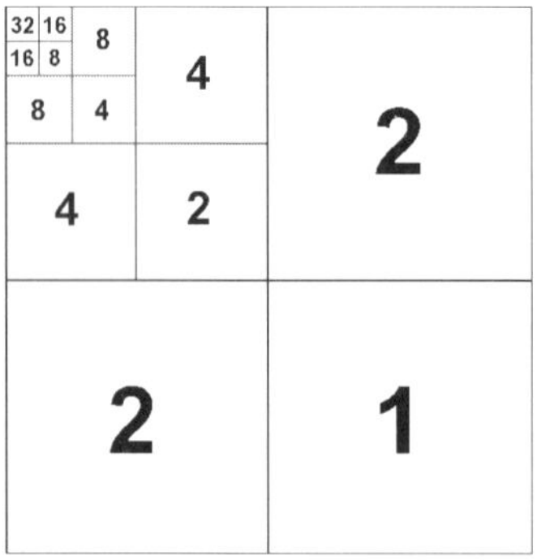

FIGURE 14.4.1 *Step numbers of a subband decomposition*

For the sake of convenience, these four screens are tiled into a single screen, say Q, of dimension $2w \times 2h$:

$$Q[x,y] = LL[x,y], \quad Q[x,y+h] = HL[x,y],$$
$$Q[x+w,y] = LH[x,y], \quad Q[x+w,y+h] = HH[x,y]. \qquad (14.4.3)$$

If the step of the image $\{P[x, y]\}$ is k, then after the transform, the step of $LL[x, y]$, $HL[x, y]$, $LH[x, y]$, and $HH[x, y]$ will be $4k$, $2k$, $2k$, and k, respectively.

To the low–low subimage $\{LL[x, y]\}$, the S transform can be applied repeatedly until either the width or the height of the new low–low subimage is an odd number. The last low–low subimage is called the root image. In Figure 14.4.1 we show the variable steps of a subband decomposition. For any pixel in the subband decomposition, the *step* of the pixel is defined to be the step of the subimage where the pixel is allocated.

14.4.2 SUBBAND SLICING

We continue to use the notation given in Chapter 9, Section 9.3.1.

Let $\mathbf{R}$ denote the set of all pixel coordinates of the root subimage, *roots* of the zero tree. Let $\mathbf{L}$ denote the set of all *leaves* of the tree consisting of coordinates of the form (i, j) such that the coordinate $(2i, 2j)$ is off the image dimension. For any coordinate $(i, j) \notin \mathbf{L}$, let $\mathbf{O}(i, j)$ denote the offspring of the node (i, j), and $\mathbf{D}(i, j)$ the descendant of the node (i, j).

For presenting the new algorithm, we use the following notation for the set of offspring branches: $\mathbf{B}(i,j) = \mathbf{D}(i,j) \setminus \mathbf{O}(i,j)$. Given any number n, a set $\mathbf{S}$ of coordinates is said to be *n-significant* if and only if there is some coordinate $(k, h) \in \mathbf{S}$ such that the absolute pixel value at (k, h), $Q(k, h)$, is significant in the nth power of 2, i.e., $|Q(k,h)| \geq 2^n$. We write $s_n(\mathbf{S}) = 1$ if $\mathbf{S}$ is n-significant and $s_n(\mathbf{S}) = 0$ otherwise.

The order in which the subsets are tested for significance is important, for both the encoder and decoder must follow the same order. Below, the exact slicing algorithm is presented.

Given $\{Q(i, j)\}$ a subband decomposed image, let

$$N = \max_{(i,j)} \left\{ \left\lfloor \log_2 \left| Q(i, j) \right| \right\rfloor \right\} \tag{14.4.4}$$

denote the highest possible bit power of all possible pixel values. Set a variable n to N, and decrement by 1 repeatedly until $n=0$. For each n, three slices will be built: the *insignificant pixel sorting slice*, the *insignificant set sorting slice*, and the *significant pixel refinement slice*. Thus, the lossless file contains a total of $3(N+1)$ slices.

Initially, we set $n=N$; the *list of significant pixels* **LSP** to empty, the *list of insignificant pixels* **LIP** to all root subimages **R** in a fixed order, e.g., scan line order; and the *list of insignificant sets* **LIS** to all root descendant sets, i.e., $\mathbf{LIS} = \left\{ \mathbf{D}(i,j) \middle| \ (i,j) \in \mathbf{R} \right\}$.

For each $n \geq 0$, first make a copy of the list of significant pixels: $\mathbf{CP} = \mathbf{LSP}$.

1. *The insignificant pixel sorting slice*. For each $(i, j) \in \mathbf{LIP}$, output $s_n(i, j)$ in the given order. If $s_n(i, j) = 1$, i.e., it is n-significant, move (i, j) to the end of **LSP** and output the sign of $Q(i, j)$.
2. *The insignificant set sorting slice*. For each set $\mathbf{S} \in \mathbf{LIP}$, it must have the form of either $\mathbf{S} = \mathbf{D}(i, j)$ or $\mathbf{S} = \mathbf{B}(i, j)$, for some (i, j). Again output $s_n(\mathbf{S})$ in the given order, and if $s_n(\mathbf{S}) = 1$, remove the set **S** from **LIS**. Then,

   a. if $\mathbf{S} = \mathbf{D}(i, j)$, first add the set $\mathbf{B}(i, j)$ to the end of **LIS** if $\mathbf{B}(i, j) \neq \emptyset$. Now, for each $(h, k) \in \mathbf{O}(i, j)$, first output $s_n(h, k)$:

      - if $s_n(h, k) = 1$, we add (h, k) to the end of **LSP** and output the sign of $Q(h, k)$;
      - if $s_n(h, k) = 0$, we add (h, k) to the end of **LIP**.

   b. if $\mathbf{S} = \mathbf{B}(i, j)$, then for each $(h, k) \in \mathbf{O}(i, j)$, we add the set $\mathbf{D}(h, k)$ to the end of **LIS**.
3. *The significant pixel refinement slice*. For each $(i, j) \in \mathbf{CP}$, a significant pixel determined by some earlier round, if $Q(i, j)$ has a step less than or equal to 2^n, we simply output the nth bit of $\left| Q(i, j) \right|$, i.e., $((\left| Q(i, j) \right| >> n) \,\&\, 1)$.

One important characteristic of the algorithm is that the entries added in the second slice to the end of **LIS** are evaluated before that same sorting pass ends. So, when we say "for each set $\mathbf{S} \in \mathbf{LIS}$," we also mean those that are being added to its end.

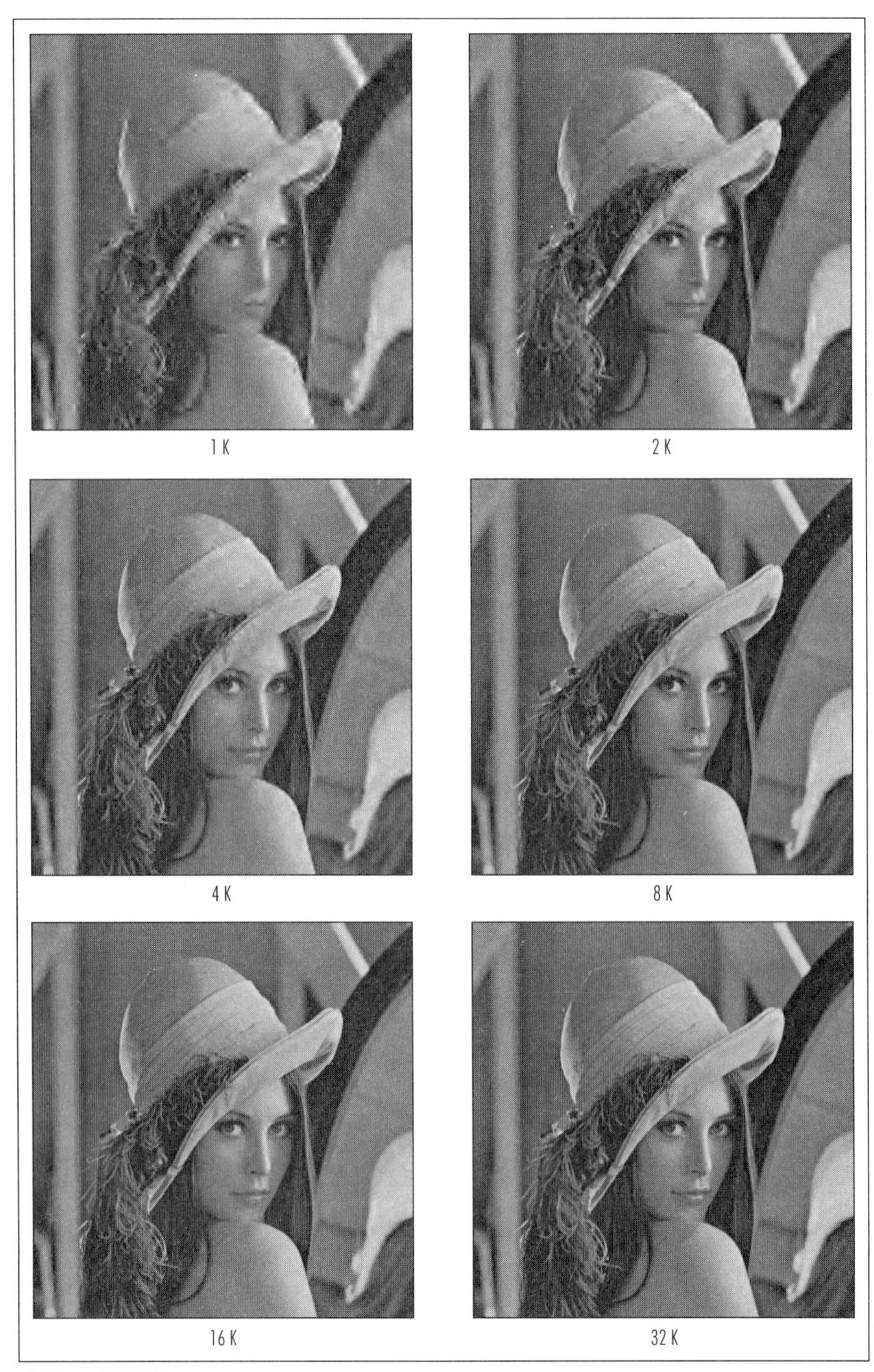

FIGURE 14.4.2 *Progressiveness of Lena*

Table 14.4.1 *Performance of* Lena

File size	1 K	2 K	4 K	8 K	16 K	32 K	55 K
bpp rate	0.031	0.062	0.125	0.250	0.500	1.000	1.710
rmse	12.80	9.89	7.30	5.27	3.77	2.69	1.60
PSNR	25.98	28.23	30.86	33.69	36.60	39.52	44.05

In a decoder, the three control lists (**LSP**, **LIP**, and **LIS**) will be reconstructed identically to the ones used by the encoder. As a consequence, in this scheme, both the encoding and the decoding have the same computational complexity.

The lossy image can be cut not only on the boundary of the slices, but anywhere in the file. Every bit of additional data will always contribute to some decoded image improvement. When a decoder reaches the end of a lossy file, it is assumed that the control lists will remain unchanged from that moment on. If the lossy file stops at the bit power n, for any (i, j) from **LSP**, if $Q(i, j)$ has a step less than or equal to 2^{n-1}, $2^{n-1}\, sign(Q(i, j))$ will be added to $Q(i, j)$ for decreasing distortion.

In Figure 14.4.2, the testing image *Lena* has been compressed using this subband technique and displayed progressively in file sizes of 1K, 2K, 4K, 8K, 16K, and 32K. The visually lossless image quality has been achieved in the compression ratio of about 7:1 to this testing image. The final lossless file has a size of 137,250 bytes, which is a 1.9:1 compression ratio, better than the DPCM methods described in Chapter 11. This statement actually holds for almost all images that have been compressed. The numerical results shown in Table 14.4.1 are as impressive as the visual ones.

The subband coding scheme inherited a natural progressive structure. Could the fractal image representation be designed in a similar fashion? This question has been explored in Chapter 9.

I feel along the edges of life
for a way
that will lead to open land.

—Rescue the Dead

David Ignatow

Epilogue: It's Real

15

Picture something out of a dream like a ruby moon hanging on the saffron sky with an emerald ocean reflecting its vibrant colors. The ability to dream is one of the most intense capacities one can possess. Beautiful dreams are flowers blooming in the heart and music singing in the soul. This book, begun with magic, will end with a dream.

During the process of writing this book, I found so many people who share the same dreams and the same passions and are excited about the technological future. Like the rapid progress of today's technology, those dreams have turned out to be more real than might have been imagined. Actually, the themes discussed in this chapter could more appropriately be labeled predictions, since they will very likely become realized in the foreseeable future—although the exact form of these realizations could be strange in relation to our current knowledge and computer experience.

This chapter will demonstrate how fractal image compression and decompression systems can be built in the optical world—the world they really belong to. In fact, in the digital world, fractal imaging systems have been crippled from the very beginning, since they have been forced to work with image digital data in the form of a finite resolution grid—which is the only digital image form that is available at the moment.

Therefore, this chapter will start from the very beginning of the image process, "image capturing," to obtain an image fractal representation directly from the natural image source. A fractal image capturer, a *fractal camera* or *fractal scanner*, is a photographic digitization machine that converts an optical or chemical image into fractal image format using a precise optical matching procedure.

The reverse process of *capturing* is *displaying*. For an output device that has no inherent resolution, a decompression system will directly output a resolution-independent picture through an optical system, such as a *fractal printer* or a *fractal television*. Consequently, a *fractal copier* is simply a combination of a fractal scanner and a fractal printer.

15.1 FRACTAL CAPTURER

The name "*fractal capturer*" is used for both a fractal camera and a fractal scanner. One design of a fractal capturer can be obtained by modifying Shih *et al.*'s *electro-optical IFS finder*, described in their patent [SCMD], which was built only for binary fractals and probabilistic grayscale fractals (see Chapter 2). In this design a fixed image partition, e.g., the *uniform square partition*, is assumed. Thus, the shape of destination regions and the number of fractal transformations are exactly as designed by the hardware of a given capturing system.

15.1.1 OPTICAL REALIZATION OF AN AFFINE TRANSFORMATION

Before drawing the design of the capturer, we shall present a component of how a reference region can be masked and mapped to a destination region. Such a

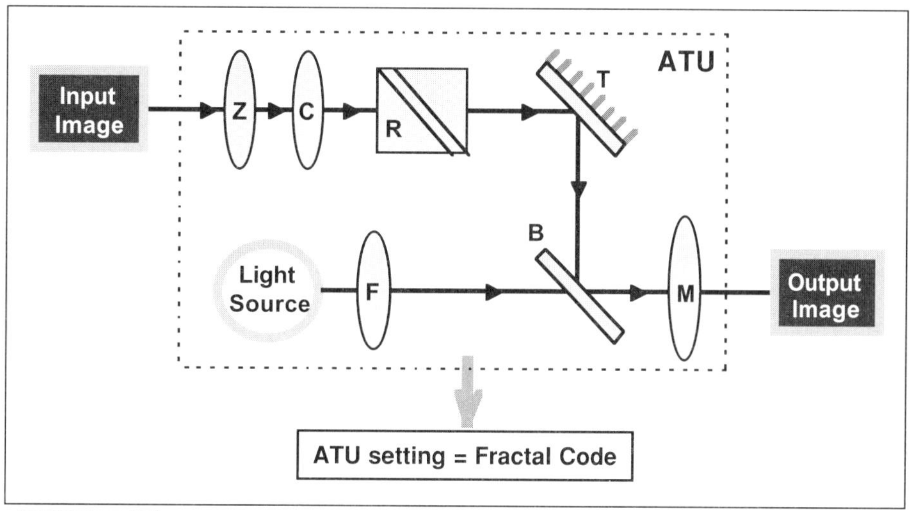

FIGURE 15.1.1 *Optical affine transformation unit*

component, corresponding to a fractal transformation element, is called an *affine transformation unit* (*ATU*), which is formed by a *contrast adjustment glass* C, a *zoom lens* Z, a *rotating prism* R, *translation mirrors* T and B, a *constant light source* with an adjustable *light filter* F, and a *masking blocking filter* M, as shown in Figure 15.1.1.

For example, an input image could be photographic film illuminated by a light source. This image light goes through a zoom lens Z first. This provides a spatial magnification or demagnification as required by the spatial scaling factor of the affine transformation; then this image light goes through a contrast adjustment lens C that reduces the intensity contrast (which is equivalent to multiplying a contrast adjustment γ-value); then the image light goes through a rotation prism R, which could be a Harting-Dove prism or a Pechan prism, that generates standard rotations and flips; then the image light goes through a translation mirror set T that performs the spatial shifting and translating; it then merges with a constant-intensity light adjustment given by a constant light source through an adjustable intensity filter F that add a constant positive intensity β-value and at last goes through a masking filter M that block out all the image except the active destination region. What comes out is exactly the desired output image as applied by the affine transformation (if the β-value of the transform is not negative).

This ATU is the main building block of our capturer. In this ATU the lenses Z, C, and M could be fixed or adjustable depending on the scheme, and the prism R, the mirror set T, and the intensity filter F must be adjustable. They are actually all controlled by a central controller in the capturer.

The magnification ratio adjustment of the lens Z, the spatial translation adjustment of the mirror set T, the setting of the rotation and flip of the prism R, and the intensity shifting of the filter F can all be mechanically mounted in respective fixtures by many known servomechanisms. The adjustment of the lens C and the exchange of the masking filter M can be obtained in the same way using a set of multiple lenses and glasses. All of these adjustments can be controlled by the capturer central controller through some associated peripheral devices.

15.1.2 DESIGN OF A FRACTAL CAPTURER

Using ATUs, a capturer can be designed. Given a fixed image partition (or weighted covering, cf. Chapter 4), for each destination region, an ATU with the masking filter exactly matching the masking function of the region is assigned. Thus, this system will use a fixed image partition. The number of ATUs is exactly the number of destination regions.

Figure 15.1.2 illustrates a case of four ATUs—the light of the original input image is optically branched into several light paths of equal intensity using some beamsplitters, *BI1*, *BI2*, *BI3*, and *BI4*, of particular branching ratios. Going through the ATUs, the light paths merge together using some mirrors, *MO1*, *MO2*, *MO3*, and *MO4*, and some beamsplitters, *BO1*, *BO2*, *BO3*, and *BO4*, and form a composed image (called the *collage image*). The collage image is compared with the *adjusted original input image* in an *image comparison box*, (described in the next subsection), which could be built either optically or digitally. The adjusted original input image is obtained from the original image by adding a constant light intensity b from the light source *LS* through a tunable filter F reflected by a beamsplitter B. The image comparison box outputs its

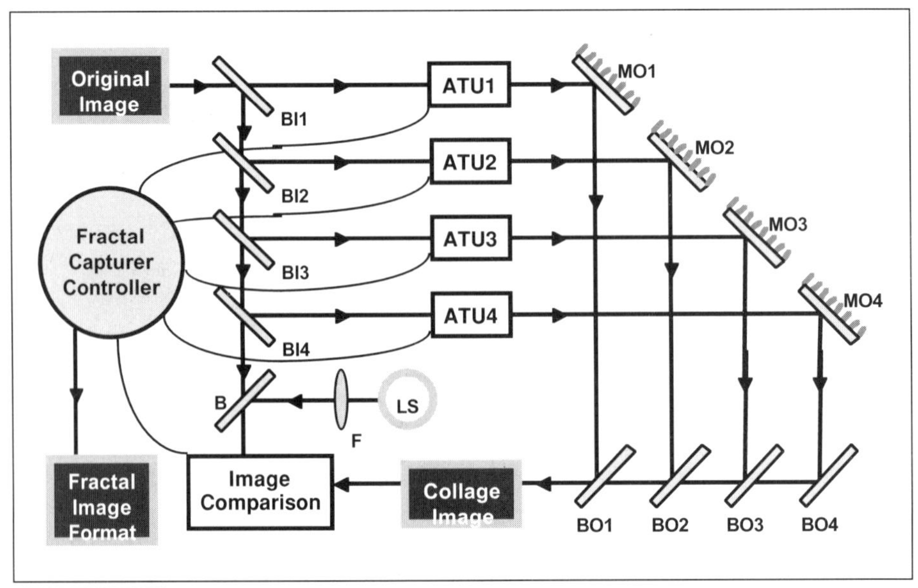

FIGURE 15.1.2 Fractal capturer

evaluation value to the *fractal capturer controller*. The controller records the best setting while controlling ATUs scanning through all possible settings by varying the parameters of the affine transforms in a systematic manner. The best setting, modulo subtracting the intensity value b from the β-value of each transformation, is exactly a fractal representation of the original image. It will finally be stored in the *fractal image format*.

Because of the high speed of the optical process, overall, the scanning process is much faster than a conventional method using a sample pixel CCD sensor for an image of similar quality.

15.1.3 OPTICAL IMAGE COMPARISON

The image comparison box has the option to use either digital or optical methods. Digital comparison is achieved by first digitizing both the original and the collage images using some image detector array (e.g., a CCD imager—which will be discussed further in a later section on the fractal camera) and then differentiating the digital data.

Instead of taking the difference of the tiled image digitally, the evaluation can also be done optically. For example, a *liquid crystal light valve* (*LCLV*) is used to convert the collage image into a coherent light source. Then the image can be correlated with the original one using traditional coherent optical processing for image subtraction.

As shown in Figure 15.1.3, the collage image and the original image are projected by respective lenses *IC* and *IO* onto the rear of the *LCLV* through a Ronchi grating *RG*—a grating with equal-width opaque and transparent stripes. The composite image of both images is obtained by a coherent light beam (a laser beam). A beamsplitter *B* directs the coherent light beam onto the front side of the *LCLV*, and the reflected light beam is transmitted through the beamsplitter *BS* to lens *L*. A filtering slit *F* is used to select out an odd order of the composite image so that the filtered image on the image plane *ID* is just the

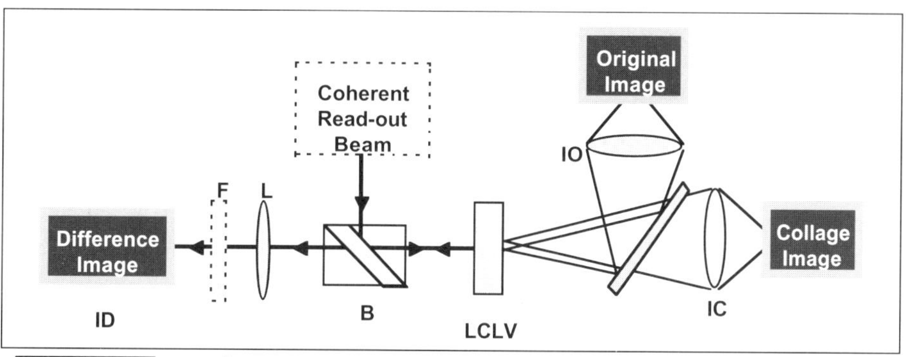

FIGURE 15.1.3 Optical image comparison

difference of the collage and the original images. The accuracy of the match is indicated by the sum of total intensity at the image plane *ID*; the higher is the sum, the poorer the match.

The use of LCLV in optical data processing, including image subtraction, has been discussed in the literature. See [B5] and [M3] for more details.

15.2 DIGITAL CAMERA

A blueprint of a fractal image capturer has been presented. It will take several more years to see a "commercially ready" fractal scanner. After this period of time, fractal cameras will become available. A camera needs to capture an image in not less than 1/10 second, but a scanner is allowed up to a number of minutes to acquire an image. This is not the only extra challenge to a camera and not to a scanner. Color interpolation is another challenge unique to a camera. This section will describe the current digital camera design and show how fractal techniques can be used to improve color image quality.

15.2.1 SINGLE-CHIP CCD CAMERA

A camera (different from a scanner, which can scan the same image three times for different color components: red, green, and blue) must capture all color information all at once in a single procedure. A single-chip *charge coupled device* (*CCD*) camera solves this problem by using a *color filter array* (*CFA*), which is formed as a layout of color filter sites, to obtain both luminance and chrominance signals. A *color filter*, corresponding to a pixel signal, can read only a color spectral filtered luminance intensity. In reality, keeping the camera cost low puts a limit on the total number of filters. So the question of getting a high-quality color image from a cheap digital camera is how to lay out the filters for maximal information and how to process the filter signal data for maximal image quality. The problem can be decomposed into three steps: *geometry of CFA layout*, *color signal interpolation*, and *image enhancement*. Here, a fractal approach to this problem by focusing on the last two steps will be presented. With additional local image analysis, we will present a new interpolation scheme. Combining fractal image enhancement, one of the best image high-resolution prediction technologies, we offer a complete solution for a near-future, single CCD digital camera.

Figure 15.2.1 gives a design sketch of a digital camera. The real-world image goes through a *CFA* and is detected by a *CCD sensor*. An *analog-to-digital converter A/D* translates analog signals into digital signals. A *γ-corrector G* does a γ-correction on the digital signals. A *color interpolator CI* interpolates missing color pixel values. Then, a *fractal image enhancement E* step is involved to enhance the image for high visual quality. Finally, the image data will be packed using some *compression chip P* and then sent to the camera's *flash memory* for

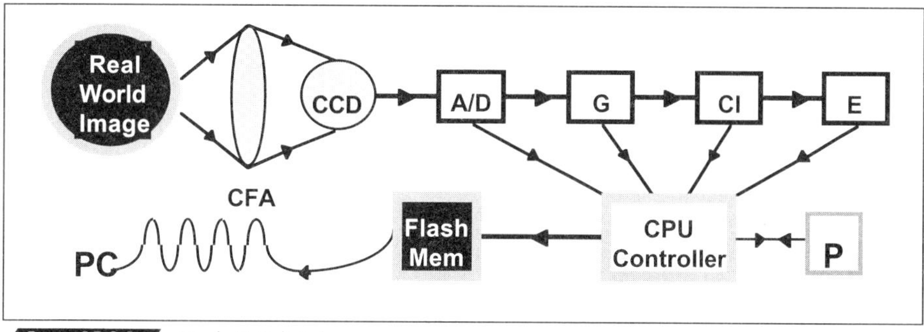

FIGURE 15.2.1 Single CCD digital camera

temporary storage. And later, the image data will be passed to a computer for printing and display. The current commercial digital cameras have a CFA resolution from about 300×200 up to 1200×800. For a 4MB flash memory camera, if the image data of a picture is packed into 200K bytes, 20 pictures can be stored in the camera. And if the compression is pushed further to 80K bytes per picture, 50 pictures can be stored in the camera.

One may notice the connection lines from the *A/D converter*, the *γ-corrector*, the *color interpolator*, and the *image enhancer* to the *camera controller*. In fact, in any of the steps of the image γ-correction, the color interpolation and the image enhancement can be switched to the main computer. In that case, the image data from the camera will be less optimized in compression, but the cost of the camera becomes far less.

15.2.2 COLOR FILTER ARRAY

The *Color filter array* (*CFA*), lying on top of a single CCD, determines the detected image data color pattern. Each color filter, corresponding to a light signal reading, gives only a monochromatic light intensity and is designed to have a specific spectral response that in most cases is not uniform for the same intensity of light falling on them. Figure 15.2.2 shows the spectral response curves of red, green, and blue filters used in a TV sensor.

Based on colorimetry theory, each color is combined from the three primaries, e.g., red (R), green (G), and blue (B). Thus, a color filter can read only a fixed monochromatic color combination. Hence, it can be assumed to have the following formula:

$$\alpha \cdot R + \beta \cdot B + \gamma \cdot G, \tag{15.2.1}$$

for any $\alpha > 0, \beta > 0, \gamma > 0$ and $\alpha + \beta + \gamma = 1$.

For example, in current television standards, e.g., *YIQ*, we may have a color filter to read the luminance component Y directly, but neither I nor Q is directly readable by a color filter.

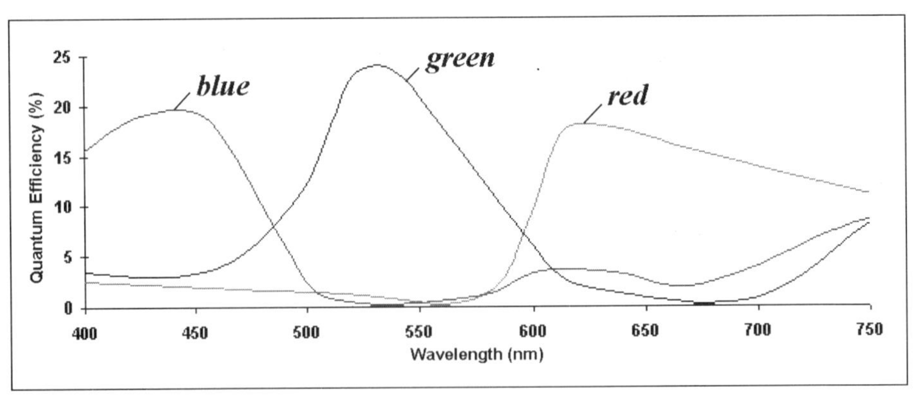

FIGURE 15.2.2 Spectral response curves of a TV image sensor

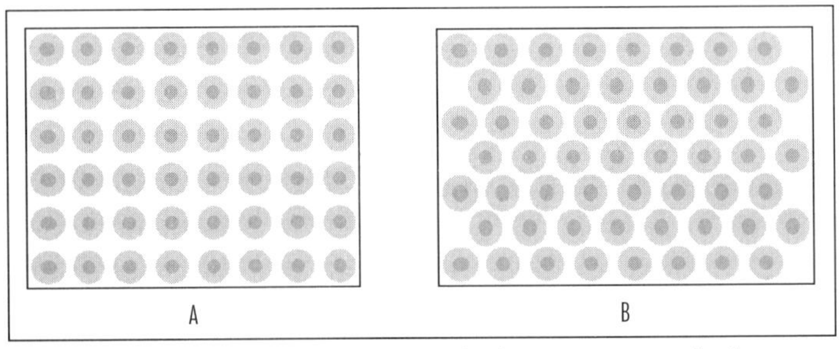

FIGURE 15.2.3 CFA layouts: (A) rectangular diagram; (B) triangular diagram

Geometrically, there are two major layouts for CFAs: the *rectangular layout* and the *triangular layout*, as shown in Figure 15.2.3.

In each geometric arrangement there are many CFA patterns that have been studied. It is widely believed that the human eye is far more sensitive to green than to red or blue. Therefore, in most popular CFA patterns there are always more G filters than other color filters.

In a rectangular diagram, the most common CFA pattern is the Bayer 2G pattern (i.e., the red–green–blue ratio is 1:2:1) (cf. Figure 15.2.4). And in a triangular diagram we can have the symmetric 1G pattern (1:1:1), the hexagon 4G pattern (1:4:1), or a 2G pattern (1:2:1) (as in Figure 15.2.5).

In converting the CFA signals into an RGB true color image, the missing pixel color values need to be filled up. Using the rectangular Bayer 2G pattern as a concrete example, we will first interpolate the missing green pixels and then interpolate the other chromatic components.

For notation, assume that there are three color screens, $\{R(i, j)\}_{i,j=0,1,\ldots}$, $\{G(i, j)\}_{i,j=0,1,\ldots}$, and $\{B(i, j)\}_{i,j=0,1,\ldots}$, where $G(i, j)$'s are given for all i and j that

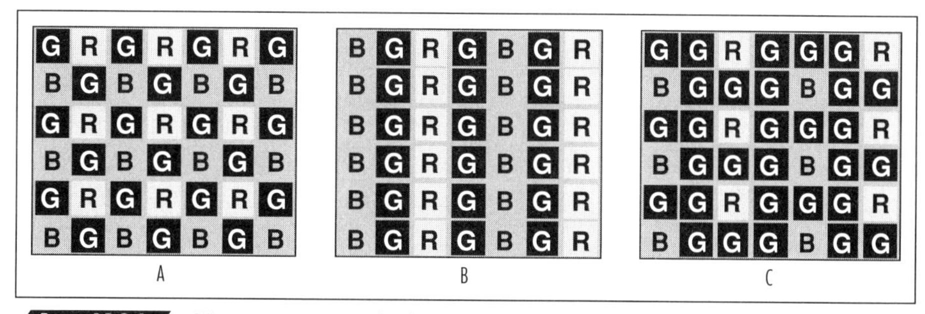

Figure 15.2.4 *CFA patterns in rectangular diagrams: (A) Bayer 2G pattern for progressive scanning; (B) Bayer 2G pattern for any type of scanning; (C) Bayer 3G pattern for progressive scanning*

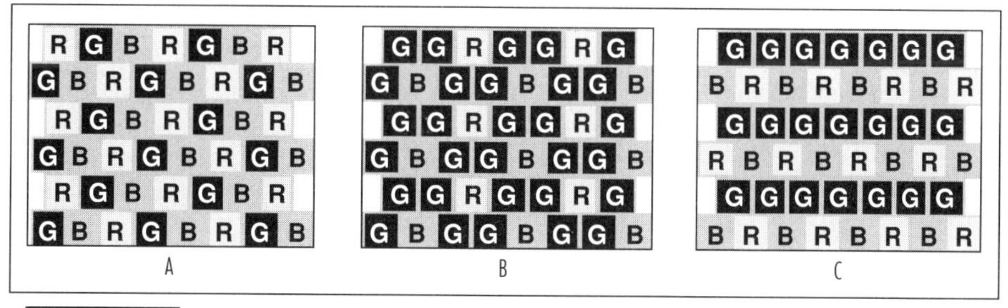

Figure 15.2.5 *CFA patterns in triangular diagram: (A) 1G CFA pattern; (B) 4G CFA pattern; (C) 2G CFA pattern*

are both odd or even, $R(i, j)$'s are given for all odd i even j, and $B(i, j)$'s are given for all even i odd j.

Better color filters are used in practice. At first we can switch green G to any luminance expression, i.e., switch from G-R-B-G to L-R-B-L. A good choice of L is given by

$$L = \frac{R + B + 2G}{4}. \tag{15.2.2}$$

Focusing further on image luminance, replace red R and blue B with yellow Ye and cyan Cy, respectively. In the formulas

$$Ye = \frac{R + G}{2}, \text{ and } Cy = \frac{B + G}{2}, \tag{15.2.3}$$

all CFA patterns described before can be alternated from R-G-B to Ye-L-Cy. In particular, the G-R-B-G can be switched to L-Ye-Cy-L.

However, in all of those cases, except color conversion, everything else will be constructed in exactly the same manner.

15.2.3 MISSING GREEN PIXEL INTERPOLATION

In the rectangular Bayer 2G CFA, pattern G is the dominant color filter. For half of the image pixels the green G values are given. What about the other half? An interpolation or prediction is required. There are many approaches that have been studied, but none of them are satisfactory. The simplest one is the bilinear interpolation, which is the average of its neighboring four G values from its left, its right, above, and below. The performance is not adequate, especially on textures and edges.

Hibbard [H4] improved the bilinear method by taking the horizontal average and the vertical average into account. For example, let us interpolate the value $G(2, 3)$. We have $G(2, 2)$ above it, $G(2, 4)$ below it, $G(1, 3)$ on its left, and $G(3, 3)$ on its right. The *neighborhood average,* the *horizontal average* and the *vertical average* are defined to be

$$A = \frac{G(1,3) + G(3,3) + G(2,2) + G(2,4)}{4}, \tag{15.2.4}$$

$$H = \frac{G(1,3) + G(3,3)}{2}, \text{ and } V = \frac{G(2,2) + G(2,4)}{2}, \tag{15.2.5}$$

respectively. The *horizontal difference* and *vertical differences* are defined to be

$$\Delta H = \left|G(1,3) - G(3,3)\right| \text{ and } \Delta V = \left|G(2,2) - G(2,4)\right|, \tag{15.2.6}$$

respectively.

Thus, for a previously fixed threshold T, the green pixel value $G(2, 3)$ is predicted as

$$G(2,3) = \begin{cases} H, & \text{if } \Delta H < T \text{ and } \Delta V > T; \\ V, & \text{if } \Delta V < T \text{ and } \Delta H > T; \\ A, & \text{otherwise.} \end{cases} \tag{15.2.7}$$

A typical artifact of the Hibbard interpolation is the annoying aliasing dots along the texture's edge (see Figure 15.2.8). Further improvement was made by Cok (cf. [C1]) by adopting "edge enhancement" to the missing green pixel interpolation. In that scheme, local geometry is involved.

Cok's method adds an additional term to two special cases based on the pixel values: the *stripe pattern,* which can be defined by a formula like

$$\min\left(G(1,3), G(3,3)\right) - \max\left(G(2,2), G(2,4)\right) > T, \tag{15.2.8}$$

and the *corner pattern,* which is given by a formula like

$$\min\left(G(1,3), G(2,4)\right) - \max\left(G(2,2), G(3,3)\right) > T, \tag{15.2.9}$$

for a some threshold T, as shown in Figure 15.2.6. For all other patterns he uses the *median predictor* instead of the average one.

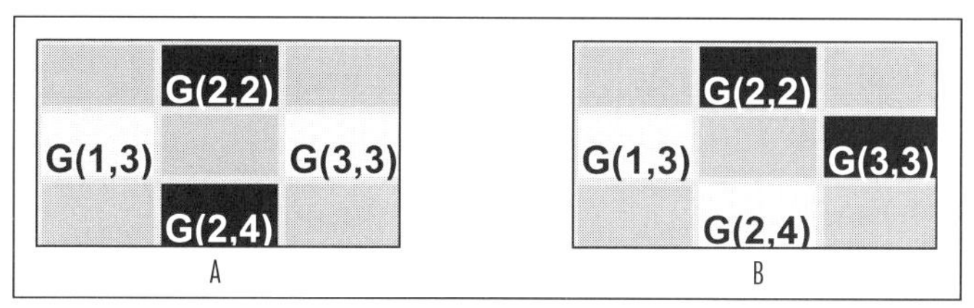

FIGURE 15.2.6 *Cok's exceptional cases: (A) the stripe pattern; (B) the corner pattern*

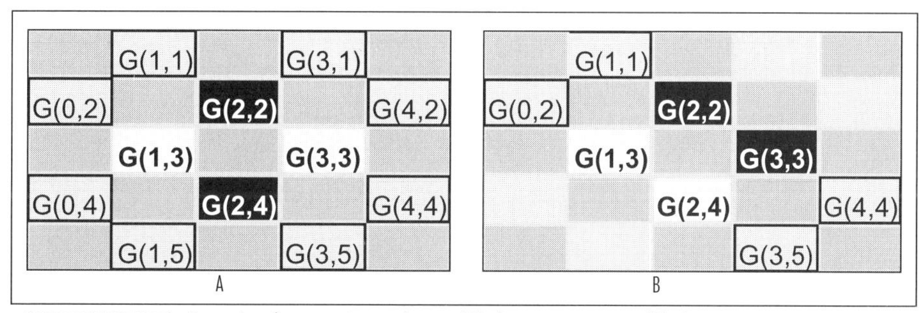

FIGURE 15.2.7 *Exceptional pattern interpolation: (A) the stripe pattern; (B) the corner pattern*

The *median* of the four values $G(2, 2)$, $G(2, 4)$, $G(1, 3)$ and $G(3, 3)$, denoted by $m(2, 3)$, is the average of the middle two value, after excluding the largest and smallest, of the four values. If we reorder these four values as $A \geq B \geq C \geq D$, then $m(2,3) = \frac{B+C}{2}$. Therefore, in most cases we will define $G(2, 3) = m(2, 3)$, excepting the stripe pattern and corner pattern cases displayed in Figure 15.2.6. For those cases, we add an extra term S:

$$G(2,3) = clip_C^B(2 \cdot m(2,3) - S), \tag{15.2.10}$$

where the S term is given as shown in Figure 15.2.7: In the stripe case,

$$S = \frac{G(1,1) + G(3,1) + G(4,2) + G(4,4) + G(3,5) + G(1,5) + G(0,4) + G(0,2)}{8}, \tag{15.2.11}$$

and in the corner case,

$$S = \frac{G(1,1) + G(4,4) + G(3,5) + G(0,2)}{4}, \tag{15.2.12}$$

while the clip function is defined to be

$$clip_C^B(X) = \max(C, \min(B, X)), \tag{15.2.13}$$

for any value X.

Instead of interpolating and predicting the missing green pixels using neighboring green pixels, our interpolation will use all neighboring color filter pixels, including *red* and *blue* ones [LL].

Given a pixel whose green value is not given, interpolate this missing green pixel. One of the blue or red values of that pixel must be given. For example, at the pixel (2, 3) the blue $B(2, 3)$ is given. The row and the column across this pixel consists of a string of pixel values in the color filter array alternating between two colors: green G and the given value color of this pixel (either red R or blue B). In the example, at the pixel (2, 3), the row pixel values are

$$\cdots, B(0,3), G(1,3), B(2,3), G(3,3), B(4,3), \cdots \tag{15.2.14}$$

and the column pixel values are:

$$\cdots, B(2,1), G(2,2), B(2,3), G(2,4), B(2,5), \cdots . \tag{15.2.15}$$

In general, both row and column have the form

$$\cdots, B(-2), G(-1), B(0), G(1), B(2), \cdots . \tag{15.2.16}$$

The pixel value of $G(0)$ is predicted from the following combined color criteria:

1. *Both the green G screen and blue B screen have the same local geometric shape,* i.e., locally, the color ratio B/G is constant.
2. *Both the green G screen and blue B screen have the same local fractal complexity*, i.e., locally, the G screen segment between $G(-1)$ and $G(1)$ is similar to the B screen segment between $B(-2)$ and $B(2)$.

The simplest prediction based on these criteria can be expressed in the following formula:

$$G(0) = \frac{G(-1)+G(1)}{2} + \sigma \cdot \frac{2 \cdot B(0) - B(-2) - B(2)}{2}, \tag{15.2.17}$$

where σ is a fixed scaling factor. Referring to our example case, at (2, 3), let $HG(2, 3)$ denote the *horizontal prediction*, which is predicted by the row, and let $VG(2, 3)$ denote the *vertical prediction*, which is predicted by the column. These predictions are actually the generalizations of horizontal and vertical average predictions. Therefore, the same criterion as Hibbard's scheme could be applied in order to determine which prediction to use. Nonetheless, a better criterion is found that again involves both green and blue screens.

Define the *green horizontal difference* and the *green vertical difference* to be

$$\Delta HG = \left|G(1,3) - G(3,3)\right| \text{ and } \Delta VG = \left|G(2,2) - G(2,4)\right|, \tag{15.2.17}$$

respectively. And define the *blue horizontal* and *vertical differences* to be

TABLE 15.2.1 Green pixel G(2, 3) interpolation

if	$\Delta HG<\tau$ and $\Delta VG<\tau$		$\Delta HG\geq\tau$ or $\Delta VG\geq\tau$	
and if	$\Delta HB<\Delta VB$	$\Delta HB\geq\Delta VB$	$\Delta HG<\Delta VG$	$\Delta HG\geq\Delta VG$
then	$HG(2, 3)$	$VG(2, 3)$	$HG(2, 3)$	$VG(2, 3)$

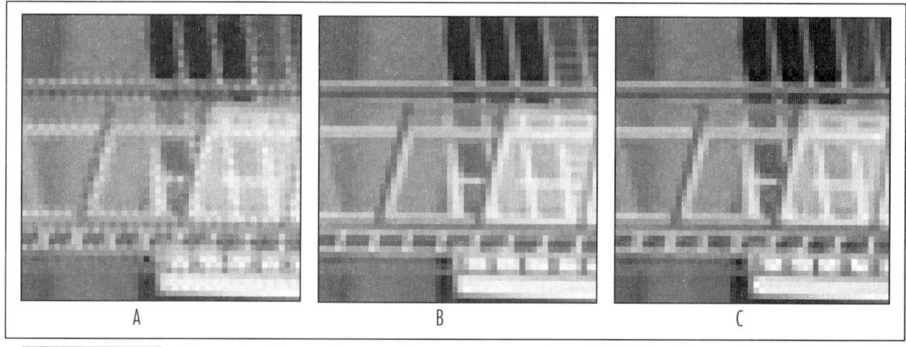

FIGURE 15.2.8 Green pixel interpolation performance comparison: (A) Hibbard Interpolation; (B) Original; (C) Our Method

$$\Delta HB = \left|B(2,3) - B(0,3)\right| + \left|B(2,3) - B(4,3)\right| \text{ and}$$
$$\Delta VB = \left|B(2,3) - B(2,1)\right| + \left|B(2,3) - B(2,5)\right|, \qquad (15.2.18)$$

respectively. Then, for a previously fixed threshold τ, the green pixel value $G(2, 3)$ will be predicted as charted in Table 15.2.1.

In Figure 15.2.8, a portion of the sample testing image *Peachtree One Building* (Color Plate 15) is displayed to demonstrate significant image quality improvement. The exact procedure is as follows: take an RGB scanned image, mask it by simulating Bayer 2G CFA, and then predict the missing pixels. This test image is typically used for digital photography.

15.2.4 CHROMATIC INTERPOLATION

Chromatic interpolation has been approached differently from the missing green pixel interpolation. The main technique is the *smooth hue transition interpolation* (introduced by Cok [C3]), assuming that a smooth hue change will reduce unpleasant color artifacts. A *red hue value* or a *blue hue value* is defined to be the ratio B/G or R/G, respectively. After the green screen is completely interpolated, the red screen (similarly, the blue screen) could be obtained from the hue approximation.

For example, as show in Figure 15.2.9*a*, $R(1, 2)$, $R(3, 2)$, $R(1, 4)$, and $R(3, 4)$ are four given red pixels next to each other. Then,

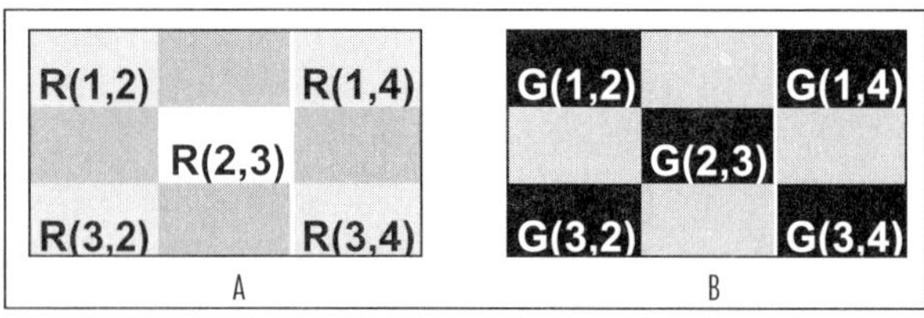

FIGURE 15.2.9 *Red pixel interpolation following green pixels: (A) Red screen R; (B) Green screen G*

$$R(2,2) = \frac{G(2,2)}{2}\left(\frac{R(1,2)}{G(1,2)} + \frac{R(3,2)}{G(3,2)}\right), \tag{15.2.19}$$

and similarly, $R(2, 4)$, $R(1, 3)$, and $R(3, 3)$ are deduced, and also

$$R(2,3) = \frac{G(2,3)}{4}\left(\frac{R(1,2)}{G(1,2)} + \frac{R(3,2)}{G(3,2)} + \frac{R(1,4)}{G(1,4)} + \frac{R(3,4)}{G(3,4)}\right). \tag{15.2.20}$$

In color interpolation applications, there are advantages in describing the pixel data in a logarithmic exposure space. The above scheme can be replaced by the next two formulas:

$$r(2,2) = g(2,2) + \frac{r(1,2) - g(1,2) + r(3,2) - g(3,2)}{2}, \tag{15.2.21}$$

$$r(2,3) = g(2,3) + \frac{r(1,2) - g(1,2) + r(3,2) - g(3,2) + r(1,4) - g(1,4) + r(3,4) - g(3,4)}{4}, \tag{15.2.22}$$

where $r(i, j) = \log R(i, j)$ and $g(i, j) = \log G(i, j)$.

These are equivalent to

$$R(2,2) = G(2,2)\sqrt{\frac{R(1,2)}{G(1,2)} \cdot \frac{R(3,2)}{G(3,2)}}, \tag{15.2.23}$$

$$R(2,3) = G(2,3) \cdot \sqrt[4]{\frac{R(1,2)}{G(1,2)} \cdot \frac{R(3,2)}{G(3,2)} \cdot \frac{R(1,4)}{G(1,4)} \cdot \frac{R(3,4)}{G(3,4)}}. \tag{15.2.24}$$

How accurate is the new "missing color interpolation method" (described in the previous subsection) when applied to red and blue screens in comparison with Cok's method? The result was excellent.

To interpolate the red screen, the first step is to predict red pixels where blue pixel values are given, i.e., the pixels whose addresses have an even i and an odd j. For example, (2, 3) is one of such pixels and $R(2, 3)$ is one of the pixel value to be predicted. For each such pixel, the four nearest given red pixel values are allocated at its four diagonal corners. For $R(2, 3)$, these four pixels are shown in Figure 15.2.9*a*; they form two diagonal lines. Together with the green values in those locations as shown in Figure 15.2.9*b*, the value $R(2, 3)$ is predicted using similar fractal interpolation on both diagonal lines:

$$\cdots, G(1,2), R(1,2), G(2,3), R(3,4), G(3,4), \cdots \tag{15.2.25}$$

and

$$\cdots, G(1,4), R(1,4), G(2,3), R(3,2), G(3,2), \cdots . \tag{15.2.26}$$

The two diagonal predictions will be

$$XR(2,3) = \frac{R(1,2)+R(3,4)}{2} + \sigma \cdot \frac{2 \cdot G(2,3) - G(1,2) - R(3,4)}{2}, \tag{15.2.27}$$

and

$$YR(2,3) = \frac{R(1,4)+R(3,2)}{2} + \sigma \cdot \frac{2 \cdot G(2,3) - G(1,4) - R(3,2)}{2}, \tag{15.2.28}$$

where σ is a fixed scaling factor.

Similarly, define the *red diagonal differences* to be

$$\Delta XR = |R(1,2) - R(3,4)| \text{ and } \Delta YR = |R(1,4) - R(3,2)| \tag{15.2.29}$$

and the *green diagonal differences* to be

$$\begin{aligned} \Delta XG &= |G(2,3) - G(1,2)| + |G(2,3) - G(3,4)| \\ \Delta YG &= |G(2,3) - G(1,4)| + |G(2,3) - G(3,2)|. \end{aligned} \tag{15.2.30}$$

Then, for a previously fixed threshold τ, the red pixel value $R(2, 3)$ will be defined in Table 15.2.2.

In the second step, the remaining red pixels are filled up using the identical method in the missing green pixel interpolation by using the green screen as reference.

The blue screen is always interpolated in the identical fashion as the red one.

TABLE 15.2.2 *Red pixel R(2, 3) interpolation*

if	$\Delta XR<\tau$ and	$\Delta YR<\tau$	$\Delta XR\geq\tau$ or	$\Delta YR\geq\tau$
and if	$\Delta XG<\Delta YG$	$\Delta XG\geq\Delta YG$	$\Delta XR<\Delta YR$	$\Delta XR\geq\Delta YR$
then	$XR(2, 3)$	$YR(2, 3)$	$XR(2, 3)$	$YR(2, 3)$

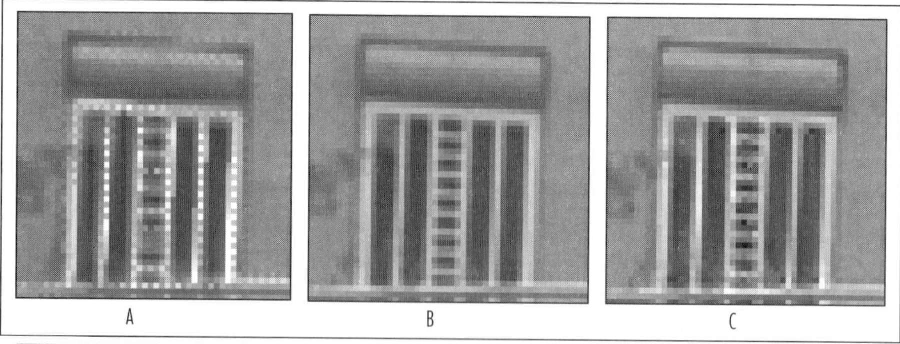

FIGURE 15.2.10 *Red pixel interpolation performance comparison: (A) Hibbard–Cok; (B) Original; (C) Combined Color*

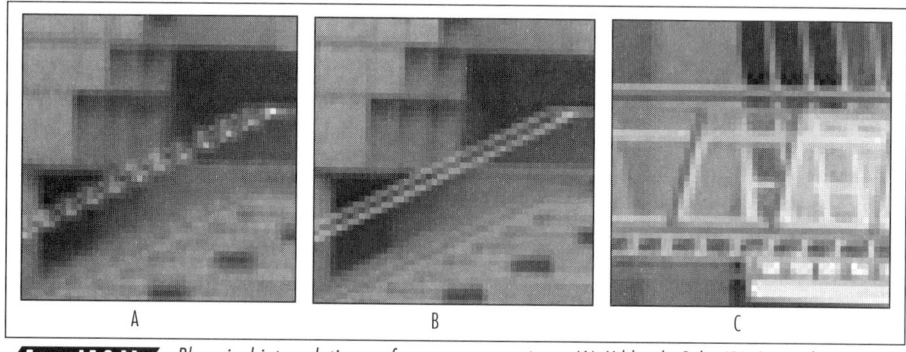

FIGURE 15.2.11 *Blue pixel interpolation performance comparison: (A) Hibbard–Cok; (B) Original; (C) Our Method*

In Figures 15.2.10 and 15.2.11 are a piece of red screen and a piece of blue screen of the same simulated sample image, Color Plate 15—*Peachtree One Building*, are shown. The high performance of the new interpolation method is clear. The combined result is illustrated in Color Plate 15.

15.3 FRACTAL DISPLAYER

It is interesting to realize that printing is actually the inverse procedure of scanning. The history of printing dates back more than 4000 years, but the industry of capturing, including Xerox copying and photographic duplicating, has less than 200 years on record.

The fractal capturer is indeed a fractal compressor. And a fractal displayer is actually a fractal decompressor. When fractal images are displayed on a screen monitor, the monitor is called a *fractal projector*, or *fractal television*. And when fractal images are printed out onto paper, the device is called a *fractal printer*. If a fractal capturer is designed for a specific image partition, a fractal displayer needs to be designed for direct output using the same image partition that the fractal capturer used. A design for the fractal displayer is obtained by modifying a dual patent of an *electro-optical IFS constructor* by Shih *et al.* [SCM].

15.3.1 Design of a Fractal Displayer

The same ATUs used for the capturer will be used in the design of the *fractal displayer* for imitating the corresponding affine mappings. The design follows the same software fractal decompression algorithm. We have seen how to perform fractal transforms using optical ATUs; the remaining link is knowing how to realize the IFS iteration, which is done by using *pulsed laser light* as the input light source.

The design given in Figure 15.3.1 illustrates a case of four fractal transforms. The input pulsed-laser-light-source generates a pulse of laser light directed through the initial input image plane *I0* and is optically branched into several light paths of equal intensity using a set of beamsplitters, *BI1*, *BI2*, *BI3*, and *BI4*, of particular branching ratios. These light paths again go through the ATUs, merge together by a set of mirrors, *MO1*, *MO2*, *MO3*, and *MO4*, and another set of beamsplitters, *BO1*, *BO2*, *BO3*, and *BO4*. Then the merged light goes through a *constant light subtract box CLS* to adjust the total intensity β-value and forms a *composite image*. The *CLS* box can use traditional coherent optical processing for image subtraction, as described in Section 15.1.3. The

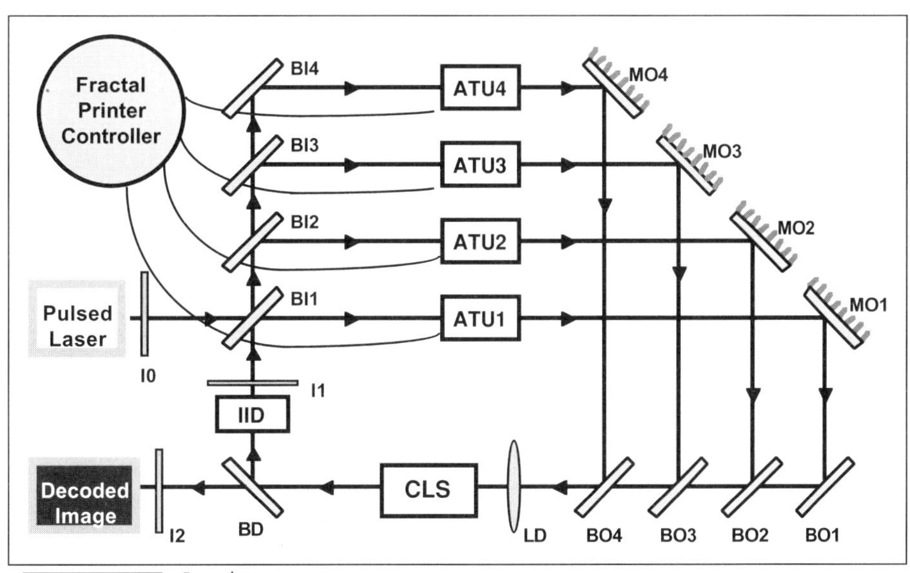

Figure 15.3.1 *Fractal printer*

composite image is now split by a beamsplitter *BD* into two paths: one, called the *iteration image*, goes through an *image intensifier device IID* and the subsequent image plane *I1* for further iteration; and the other, called the *decoded image*, goes through the image plane *I2* and serves as the output image for either a *fractal television* screen display or a direct *fractal printer* printing.

The image intensifier device *IID* amplifies the incident light energy and passes the amplified light energy back as the input image.

The input image source should be capable of generating a light pulse of very short duration—much shorter than the time it takes for the light to traverse the optical loop once. This loop traverse time is typically on the order of nanoseconds. This prevents the input laser light from overlapping the iteration image light. There are light sources commercially available that meet these requirements.

15.3.2 HDTV AND DETV

Television (which modern civilization seems unable to live without), invented in 1927, was initially intended as an extension of radio and was distributed entirely by terrestrial broadcasts. In North America, the black-and-white TV standard was adopted by NTSC in 1941; color was added in 1953, and stereo sound was appended in 1984. In this standard, images are presented in 525 interlaced lines (i.e., a 525 horizontal line frame is presented as two 262.5 interlaced fields as shown in Figure 15.3.2) at a frame rate of 30Hz interlaced, which no longer suits today's consumer expectations. Many proposals for *high definition television* (*HDTV*) were introduced to increase the resolution up to 1920×1080 at a progressive frame rate 60Hz. So what is the problem that prevents the proposals from being widely accepted? It is the limited broadcasting bandwidth. Higher resolution requires wider bandwidths, while more channels are demanded to fit the limited bandwidth.

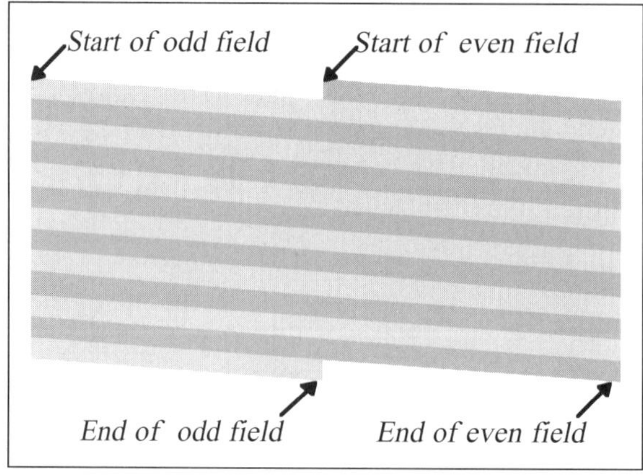

FIGURE 15.3.2 Interlaced Scanning

Our design of fractal television gives an alternative solution to this problem. Instead of increasing image resolution by sending more pixel data, sending image data in fractal format that fits within a given bandwidth is the better option. Where television is concerned, the image will be displayed according to a particular TV's resolution. A 40" TV may have much higher resolution than a 14" one, while still receiving the same broadcasting data. Actually we separated the TV manufacturers' problem of a TV resolution standard from the TV broadcast companies' problem of TV signal broadcasting and transmission bandwidth standards.

Before a TV signal will switch to transmit in fractal form, we still can build "*digitally enhanced TV*" (*DETV*) that receives today's TV broadcasting signals—it processes digitally through a fractal enhancement chip and outputs real-time, high-resolution TV with possibly a few seconds constant delay. The design of the fractal enhanced chip can be easily derived from the algorithms discussed in Chapter 8.

While writing this book, a new revolution in television is happening. PC-TV is taking over and turning the industry from network broadcasting into digital delivery on-demand [P2]. In the 50s, you needed properly pointed rabbit ears to pick up just a few stations; today, thanks to new technological marvels, you have up to 200 cable channels. With a billion users logged onto the Internet, the one-way broadcasting will soon be replaced by user-interacted transmission of millions of online real-time programs. Consequently, HDTV, a dream for many years could soon become history and never become commercially adopted. However, as PC becomes a part of TV, DETV will find itself in an even wider market.

15.4 IMAGE CAPTURING

From the cave dwellers that drew ancient symbols to Internet surfers that design their own web pages, imaging attempts to preserve our history, our experience, passing down to future generations "our story." Our existence pales in comparison to Nature, though we are a significant part. Our motivation for living seems to stem from our thought processes—in fighting for honors and in pursuing memorials—and is always connected with the dream of prolonging our own existence: ultimately, *immortality*. It is known that everything will pass, sooner or later, in this transient universe.

Then what is so exciting about digital imaging? The answer is simple:

> *The digital image, unlike photo prints or magnetic tapes, can be duplicated identically and preserved forever without fading away by aging.*

The true implication is the empowerment to fulfill our dreams. Now, for the first time our images can be preserved, perfectly vivid, until the end of time (as we can imagine it).

Rapid technological progress enables us to have increasingly powerful tools to capture extremely realistic still and video images. However, true image capturing is far more than the design of a camera or other hardware devices. An image that looks perceptually real to our eyes reflects only an illusion that has little to do with our internal instincts. A photographic image or movie scene helps us recall our memories and amplifies the beauty within our souls. But only the real experience (Nature) can originate and plant a seed of beauty. The spirit, energy, touch and penetration — will we someday have a device to capture all of these? As a matter of fact, we ourselves are these capturing devices.

The book pauses with a final question:

> *A scenic picture from the Alps, and a fresh leaf from one of its valleys, which of them tells you more about these mountains' legendary past and mystic future?*

The memory,
like an enchanted melody unfolds, unrolls,
expands . . . late in life.

At the suspension bridge, we spent many
long hours watching boats and passers-by
remembering our aspirations
which are carried off with the current.
The wooden bench has seen a generation pass
a school child rambling
in the drop of a nightfall curtain.
The spark of a familiar tune
pricks my aged ear.
The rebirth of freshly-cut grass.
The marquee of the old theater now
flashes in neon.
I have traced memory lane and found sportsters
instead of roadsters.

. . . Life, a coded book, a reconnaissance,
. . . something new every time you open it.

—Timeless Home Town

Anonymous

[A] J. E. Adams, Jr., *Interactions between color plane interpolation and other image processing functions in electronic photography,* SPIE Vol. 2416, (1995), 144–151.

[AB] A. J. Ahumada, Jr., and B. L. Beard, *Image Discrimination Models: Detection in Fixed and Random Noise,* in *Human Vision and Electronic Imaging*—SPIE Proceeding. 3016 (1997).

[AH] A. N. Akansu and R. A. Haddad, *Multiresolution Signal Decomposition—Transforms, Subbands, Wavelets,* Academic Press, 1992.

[A2] J. Arvo, *Graphics Gems II,* Academic Press, 1991.

[B] M. F. Barnsley, *Fractals Everywhere,* Academic Press, 1988, 1992.

[B2] M. F. Barnsley, *Fractal Functions and Interpolation,* Constr.Approx. 2 (1986) 303–329.

[BDX] M. F. Barnsley, A. Deliu, and R. Xie, *Stationary Stochastic Processes and Fractal Data Compressiony,* Int. J. of Bifurcation and Chaos in Appl. Sci. and Eng. 7 no. 2 (1997).

[BD] M. F. Barnsley and S. G. Demko, *Iterated Function Systems and the Global Construction of Fractals,* Proc. Royal Soc. London A 399 (1985) 245–275.

[BEH] M. F. Barnsley, J. H. Elton, and D. P. Hardin, *Recurrent Iterated Function Systems,* Constr.Approx. 5 (1989) 3–48.

[BEHL] M. Barnsley, V. Ervin, D. Hardin, and J. Lancaster, *Solution of an Inverse Problem for Fractals and other Sets,* Proc. Natl. Acad. Sci. USA 83 (1986) 1975–1977.

[BH] M. F. Barnsley and L. P. Hurd, *Fractal Image Compression,* AK Peters, Ltd., Wellesley, 1993.

[BJ] M. F. Barnsley and A. Jacquin, *Applications of Recurrent Iterated Function Systems to Images,* SPIE Visual Com. & Image Proc. 1001 (1988) 122–131.

[BL] M. F. Barnsley and N. Lu, *Method and System for Camera Data Processing,* U.S. Patent, filed 1997.

[BLH] M. F. Barnsley, N. Lu, and D. R. Howard, *Method for Transmitting Fractal Transform Data to Support Different Compression-Decompression Designs,* U.S. Patent, filed 1995.

[BS] M. F. Barnsley and A. D. Sloan, *A Better Way to Compress Images,* Byte, January 1988.

[BS2] M. F. Barnsley and A. D. Sloan, *Method and Apparatus for Processing Digital Data,* U.S. Patent No. 5065447, 1991.

[BX] M. F. Barnsley and R. Xie, *Fractal Lossless Data Compression,* Preprint, 1997.

[B3] B. E. Bayer, *Color Imaging Array,* U.S. Patent No. 3971065, 1976.

[B4] M. A. Berger, *Apparatus and Method for Encoding Digital Signals,* U.S. Patent No. 5497435, 1996.

[B5] W. P. Bleha *et al., Application of the Liquid Crystal Light Valve to Real-Time Optical Data Processing,* Optical Engineering, 17 (1978) 371–384.

[BS3] C. W. Brown and B. J. Shepherd, *Graphics File Formats—Reference and Guide,* Manning Publications Co., 1995.

[BW] M. Burrows and D. J. Wheeler, *A Block-sorting Lossless Data Compression Algorithm,* Digital Systems Research Center Research Report 124, 1994.

[C-R] S. Calzone *et al., Video Compression by Mean-corrected Motion Compensation of Partial Quad-trees,* IEEE Trans. on Circ. & Sys. for Video Tech., (accepted).

[CM] F. C. Cesbron and F. J. Malassenet, *Wavelet and Fractal Transforms for Image Compression,* Fractal in Engineering, Arcachon, France, June 1997.

[C] C. K. Chui, *An Introduction to Wavelets,* Academic Press, 1992.

[C2] D. R. Cok, *Signal Processing Method and Apparatus for Sampled Image Signals,* U.S. Patent No. 4630307, 1987.

[C3] D. R. Cok, *Signal Processing Method and Apparatus for Producing Interpolated Chrominance Values in a Sampled Color Image Signal,* U.S. Patent No. 4642678, 1987.

[D] I. Daubechies, *Ten Lectures on Wavelets,* CBMS-NSF Reginal Conf. Series in Appl. Math. vol. 61, SIAM Press, 1992.

[D2] G. M. Davis, *A Wavelet-Based Analysis of Fractal Image Compression,* IEEE Trans. On Image Processing, (1996) 100–116.

[DC] F. Davoine, and J.-M. Chassery, *Adaptive Delaunay Triangulation for Attractor Image Coding,* in *Proc. of 12th Int'l Conf. on Pattern Recognition*, Jerusalem, Oct. 1994.

[DHN] S. Demko, L. Hodges, and B. Naylor, *Construction of Fractal Objects with Iterated Function Systems,* Computer Graphics 19, July (1985), 271–278.

[D3] F. Dudbridge, *Image Approximation by Self-affine Fractals,* Ph.D. thesis, University of London, 1992.

[EAW] M. P. Eckstein, A. J. Ahumada, Jr., and A. B. Watson, *Image Discrimination Models Predict Signal Detection in Natural Medical Image Background,* in *Human Vision and Electronic Imaging*—SPIE Proceeding. 3016 (1997).

[E2] W. H. Equitz, *Fast Algorithms for Vector Quantization Picture Coding,* M.S. thesis, MIT, June 1984.

[E] M. C. Escher, *The World of M.C. Escher,* H. N. Abrams, 1971.

[F2] R. Fano, Ph.D. thesis, MIT, 1949.

[F3] P. Fatou, *Sur les Équations fonctionelles,* Bull. Soc. Math. Fr. 47 (1919) 161–271; 48 (1920) 33–94, 208–314.

[F] Y. Fisher (ed.), *Fractal Image Compression — Theory and Application,* Springer-Verlag, 1995.

[FPX] *FlashPix Format and Architecture — White Paper,* Eastman Kodak Company, 1996.

[FS] R. Floyd, and L. Steinberg, *An Adaptive Algorithm for Spatial Gray Scale,* in *Soc. for Info. Display 1975 Symposium Digest of Tech. Papers* 36 (1975).

[FDFH] J. D. Foley, A. VanDam, S. K. Feiner, and J. F. Hughes, *Computer Graphics — Principles and Practice,* Addison-Wesley, 1990, 1996.

[FV] B. Forte and E. Vrscay, *Inverse Problem Methods for Generalized Fractal Transforms,* in *Fractal Image Encoding and Analysis* (ed. Y. Fisher), Springer-Verlag, 1996 (to appear).

[F4] J. Fourier, *Théorie Analytique de la Chaleur,* Gauthiers-Villars, 1888.

[GS] I. M. Gelfand and G. E. Shilov, *Generalized Functions,* Physico-Mathematical Literature, 1959.

[GL] J. Geronimo and N. Lu, *Lectures on Fractal Techniques in Image Compression,* ImageTech Conference, Atlanta, March 1996.

[GG] A. Gersho, and R. M. Gray, *Vector Quantization and Signal Compression,* Kluwer Academic Publishers, 1991.

[GW] R. C. Gonzalez and R. E. Woods, *Digital Image Processing,* Addison-Wesley, 1992.

[G] H. Grassman, *On the Theory of Compound Colours,* Phil. Mag. Ser. 4–7 (1854) 254–264.

[HPN] B. G. Haskell, A. Puri, and A. N. Netravali, *Digital Video: An Introduction to MPEG-2,* Chapman and Hall, 1997.

[H2] S. Hecht, *The Visual Discrimination of Intensity and Weber–Fechner Law,* J. Gen. Physiol. 7 (1924) 241.

[H3] P. S. Heckbert, *Color Image Quantization for Frame Buffer Display,* Computer Graphics 16 (1982) 297–307.

[H4] R. H. Hibbard, *Apparatus and Method for Adaptively Interpolating a Full Color Image Utilizing Luminance Gradients,* U.S. Patent No. 5382976, 1995.

[H] D. Hilbert, *Über die stetige Abbildung einer Linie auf ein Flächenstück,* Math. Ann. 38 (1891) 459–460.

[H5] D. A. Huffman, *A Method for the Construction of Minimum-Redundancy Codes,* Proceedings of the IRE 40 (1952) 1098–1101.

[HR] R. Hunter and A. H. Robinson, *International Digital Facsimile Coding Standards,* Proc. IEEE, July (1980) 854–867.

[H6] J. E. Hutchinson, *Fractals and Self-similarity,* Indiana Univ. Math. J. 3 (1981) 713–747.

[IS] T. Ida and Y. Sambonsugi, *Image Segmentation Using Fractal Coding,* IEEE Trans, on Circ. & Sys. For Video Tech. 5 (1995) 567–570.

[ITU] ITU-T, *Video Coding for Low Bitrate Communication—Draft ITU-T Recommendation H.263,* May, 1996.

[J2] A. E. Jacquin, *A Fractal Theory of Iterated Markov Operators with Applications to Digital Image Coding,* Ph.D. thesis, Georgia Tech, 1989.

[J3] A. E. Jacquin, *Image Coding Based on a Fractal Theory of Iterated Contractive Image Transformations,* IEEE Trans. on Image Proc. 1. (1992) 18–30.

[J] A. K. Jain, *Fundamentals of Digital Image Processing,* Prentice Hall, 1989.

[JN] N. S. Jayant and Peter Noll, *Digital Coding of Waveforms—Principles and Applications to Speech and Video,* Prentice-Hall, 1984.

[JJS] N. Jayant, J. Johnston, and R. Safranek, *Signal Compression Based on Models of Human Perception,* Proc. of IEEE 81 (1993) 1385–1422.

[J4] G. Julia, *Mémoirs sur L'itération des Fonctions Rationnelles,* J. Math. 30 (1918) 47–245.

[JPS] H. Jürgens, H.-O. Peitgen, and D. Saupe, *The Language of Fractals,* Scientific American, August (1990) 60–67.

[K] R. Kannegundla, *Color Balancing in CCD Color Cameras Using Analog Signal Processors Made by Kodak,* SPIE vol. 2416, (1995), 152–157.

[L2] S. Lepsøy, *Attractor Image Compression—Fast Algorithms and Comparison to Related Techniques,* Ph.D. thesis, Norwegian Inst. of Tech., Trondheim, Norway, 1993.

[LM] J. Levy-Vehel and P. Mignot, *Multifractal Segmentation of Images,* Fractals, 2 (1994) 371–378.

[L] J. S. Lim, *Two-Dimensional Signal and Image Processing.* Prentice Hall, 1990.

[LBG] Y. Linde, A. Buzo, and R. Gray, *An Algorithm for Vector Quantizer Designs,* IEEE Trans. Commun., COM-28 (1980), 84–95.

[L3] N. Lu, *Encremental Entropy Coding,* Preprint, 1996.

[L4] N. Lu, *Run-length Decoding Using Huffman Inverse Table,* Preprint, 1995.

[L5] N. Lu, *Video Decompression Using Help Image Panels,* Preprint, 1996.

[LL] N. Lu, and Z. Lu, *Method and System for Interpolating Missing Picture Elements in a Single Color Component Array Obtained from a Single Color Sensor,* U.S. Patent, filed.

[LP] L. J. D'Luna and K. A. Parulski, *A System Approach to Custom VLSI for a Digital Color Imaging System,* IEEE J. of Solid-State Circuits 26 (1991), 727–737.

[M2] S. G. Mallat, *A Theory for Multiresolution decomposition: the wavelet representation,* IEEE Trans. on Pattern Analysis and Machine Intelligence, 11(1989) 674–693.

[M] B. B. Mandelbrot, *The Fractal Geometry of Nature,* W. H. Freeman and Company, 1977, 1982, 1983.

[M3] E. Marom, *Real-Time Image Subtraction Using a Liquid Crystal Light Valve,* Optical Engineering 25 (1986) 274–276.

[M4] D. Marr, *Vision,* W. H. Freeman and Company, 1982.

[M5] P. R. Massopust, *Fractal Functions, Fractal Surfaces, and Wavelets,* Academic Press, 1994.

[M6] J. Munkres, *Topology — A First Course,* Prentice Hall, 1975.

[M7] D. B. Mumford, *Neuronal Architectures for Pattern-theoretic Problems,* in "*Large-scale Neuronal Theories of the Brain*" (C. Koch and J. Davis, eds.), MIT Press, 1994, 125–152.

[MPG] MPEG-4 video VM editing ad hoc group (ISO/IEC JTC1/SC29/WG11), *MPEG-4 Video Verification Model Version 3.1,* August 1996.

[NK] N. M. Nasrabadi and R. A. King, *Image Coding Using Vector Quantization: A Review,* IEEE Trans. on Communications 36 (1988) 957–971.

[N] M. Nelson, *The Data Compression Book,* M&T Books, 1992.

[N2] M. Nelson, *Data Compression with the Burrows–Wheeler Transform,* Dr. Dobb's Journal, September 1996.

[NH] A. N. Netravali and B. G. Haskell, *Digital Pictures — Representation, Compression, and Standards,* Plenum Press, 1988, 1995.

[O] G. E. Øien, *L_2-Optimal Attractor Image Coding with Fast Decoder Convergence,* Ph.D. thesis, Norwegian Inst. of Tech., Trondheim, Norway, 1993.

[P] G. Peano, *Sur une Courbe, qui Rempli Toute une Aire Plane,* Math. Ann. 36 (1890) 157–160.

[PS] H.-O. Peitgen and D. Saupe, *The Science of Fractal Images,* Springer-Verlag, 1988.

[PJS] H.-O. Peitgen, H. Jürgens, and D. Saupe, *Chaos and Fractals — New Frontiers of Science, of Fractals,* Springer-Verlag, 1992.

[PFP] E. Peli, E. M. Fine, and K. Pisano, *Video Enhancement of Text and Movies for the Visually Impaired,* in *Low Vision: Research and New Developments in Rehabilitation* (Kooijman AD et al., eds). IOS Press, Amsterdam (1994), 191–198.

[PM] W. B. Pennebaker and J. L. Mitchell, *JPEG—Still Image Data Compression Standard,* Van Nostrand Reinhold, 1993.

[PMLA] W. B. Pennebaker, J. L. Mitchell, G. G. Landgon, Jr., and R. B. Arps, *An Overview of the Basic Principles of the Q-Coder Adaptive Binary Arithmetic Coder,* IBM J. Res. Develop. 32 (1988) 717–726.

[P2] K. C. Pohlmann, *As Webcasting Hits the Mainstream—Are You Ready for 10 Million Channels?* Video Magazine, 1996.

[PZ] M. Porat and Y. Y. Zeevi, *The Generalized Gabor Scheme of Image Representation in Biological and Machine Vision,* IEEE Trans. on Pat. Ana. & Mac. Int. 10-4 (1988) 452–468.

[P3] W. K. Pratt, *Digital Image Processing,* Wiley-Interscience Publication, 1991.

[PTVF] W. H. Press, S. A. Teukolsky, W. T. Vetterling, and B. R. Flannery, *Numerical Recipes in C—The Art of Scientific Computing,* Cambridge University Press, 1988, 1992.

[RJ] M. Rabbani, and P. W. Jones, *Digital Image Compression Techniques,* SPIE Press, 1991.

[RG] B. Ramamurthi and A. Gersho, *Classified Vector Quantization of Images,* IEEE Trans. on Communications, 34 (1986), 1105–1115.

[RH] K. R. Rao and J. J. Hwang, *Techniques and Standards for Images, Video, and Audio Coding,* Prentice Hall, 1996.

[R2] L.-M. Reissell, *Multiresolution and Wavelets,* Chapter 2 of *Wavelets and Their Applications to Computer Graphics.* SIGGRAPH 95, 26 (1995), 37–70.

[RRK] B. E. Rogowitz, D. A. Rabenhorst, and E. B. Kalin, *Perceptual Techniques for Visual Metadata,* in *Human Vision and Electronic Imaging*—SPIE Proceeding 3016 (1997).

[R] W. Rudin, *Functional Analysis,* McGraw Hill, 1973.

[SP] A. Said and W. Pearlman, *A New Fast and Efficient Image Codec based on Set Partitioning in Hierarchical Trees,* IEEE Trans. on Circuits & Sys. for Video Tech. 6 (1996) 243–250.

[SP2] A. Said and W. Pearlman, *An Image Multiresolution Representation for Lossless and Lossy Compression,* IEEE Trans. on Image Processing, 5 (1996) 1303–1310.

[SHH] D. Saupe, R. Hamzaoui, and H. Hartenstein, *Fractal Image Compression—An Introductory Overview,* Chapter 2 of *Fractal Models for Image Synthesis, Compression, and Analysis,* SIGGRAPH 96, 27 (1996), 47–112.

[SR] D. Saupe and M. Rulh, *Evolutionary Fractal Image Compression,* Proc. ICIP-96 IEEE Int'l Conf. on Image Processing, Lausanne, 1996.

[S] L. Schwartz, *Théorie des distributions,* Hermann, 1973.

[SW] C. E. Shannon and W. Weaver, *The Mathematical Theory of Communication,* University of Illinois Press, 1949.

[S2] J. Shapiro, *Embedded Image Coding Using Zerotrees of Wavelet Coefficients,* IEEE Trans. On Signal Processing 41 (1993) 3445–3462.

[SCM] I-F. Shih, D. B. Chang, and N. L. Moise, *Analog Optical Processing for the Construction of Fractal Objects,* U.S. Patent 5132831, 1992.

[SCMD] I-F. Shih, D. B. Chang, N. L. Moise, and J. E. Drummond, *Electro-Optical IFS Finder,* U.S. Patent 5076662, 1991.

[S3] M. A. Sid-Ahmed, *Image Processing—Theory, Algorithms, & Architectures,* McGraw-Hill, Inc., 1994.

[S4] W. J. Smith, *Modern Optical Engineering,* McGraw-Hill, Inc., 1966.

[S5] K. A. Stevens, *Computation of Locally Parallel Structure,* Biol. Cybernetics 29 (1978) 19–28.

[T] J. Van Tassel, *Legend of Lena: Image Processing's Main Asqueeze,* Advanced Imaging, (May 1996), 56–60.

[TD] L. Thomas and F. Deravi, *Region-based Fractal Image Compression Using Heuristic Search,* IEEE Trans. of Image Processing, 4 (1995), 832–838.

[T2] Y. T. Tsai, *Color Image Compression for Single-Chip Cameras,* IEEE Trans. of Electron. Devices 38, (1991), 1226–1232.

[VK] M. Vetterli, and J. Kovačevic, *Wavelets and Subband Coding,* Prentice Hall PTR, 1995.

[VR] E. R. Vrscay, and C. J. Roehrig, *Iterated Function Systems and the Inverse Problem of Fractal Construction Using Moments,* in *Computers and Mathematics* (E. Kaltofen and S. M. Watt, eds.), Springer-Verlag, 1989, 250–259 .

[W] A. B. Watson (ed.), *Digital Images and Human Vision,* A Bradford Book, The MIT Press, 1993.

[W2] A. B. Watson, *Image Data Compression Having Minimum Perceptual Error,* U.S. Patent 5426512, 1995.

[WK] J. A. Weldy and S. H. Kristy, *Single Sensor Color Video Camera with Blurring Filter,* U.S. Patent No. 4663661, 1987.

[W3] M. V. Wickerhauser, *Adapted Wavelet Analysis from Theory to Software,* A K Peters, 1994.

[WS] G. Wyszecki and W. S. Stiles, *Color Science—Concepts and Methods, Quantitative Data and Formulae,* 2nd Ed., John Wiley & Sons, 1982.

[ZASB] A. Zandi, J. Allen, E. Schwarts, and M. Boliek, *CREW: Compression with Reversible Embedded Wavelets,* ISO/X3L3.2/95-003 JPEG submission, 1995.

[Z2] S. Zhong, *Image Compression by Optimal Reconstruction,* U.S. Patent No. 5534925, 1996.

[Z] A. Zygmund, *Notes on the History of Fourier Series,* in *Studies in Harmonic Analysis* (ed. J. M. Ash), The Mathematical Association of America, 1976.

Tao,
Could be charted and taken,
Moving constantly in eternity;

Name,
Could be honored and given,
Evolving endlessly in perpetuity.

From Nothing,
Heaven and Earth originate;
Through Everything,
All Spirits eventuate.

— Tao-Te-Ching

Lao Tzu

Index

E

F

G

H

I

Credits for Figures and Plates

Picture	*Figure*	*Author and Copyright*
Fractal Sunflowers	Figure 1.2.5(a)	by Michael Barnsley, Arnaud Jacquin, Laurie Reuter, and Alan Sloan. Photograph © Annie Griffiths Belt
Fractal Wolf	Figure 1.2.5(b)	by Michael Barnsley, Arnaud Jacquin, Laurie Reuter, and Alan Sloan. Photograph by Jim Brandenburg © National Geographic Society
Birds and Trees	Figure 2.5.6	by Michael Barnsley, Steve Crawford © 1993 "Fractals Everywhere", M. Barnsley, Academic Press, 1993
Fish and Birds	Figure 4.2.7(a)	by M. C. Ecsher "The World of M. C. Ecsher" Abradale Press, New York, 1971. © 1997 Cordon Art, Baarn, Holland
Reptiles	Figure 4.2.7(b)	by M. C. Ecsher "The World of M. C. Ecsher" Abradale Press, New York, 1971 © 1997 Cordon Art, Baarn, Holland
Mosaic II	Figure 4.2.7(c)	by M. C. Ecsher "The World of M. C. Ecsher" Abradale Press, New York, 1971 © 1997 Cordon Art, Baarn, Holland
Mandelbrot Set Landscape	Figure 1.2.4	Picture by H.Jurgens, H.-O. Peitgen, D. Saupe © 1988 "The Science of Fractal Images", H.-O. Peitgen, D. Saupe (eds.), Springer-Verlag, New York, 1988
Lena	Color Plate 2-a, Figures 1.2.6, 4.2.3-4, 4.2.6, 4.3.4, 4.4.1, 4.4.4, 5.1.4, 5.2.3, 6.2.1., 6.2.4, 6.3.6, 8.3.3, 9.2.2, 11.1.4, 11.1.5, 11.3.2., 14.3.2, 14.4.2	Reproduced by special permission of Playboy Magazine © 1972 Playboy Enterprises Corp.
Mandrill Baboon	Color Plate 2-b, Figure 13.1.9	ISO JPEG standard image, (scanned by IBA)
Plava Laguna Hotel	Color Plate 3-a, Figures 3.5.4, 4.3.8, 8.2.2, 8.3.2, 9.2.2, 9.2.4.	ISO JPEG standard image, (scanned by IBA)
Golden Hill Village	Color Plate 3-b, Figures 4.3.5, 5.1.5, 5.2.4, 8.3.5	ISO JPEG standard image, (scanned by IBA)

PICTURE	FIGURE	AUTHOR AND COPYRIGHT
PARIS	Fig 10.1.3.	ISO MPEG standard video clip
HAWAIIAN ORCHID	Color Plate 5, Figures 8.1.5, 9.1.1	© 1996 Philip Greenspun, http://photo.net/philg
BOBCAT	Figure 8.1.6	© 1996 Philip Greenspun, http://photo.net/philg
CHURCHES -24043	Figure 1.3.1	© Corel Professional Photos
EXOTIC CARS -29097	Figure 1.3.2	© Corel Professional Photos
BIRDS -8093	Figures 2.2.2-3	© Corel Professional Photos
CARIBBEAN -68048	Figure 3.1.1	© Corel Professional Photos
PEOPLE -23021	Figure 13.1.2	© Corel Professional Photos
YOSEMITE -167014	Cover	© Corel Professional Photos
A RESTAURANT IN FIVE POINT	Figures 13.2.1-4	© 1992 Ning Lu
A BRIDGE IN TRONDHEIM	Figure 5.3.3	© 1995 Ning Lu
RED AND GREEN LEAVES	Color Plate 1	© 1996 Ning Lu
FLOWERS	Color Plate 4	© 1993 Ning Lu
A CITY PLAZA	Color Plate 6	© 1984 Ning Lu
CHRYSANTHEMUM	Color Plate 7	© 1993 Ning Lu
EIFFEL TOWER	Color Plate 8	© 1985 Ning Lu
IN THE CATHEDRAL OF PISA	Color Plate 9	© 1985 Ning Lu
BENCHES AND LAMPS	Color Plate 10	© 1996 Ning Lu
SAN JOSE LIVE	Color Plate 11	© 1995 Ning Lu
FACE PAINTING	Color Plate 12	© 1996 Ning Lu
A STREET CORNER WITH ATM	Color Plate 13	© 1996 Ning Lu
BOSTON HARBOR	Color Plate 14	© 1992 Ning Lu
PEACHTREE ONE BUILDING	Color Plate 15	© 1996 Ning Lu
A TEMPLE IN HANGZHOU	Color Plate 16	© 1994 Ning Lu